Principles of
Highway Engineering and
Traffic Analysis

Principles of Highway Engineering and Traffic Analysis

Seventh Edition

Fred L. Mannering
University of South Florida

Scott S. Washburn
University of Florida

WILEY

VP AND EDITORIAL DIRECTOR	Laurie Rosatone
SENIOR DIRECTOR	Don Fowley
EDITOR	Jennifer Brady
EDITORIAL MANAGER	Judy Howarth
CONTENT MANAGEMENT DIRECTOR	Lisa Wojcik
CONTENT MANAGER	Nichole Urban
SENIOR CONTENT SPECIALIST	Nicole Repasky
PRODUCTION EDITOR	Mathangi Balasubramanian
COVER PHOTO CREDIT	© Fred L. Mannering

Photos are courtesy of Scott S. Washburn, unless otherwise noted.

This book was set in 11/13 Times New Roman MT Std Roman by SPi Global and printed and bound by Quad Graphics.

Founded in 1807, John Wiley & Sons, Inc. has been a valued source of knowledge and understanding for more than 200 years, helping people around the world meet their needs and fulfill their aspirations. Our company is built on a foundation of principles that include responsibility to the communities we serve and where we live and work. In 2008, we launched a Corporate Citizenship Initiative, a global effort to address the environmental, social, economic, and ethical challenges we face in our business. Among the issues we are addressing are carbon impact, paper specifications and procurement, ethical conduct within our business and among our vendors, and community and charitable support. For more information, please visit our website: www.wiley.com/go/citizenship.

ISBN: 978-1-119-72319-6 (PBK)
ISBN: 978-1-119-49414-0 (EVAL)

Library of Congress Cataloging-in-Publication Data

Names: Mannering, Fred L., author. | Washburn, Scott S., author.
Title: Principles of highway engineering and traffic analysis / Fred L.
   Mannering, University of South Florida, Scott S. Washburn, University of
   Florida.
Description: Seventh edition. | Hoboken, NJ : Wiley, 2020. | Includes
   index.
Identifiers: LCCN 2019039405 (print) | LCCN 2019039406 (ebook) | ISBN
   9781119494133 (hardback) | ISBN 9781119494102 (adobe pdf) | ISBN
   9781119493969 (epub)
Subjects: LCSH: Highway engineering. | Traffic engineering.
Classification: LCC TE147 .M28 2020 (print) | LCC TE147 (ebook) | DDC
   625.7—dc23
LC record available at https://lccn.loc.gov/2019039405
LC ebook record available at https://lccn.loc.gov/2019039406

The inside back cover will contain printing identification and country of origin if omitted from this page. In addition, if the ISBN on the back cover differs from the ISBN on this page, the one on the back cover is correct.

SKY10059964_111523

Preface

INTRODUCTION

With every edition of *Principles of Highway Engineering and Traffic Analysis*, we have always sought to set the pedagogical standard for instruction in entry-level transportation engineering courses. When the first edition was published three decades ago, there was a need for an entry-level transportation engineering book that focused exclusively on highway transportation and provided the depth of coverage needed to serve as a basis for future transportation courses as well as the material needed to answer questions likely to appear on the Fundamentals of Engineering (FE) and/or Principles and Practice of Engineering (PE) exams in civil engineering. The subsequent use of the various editions of this book, over the years, at some of the largest and most prestigious schools in the United States and throughout the world suggests that a vision of a concise, highly focused, and well-written entry-level book is shared by many educators.

APPROACH

This seventh edition of *Principles of Highway Engineering and Traffic Analysis* continues the spirit of the previous six editions by again focusing exclusively on highway transportation and providing the depth of coverage necessary to solve the highway-related problems that are most likely to be encountered in engineering practice. The focus on highway transportation is a natural one given the dominance of highway transportation for people and freight movement in the United States and throughout the world. While the focus on highway transportation is easily accomplished, identifying the highway-related problems most likely to be encountered in practice and providing an appropriate depth of coverage of them is a more challenging task. Using the first six editions as a basis, along with the comments of other instructors and students who have used previous editions of the book, topics that are fundamental to highway engineering and traffic analysis have been carefully selected. The material provided in this book ensures that students learn the fundamentals needed to undertake upper-level transportation courses, enter transportation employment with a basic knowledge of highway engineering and traffic analysis, and have the knowledge necessary to answer transportation-related questions on the civil engineering Fundamentals of Engineering (FE) exam and the Principles and Practice of Engineering (PE) exam.

MATHEMATICAL RIGOR

Within the basic philosophical approach described above, this book addresses the concern of some that traditional highway transportation courses are not as mathematically challenging or rigorous as other entry-level civil engineering courses, and that this may affect student interest relative to other civil-engineering fields of study. This concern is not easily addressed because there is a dichotomy with regard to mathematical rigor in highway transportation, with relatively simple mathematics used in practice-oriented material and complex mathematics used in research. Thus it is common for instructors to either insult students' mathematical knowledge or vastly exceed it. This book strives for that elusive middle ground of mathematical rigor that matches junior and senior engineering students' mathematical abilities.

CHAPTER TOPICS AND ORGANIZATION

The seventh edition of *Principles of Highway Engineering and Traffic Analysis* has evolved from over three decades of teaching introductory transportation engineering classes at the University of Washington, University of Florida, Purdue University, University of South Florida, and the Pennsylvania State University, feedback from users of the first six editions, and experiences in teaching civil-engineering licensure exam review courses. The book's material and presentation style (which is characterized by the liberal use of example and practice problems) are largely responsible for transforming much-maligned introductory transportation engineering courses into courses that students consistently rate among the best civil engineering courses.

The book begins with a short introductory chapter that stresses the significance of highway transportation to the social and economic underpinnings of society. Also discussed are environmental impacts including climate change and emerging technologies including connected and automated vehicles. This chapter provides students a basic overview of the problems facing the field of highway engineering and traffic analysis. The chapters that follow are arranged in sequences that focus on highway engineering (Chapters 2, 3, and 4) and traffic analysis (Chapters 5, 6, 7, and 8).

Chapter 2 introduces the basic elements of road vehicle performance. This chapter represents a major departure from the vehicle performance material presented in all other transportation and highway engineering books, in that it is far more involved and detailed. The additional level of detail is justified on two grounds. First, because students own and drive automobiles, they have a basic interest that can be linked to their freshman and sophomore coursework in physics, statics, and dynamics. Traditionally, the absence of such a link has been a common criticism of introductory transportation and highway engineering courses. Second, it is important that engineering students understand the principals involved in vehicle technologies and the effect that continuing advances in vehicle technologies will have on engineering practice.

Chapter 3 presents current design practices for the geometric alignment of highways. This chapter provides details on vertical-curve design and the basic elements of horizontal-curve design. This edition of the book includes the latest design guidelines (*Policy on Geometric Design of Highways and Streets*, American Association of State Highway and Transportation Officials, Washington, DC, 2018).

Chapter 4 provides a detailed overview of traditional pavement design, covering both flexible and rigid pavements in a thorough and consistent manner. A brief overview of the topics of pavement distresses and mechanistic-empirical approaches to pavement design are also provided. The material in this chapter also links well with the geotechnical and materials courses that are likely to be part of the student's curriculum.

Chapter 5 presents the fundamentals of traffic flow and queuing theory, which provide the basic tools of traffic analysis. Relationships and models of basic traffic-stream parameters are introduced, as well as queuing analysis models for deterministic and stochastic processes. Considerable effort was expended to make the material in this chapter accessible to junior and senior engineering students.

Chapter 6 presents some of the current methods used to assess highway levels of performance. Fundamentals and concepts are discussed along with the complexities involved in measuring and/or calculating highway level of service. This edition of the book has been updated to the latest analysis standards (*Highway Capacity Manual*, Transportation Research Board, National Academy of Sciences, Washington, D.C., 2016).

Chapter 7 introduces the basic elements of traffic control at a signalized intersection and applies the traffic analysis tools introduced in Chapter 5 to signalized intersections. The chapter not only focuses on pretimed, isolated signals but also introduces the reader to the fundamentals of actuated and coordinated signal systems. Both theoretical and practical elements associated with traffic signal timing are presented. This edition of the book has been updated to the latest analysis standards (*Highway Capacity Manual*, Transportation Research Board, National Academy of Sciences, Washington, D.C., 2016).

Chapter 8, the final chapter, provides an overview of travel demand and traffic forecasting. This chapter concentrates on a theoretically and mathematically consistent approach to travel demand and traffic forecasting that closely follows the approach most commonly used in practice, and contains a section on the traditional four-step travel-demand forecasting process. This chapter provides the student with an important understanding of the current state of travel demand and traffic forecasting, and some critical insight into the deficiencies of forecasting methods currently used.

NEW AND REVISED PROBLEMS

This edition, like the sixth edition, includes practice problems as a new pedagogical tool. At the end of chapters that require mathematical solutions to problems (Chapters 2 through 8 inclusive), several partially solved problems are provided for students to practice their problem-solving techniques and more fully understand the material presented in the text. This enables students to follow the thought process involved in solving problems in the chapter, while also engaging them in the solution (as opposed to the traditional example problems, also provided, which present the complete solution). There are also several new and revised chapter homework problems relative to the sixth edition.

Users of the book will find the practice chapter homework problems to be extremely useful in supporting the material presented in the book. These problems are precise and challenging, a combination rarely found in transportation/highway engineering books.

NEW TO THIS EDITION

In this edition we have once again made several enhancements to the content and visual presentation, based on suggestions from instructors. Some new features in this edition of the book include:

New chapter homework problems. Several new homework problems have been added to further improve the pedagogical effectiveness of the book.

New Enhanced E-Text makes your study time more effective with show/hide solutions and answers for select practice problems.

WEBSITE

The website for this book is www.wiley.com/go/mannering/highwayandtraffic7e and contains the following resources for instructors:

Solutions Manual. An example solutions manual (several problems for each chapter). Instructions for obtaining the complete solutions manual—that is, all the problems in the book.

In-Class Design Problems. Design problems developed by the authors for in-class use by students in a cooperative-learning context. The problems support the material presented in the chapters and the chapter homework problems.

In-Class Design Problems with Solutions. A complete solutions manual for all of the In-class design problems.

Sample Exams. Sample midterm and final exams are provided to give instructors class-proven ideas relating to successful exam format and problems.

A Sample Course Syllabus. Provided to assist instructor's in structuring their course.

Online Metric Conversion Appendix is available to instructors for distribution to students.

Image Gallery. All of the text's figures and tables are provided to instructors to assist them in creating teaching slides and materials.

Visit the Instruction Companion Site section of the book website to register for a password to download these resources.

Fred L. Mannering
Scott S. Washburn

Contents

Chapter 1

Introduction to Highway Engineering and Traffic Analysis

1.1 INTRODUCTION

In many industrialized nations today, highways present engineers and governments with formidable challenges relating to safety, sustainability, environmental impacts, congestion mitigation, and deteriorating infrastructure. As a result, highways are often viewed from the perspective of the many challenges they present as opposed to the benefits they provide. Historically, highways have always played a key role in the development and sustainability of human civilization. Today, in the United States and throughout the world, highways continue to dominate the transportation system, by providing critical access for the acquisition of natural resources, industrial production, retail marketing, and population mobility. The influence of highway transportation on the economic, social, and political fabric of nations is far-reaching and, as a consequence, highways have been studied for decades as a cultural, political, and economic phenomenon. While industrial needs and economic forces have clearly played an important part in shaping highway networks, societies' fundamental desire for access to activities and affordable land has generated significant highway demand, which has helped define and shape highway networks.

Without doubt, highways have had a dramatic impact on the environment in terms of the consumption of nonrenewable resources, air pollution, and the generation of greenhouse gases. In addition, vehicle crashes result in well over a million deaths worldwide every year and are the leading cause of death among people 15 to 29 years old [World Health Organization 2015]. As with other critical infrastructures (such as electrical power generation and distributions systems, water distribution systems, and storm-water and sewage systems), highway systems are costly to build, manage, and maintain, and inadequate management and maintenance can result in additional costs with regard to congestion, safety, and a variety of adverse economic impacts.

Given the above, the focus of highway engineering has gone from one of network expansion to one that addresses issues relating to infrastructure maintenance and rehabilitation, improvements in operational efficiency, various traffic-congestion relief measures, energy conservation, improved safety, and environmental mitigation. This shift has forced a new emphasis in highway engineering and traffic analysis and is one that requires a new skill set and a deeper understanding of the impact of highway decisions than has historically been the case.

1.2   HIGHWAYS AND THE ECONOMY

It is difficult to overstate the influence that highway transportation has on the world economy. Highway systems have a direct effect on industries that supply vehicles and equipment to support highway transportation and the industries that are involved in highway construction and maintenance. Highway systems are also vital to manufacturing and retail supply chains and distribution systems and serve as regional, national, and international economic engines.

1.2.1   The Highway Economy

In the United States, more than 15% of average household income is spent on highway vehicle purchases, maintenance, and other vehicle expenditures. As a consequence, the industries providing vehicles and vehicle services for highway transportation have an enormous economic influence. In the United States alone, in the light-vehicle market (cars, vans, pickup trucks, and so on), as many as 17 million or more new vehicles can be sold annually (depending on economic conditions), which translates to roughly a half a trillion dollars in sales and more than a million jobs in manufacturing and manufacturing-supplier industries. Add to this the additional employment associated with vehicle maintenance and servicing, and more than seven million U.S. jobs can be tied directly to highway vehicles. The influence of the highway economy extends further to the heavy-vehicle sector as well, with more than 1.3 million jobs and trucking industry revenue of roughly three-quarters of a trillion dollars annually in the United States.

The direct influence that highways have also includes the construction and maintenance of highways, with over 100 billion dollars in annual expenditures in the United States alone. This too has an enormous impact on employment and other aspects of the economy.

1.2.2   Supply Chains

The survival of modern economies is predicated on efficient, reliable, and resilient supply chains. Industries have become increasingly dependent on their supply chains to reduce costs and remain competitive. As an example, most manufacturing industries today rely on just-in-time delivery to reduce inventory-related costs, which can be a substantial percentage of total costs in many industries. The idea of just-in-time delivery is that the materials required for production are supplied just before they are needed. While such a strategy significantly reduces inventory costs, it requires a very high degree of certainty that the required materials will be delivered on time. If not, the entire production process could be adversely affected and costs could rise dramatically.

In retail applications, effective supply chains can significantly reduce consumer costs and ensure that a sufficient quantity of goods is available to satisfy consumer demand. The ability of highways to provide reliable service for just-in-time inventory control and other supply chain–related industrial and retail applications has made highways critical to the function of modern economies.

1.2.3   Economic Development

It has long been recognized that highway construction and improvements to the highway network can positively influence economic development. Such improvements can increase accessibility and thus attract new industries and spur local economies. To be sure, measuring the economic development impacts of specific highway projects is not an easy task because such measurements must be made in the context of regional and national economic trends. Still, the effect that highways can have on economic development is yet another example of the far-reaching economic influences of highway transportation.

1.3   HIGHWAYS, ENERGY, THE ENVIRONMENT, AND CLIMATE CHANGE

As energy demands fluctuate and supplies vary, and nations become increasingly concerned about environmental impacts, the role that highway transportation plays has come under close scrutiny. As a primary consumer of fossil fuels and a major contributor to air-borne pollution and greenhouse gas emissions, highway transportation is an obvious target for energy conservation and environmental impact mitigation efforts.

In the United States, highway transportation is responsible for roughly 60% of all petroleum consumption. This translates into about 12 million barrels of oil a day. In light of the limitations of oil reserves, this is an astonishing rate of consumption. Highway transportation's contribution to other pollutants is also substantial. Highway travel is responsible for about 35% of all nitrous oxide emissions and 25% of volatile organic compound emissions, both major contributors to the formation of ozone. Highway travel also contributes more than 50% of all carbon monoxide emissions in the United States and is a major source of fine particulate matter (2.5 microns or smaller), which is a known carcinogen.

The effect of highways on climate change is also formidable. Highway transportation is responsible for roughly 25% of U.S. greenhouse gas emissions (including over 30% of carbon dioxide emissions). While highways affect climate change, the effect that climate change will ultimately have on highways is an issue that has just begun to be addressed [National Cooperative Research Program 2014]. The effect of climate change on extreme weather events, geographic shifts in temperature and moisture, and rising sea levels present highway engineers with extraordinary challenges with regard to highway design, infrastructure maintenance, highway operations, and highway safety.

1.4   HIGHWAYS AS PART OF THE TRANSPORTATION SYSTEM

It is important to keep in mind that highway transportation is part of a larger transportation system that includes air, rail, water, and pipeline transportation. In this system, highways are the dominant mode of most passenger and freight movements. For passenger travel, highways account for about 90% of all passenger-miles. On the freight side, commercial trucks account for about 37% of the freight ton-miles and, because commercial trucks transport higher-valued goods than other modes of transportation (with the exception of air transportation), nearly 80% of the dollar value of all goods is transported by commercial trucks.

While highways play a dominant role in both passenger and freight movement, in many applications there are critical interfaces among the various transportation modes. For example, many air, rail, water, and pipeline freight movements involve highway transportation at some point for their initial collection and final distribution. Interfaces between modes, such as those at water ports, airports, and rail terminals, create interesting transportation problems but, if handled correctly, can greatly improve the efficiency of the overall transportation system. However, inter-modal coordination can be problematic because of institutional, regulatory, and other barriers.

1.5 HIGHWAY TRANSPORTATION AND THE HUMAN ELEMENT

Within the highway transportation system, traveler options include single-occupant private vehicles, multi-occupant private vehicles, and public transportation modes (such as bus). It is critical to develop a basic understanding of the effect that highway-related projects and policies may have on the individual highway modes of travel (single-occupant private vehicles, bus, and so on) because the distribution of travel among modes will strongly influence overall highway-system performance. In addition, highway safety and the changing demographics of highway users are important considerations.

1.5.1 Passenger Transportation Modes and Traffic Congestion

Of the available urban transportation modes (bus, commuter train, subway, private vehicle, and others), private vehicles, and single-occupant private vehicles in particular, offer an unequaled level of mobility. The single-occupant private vehicle has been such a dominant choice that travelers have been willing to pay substantial capital and operating costs, confront high levels of congestion, and struggle with parking-related problems just to have the flexibility in travel departure time and destination choices that is uniquely provided by private vehicles. In the last 50 years, the percentage of trips taken in private vehicles has risen from slightly less than 70% to over 90% (public transit and other modes make up the balance). Over this same period, the average private-vehicle occupancy has dropped from 1.22 to 1.09 persons per vehicle, reflecting the fact that the single-occupant vehicle has become an increasingly dominant mode of travel.

Traffic congestion that has arisen as a result of extensive private-vehicle use and low-vehicle occupancy presents a perplexing problem. The high cost of new highway construction (including monetary, environmental, and social costs) often makes building new highways or adding additional highway capacity an unattractive option. Trying to manage the demand for highways also has its problems. For example, programs aimed at reducing congestion by encouraging travelers to take alternate modes of transportation (bus-fare incentives, increases in private-vehicle parking fees, tolls and traffic-congestion pricing, rail- and bus-transit incentives) or increasing vehicle occupancy (high-occupancy vehicle lanes and employer-based ridesharing programs) can be considered as viable options.

However, such programs have the adverse effect of directing people toward travel modes that inherently provide lower levels of mobility because no other mode offers the departure-time and destination-choice flexibility provided by private, single-occupant vehicles. Managing traffic congestion is an extremely complex problem with significant economic, social, environmental, and political implications.

1.5.2   Highway Safety

The mobility and opportunities that highway infrastructure provides also have a human cost. Although safety has always been a primary consideration in highway design and operation, highways continue to exact a terrible toll in loss of life, injuries, property damage, and reduced productivity as a result of vehicle accidents. Highway safety involves technical and behavioral components and the complexities of the human/machine interface. Because of the high costs of highway accidents, efforts to improve highway safety have been intensified dramatically in recent decades. This has resulted in the implementation of new highway design guidelines and countermeasures (some technical and some behavioral) aimed at reducing the frequency and severity of highway accidents. Fortunately, efforts to improve highway design (such as more stringent design guidelines, breakaway signs, and so on), vehicle occupant protection (safety belts, padded dashboards, collapsible steering columns, driver- and passenger-side airbags, improved bumper design), as well as advances in vehicle technologies (antilock braking, traction control systems, electronic stability control) and new accident countermeasures (campaigns to reduce drunk driving), have managed to gradually reduce the fatality rate (the number of fatalities per mile driven). However, in spite of continuing efforts and unprecedented advancements in vehicle safety technologies, the total number of fatalities per year has remained unacceptably high worldwide (in the United States, the fatality rate has remained at more than 30,000 per year).

To understand why highway fatality numbers have not dramatically decreased or why fatality rates (fatalities per distance driven) have not dropped more than they have as a result of all the safety efforts, a number of possible explanations arise including: an increase in the overall level of aggressive driving; increasing levels of disrespect for traffic control devices (red light and stop-sign running being two of the more notable examples); in-vehicle driving distractions (such as cell phones); and poor driving skills in the younger and older driving populations. Two other phenomena are being observed that may be contributing to the persistently stable number of fatalities. One is that some people drive more aggressively (speeding, following too closely, frequent lane changing) in vehicles with advanced safety features, thus offsetting some or all of the benefits of new safety technologies [Winston et al., 2006]. Another possibility is that many people are more influenced by style and function than safety features when making vehicle purchase decisions. This is evidenced by the growing popularity of vehicles such as sport utility vehicles, mini-vans, and pickup trucks, despite their consistently overall lower rankings in certain safety categories, such as roll-over probability, relative to traditional passenger cars. These issues underscore the overall complexity of the highway safety problem and the trade-offs that must be made with regard to cost, safety, and mobility (speed).

1.5.3 Demographic Trends

Travelers' commuting patterns (which lead to traffic congestion) are inextricably intertwined with socioeconomic characteristics such as age, income, household size, education, and job type, as well as the distribution of residential, commercial, and industrial developments within the region. Many American metropolitan areas have experienced population declines in central cities accompanied by a growth in suburban areas. One could argue that the population shift from the central cities to the suburbs has been made possible by the increased mobility provided by the major highway projects undertaken during the 1960s and 1970s. This mobility enabled people to improve their quality of life by gaining access to affordable housing and land, while still being able to get to jobs in the central city with acceptable travel times. Conventional wisdom suggested that as overall metropolitan traffic congestion grew (making the suburb-to-city commuting pattern much less attractive), commuters would seek to avoid traffic congestion by reverting back to public transport modes and/or once again choosing to reside in the central city. This has certainly happened to some extent, but a different trend has also emerged. Employment centers have developed in the suburbs and now provide a viable alternative to the suburb-to-city commute (the suburb-to-suburb commute). The result is a continuing tendency toward low-density, private vehicle–based development as people seek to retain the high quality of life associated with such development.

Ongoing demographic trends also present engineers with an ever-moving target that further complicates the problem of providing mobility and safety. An example is the rising average age of the U.S. population that has resulted from population cohorts and advances in medical technology that prolong life. Because older people tend to have slower reaction times, taking longer to respond to driving situations that require action, engineers must confront the possibility of changing highway design guidelines and practices to accommodate slower reaction times and the potentially higher variance of reaction times among highway users.

1.6 HIGHWAYS AND EVOLVING TECHNOLOGIES

As in all fields, technological advances at least offer the promise of solving complex problems. For highways, technologies can be classified into those impacting infrastructure, vehicles, and traffic control.

1.6.1 Infrastructure Technologies

Investments in highway infrastructure have been made continuously throughout the twentieth and twenty-first centuries. Such investments have understandably varied over the years in response to need and political and national priorities. For example, in the United States, an extraordinary capital investment in highways during the 1960s and 1970s was undertaken by constructing the interstate highway system and upgrading and constructing many other highways. The economic and political climate that permitted such an ambitious construction program has not been replicated before or since. It is difficult to imagine, in today's economic and political environment, that a project of the magnitude of the interstate highway system would ever be seriously considered. This is because of the prohibitive costs associated with land acquisition and construction and the community and environmental impacts that would result.

It is also important to realize that highways are long-lasting investments that require maintenance and rehabilitation at regular intervals. The legacy of a major capital investment in highway infrastructure is the proportionate maintenance and rehabilitation schedules that will follow. Although there are sometimes compelling reasons to defer maintenance and rehabilitation (including the associated construction costs and the impact of the reconstruction on traffic), such deferral can result in unacceptable losses in mobility and safety as well as more costly rehabilitation later.

As a consequence of past capital investments in highway infrastructure and the current high cost of highway construction and rehabilitation, there is a strong emphasis on developing and applying new technologies to more economically construct and extend the life of new facilities and to effectively combat an aging highway infrastructure. Included in this effort are the extensive development and application of new sensing technologies in the emerging field of structural health monitoring. There are also opportunities to extend the life expectancy of new infrastructure with the ongoing advances in material science. Such technological advances are essential elements in the future of highway infrastructure.

1.6.2   Traffic Control Technologies

Traffic signals at highway intersections are a familiar traffic control technology. At signalized intersections, the trade-off between mobility and safety is brought into sharp focus. Procedures for developing traffic-signal control plans (allocating green time to conflicting traffic movements) have made significant advances over the years. Today, signals at critical intersections can be designed to respond quickly to prevailing traffic flows, groups of signals can be coordinated to provide a smooth through-flow of traffic, and, in some cases, computers control entire networks of signals. Still, at some level, the effectiveness of improvements in signal control is fundamentally limited by the reaction times and the driving behavior of the motoring public as well as the braking and acceleration performance of the vehicles they drive. This presents highway engineers with a formidable barrier.

In addition to traffic signal controls, numerous safety, navigational, and congestion-mitigation technologies are reaching the market under the broad heading of Intelligent Transportation Systems (ITS). Such technological efforts offer the potential to reduce traffic congestion and improve safety on highways by providing an unprecedented level of traffic control. There are, however, many obstacles associated with ITS implementation, including system reliability, human response, and the human/machine interface. Numerous traffic control technologies offer the potential for considerable improvement in the efficient use of the highway infrastructure, but there remain limits defined by vehicle performance characteristics and their human operators.

1.6.3   Vehicle and Autonomous Vehicle Technologies

Until the 1970s, vehicle technologies evolved slowly and often in response to mild trends in the vehicle market as opposed to an underlying trend toward technological development. Beginning in the 1970s, however, three factors began a cycle of unparalleled advances in vehicle technology that continues to this day: (1) government regulations on air quality, fuel efficiency, and vehicle-occupant safety; (2) energy shortages and fuel-price increases; and (3) intense competition among vehicle manufacturers (foreign and domestic). The aggregate effect of these factors has resulted in vehicle consumers that demand new technology at highly competitive prices. Vehicle manufacturers have found it necessary to reallocate resources and to restructure manufacturing and inventory control processes to meet this demand. In recent years, consumer demand and competition among vehicle manufacturers has resulted in the widespread implementation of new technologies including supplemental restraint systems, antilock brake systems, traction control systems, electronic stability control, and a host of other applications of new technologies to improve the safety and comfort in highway vehicles. There is little doubt that the combination of consumer demand and intense competition in the vehicle industry will continue to spur vehicle technological innovations.

As vehicle technologies have progressed, the highway engineering profession is approaching a potential breakthrough in how traffic can be managed and how future highways may be designed. Specifically, the prospect of connected vehicles (being able to receive and send information to other vehicles and central controls and potentially from infrastructure sensors that may transmit pavement conditions and other factors relevant to vehicle operation) and fully automated vehicles has the potential to reshape the landscape of highway engineering. With connected and fully automated vehicle operation, one can imagine immense changes in highway design because the need to account for variation in human abilities (reaction times, sight distances, detection of changing road conditions with regard to weather, ability to predict the actions of other drivers, etc.) and human behavior (choice of car-following distances, speed choice, route choice, lane choice on multilane highways, choices at traffic signals, etc.) would now be replaced simply by the need to assess and account for variation in vehicle characteristics (vehicle dimensions, acceleration, deceleration, etc.), system operating software, and possible component and system failures, all of which are much more "predictable" than the human-related factors that currently dominate highway design. The eventual result is that, with connected and autonomous vehicles, there could be substantial increases in highway capacity for some situations, vast improvements in safety, and essentially a complete change in how highway transportation affects the economy, energy consumption, the environment, the spatial structure of cities, the impact on human activities, and so on.

As autonomous vehicles begin to enter the highway traffic mix, presumably in the coming decade, there will be enormous challenges. While operating a completely connected and autonomous vehicle fleet would be relatively easy because of the complete control available, having autonomous vehicles mixed with human drivers creates a serious challenge because sensor systems must be able to detect, and software systems must be able to quickly respond to the incredible range of human-driver behaviors.

To better understand the range of vehicle technologies that are likely to be present in highway traffic of the future, the U.S. Department of Transportation's National Highway Traffic Safety Administration has chosen to define levels of vehicle automation as follows (NHTSA, 2018):

No-Automation (Level 0): The driver is in complete and sole control of the primary vehicle controls (brake, steering, throttle, and motive power) at all times.

Driver Assistance (Level 1): Automation at this level involves the driver with driver assistance features in the design of the vehicle.

Partial Automation (Level 2): This level involves engagement from the driver, but the vehicle has combined automated functions controlling both the steering and acceleration/deceleration capabilities.

Conditional Automation (Level 3): Vehicles at this level of automation enable the driver to cede full control of all safety-critical functions under certain traffic or environmental conditions and in those conditions to rely heavily on the vehicle to monitor for changes in those conditions requiring transition back to driver control. The driver is expected to be available for occasional control, but with sufficiently comfortable transition time.

High Automation (Level 4): The vehicle is designed to perform all safety-critical driving functions and monitor roadway conditions for an entire trip. Such a design anticipates that the driver will provide destination or navigation input, but is not expected to be available for control at any time during the trip. The driver has the option to control the vehicle.

Full Automation (Level 5): The vehicle functions under all conditions without a driver and does not require a steering wheel, throttle, or brake pedals. Level 5 vehicles have the most advanced environment detection system.

In today's vehicle fleet, the vast majority of vehicles are at Levels 0 and 1. Many newer cars can be purchased at automation Level 2, and it is quite likely that within the coming decade, that substantial numbers of Level 3 and Level 4 vehicles will be on highways worldwide. While the presence of these vehicles will inevitably change highway design, the fundamental principles covered in this text will still apply with appropriate modification.

1.7   SCOPE OF STUDY

Highway engineering and traffic analysis involves an extremely complex interaction of economic, behavioral, social, political, environmental, and technological factors. This complexity makes highway engineering and traffic analysis far more challenging than many typical engineering disciplines that tend to have an overriding focus on only the technical aspects of the problem. To be sure, the technical challenges encountered in highway engineering and traffic analysis easily rival the most complex technical problems encountered in any other engineering discipline (and this will be even more pronounced with the potential introduction of autonomous vehicles). However, it is the economic, behavioral, social, political, and environmental elements that introduce a level of complexity unequalled in any other engineering discipline.

The remaining chapters in this book do not intend to provide a comprehensive assessment of the many factors that influence highway engineering and traffic analysis. Instead, Chapters 2 through 8 seek to provide readers with the fundamental elements and methodological approaches that are used to design and maintain highways and assess their operating performance. This material constitutes the fundamental principles of highway engineering and traffic analysis that are needed to begin to grasp the many complex elements and considerations that come into play (now and in the future) during the design, construction, maintenance, and operation of highways.

REFERENCES

National Cooperative Research Program. *Climate Change, Extreme Weather Events, and the Highway System: Practitioner's Guide and Research Report*. Strategic Issues Facing Transportation, Volume 2, NCHRP Report 750, Transportation Research Board, Washington, D.C., 2014.

National Highway Traffic Safety Administration (NHTSA). https://www.nhtsa.gov/technology-innovation/automated-vehicles-safety

Winston, C., V. Maheshri, and F. Mannering. An exploration of the offset hypothesis using disaggregate data: The case of airbags and antilock brakes. *Journal of Risk and Uncertainty*, vol. 32, no. 2, 2006.

World Health Organization. *Global Status Report on Road Safety 2015*. Geneva, Switzerland, 2015.

Chapter 2

Road Vehicle Performance

2.1 INTRODUCTION

Road vehicle performance forms the basis for highway design guidelines and traffic analysis. As an example, in highway design, the determination of the length required for freeway acceleration and deceleration lanes, maximum highway grades, stopping-sight distances, passing-sight distances, and numerous crash mitigation highway features all rely on a basic understanding of vehicle performance. Vehicle performance is also a major consideration in the selection and design of traffic control devices, the determination of speed limits, and the timing and control of traffic signal systems.

Studying vehicle performance serves two important functions. First, it provides insight into highway design and traffic operations and the compromises that are necessary to accommodate the wide variety of vehicles that use highways (from high-powered sports cars to heavily laden trucks). Second, vehicle performance is critical to the assessment of the impact that new vehicle technologies will have on existing highway design guidelines. This second function is particularly important in light of the ongoing unprecedented advances in vehicle technologies, including autonomous vehicles. Such advances will necessitate more frequent updating of highway design guidelines as well as engineers who have a better understanding of the fundamental principles underlying vehicle performance.

The objective of this chapter is to introduce the basic principles of road vehicle performance. Primary attention will be given to the straight-line performance of vehicles (acceleration, deceleration, top speed, and the ability to ascend grades). Cornering performance of vehicles is overviewed in Chapter 3, but detailed presentations of this material are better suited to more specialized sources [Wong, 2008].

2.2 TRACTIVE EFFORT AND RESISTANCE

Tractive effort (also referred to as thrust) and resistance are the two primary opposing forces that determine the straight-line performance of road vehicles. Tractive effort is simply the force available, at the roadway surface, to perform work and is expressed in lb. Resistance, also expressed in lb, is defined as the force impeding vehicle motion. The three major sources of vehicle resistance are (1) aerodynamic resistance, (2) rolling resistance (which originates from the roadway surface and tire interface), and (3) grade or gravitational resistance. To illustrate these forces, consider the vehicle force diagram shown in Fig. 2.1.

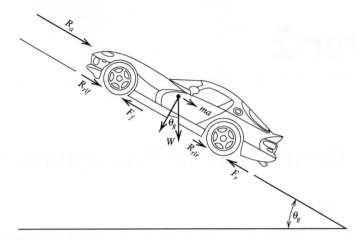

Figure 2.1 Forces acting on a road vehicle.

R_a = aerodynamic resistance in lb,
R_{rlf} = rolling resistance of the front tires in lb,
R_{rlr} = rolling resistance of the rear tires in lb,
F_f = available tractive effort of the front tires in lb,
F_r = available tractive effort of the rear tires in lb,

W = total vehicle weight in lb,
θ_g = angle of the grade in degrees,
m = vehicle mass in slugs, and
a = acceleration in ft/s^2.

Summing the forces along the vehicle's longitudinal axis provides the basic equation of vehicle motion:

$$F_f + F_r = ma + R_a + R_{rlf} + R_{rlr} + R_g \qquad (2.1)$$

where R_g is the grade resistance and is equal to $W \sin \theta_g$. For exposition purposes, it is convenient to let F be the sum of available tractive effort delivered by the front and rear tires ($F_f + F_r$) and similarly to let R_{rl} be the sum of rolling resistance ($R_{rlf} + R_{rlr}$). This notation allows Eq. 2.1 to be written as

$$F = ma + R_a + R_{rl} + R_g \qquad (2.2)$$

Sections 2.3 to 2.8 present a thorough discussion of the components and implications of Eq. 2.2.

2.3   AERODYNAMIC RESISTANCE

Aerodynamic resistance is a resistive force that can have significant impacts on vehicle performance. At high speeds, where this component of resistance can become overwhelming, proper vehicle aerodynamic design is essential. Aerodynamic efficiency has long been given primary consideration in the design of racing and sports cars. More recently, concerns over fuel efficiency and overall vehicle performance have resulted in more efficient aerodynamic designs in common passenger cars, although not necessarily in pickup trucks or sport utility vehicles where the need for cargo space often limits what can be done with aerodynamic efficiency.

Aerodynamic resistance originates from a number of sources. The primary source (typically accounting for over 85% of total aerodynamic resistance) is the turbulent flow of air around the vehicle body. This turbulence is a function of the shape of the vehicle, particularly the rear portion, which has been shown to be a major source of air turbulence. To a much lesser extent (on the order of 12% of total aerodynamic resistance), the friction of the air passing over the body of the vehicle contributes to resistance. Finally, approximately 3% of the total aerodynamic resistance can be attributed to air flow through vehicle components such as radiators and air vents.

Based on these sources, the equation for determining aerodynamic resistance is

$$R_a = \frac{\rho}{2} C_D A_f V^2 \qquad (2.3)$$

where

R_a = aerodynamic resistance in lb,

ρ = air density in slugs/ft^3,

C_D = coefficient of drag (unitless),

A_f = frontal area of the vehicle (projected area of the vehicle in the direction of travel) in ft^2, and

V = speed of the vehicle in ft/s.

To be truly accurate, for aerodynamic resistance computations, V is actually the speed of the vehicle relative to the prevailing wind speed. To simplify the exposition of concepts soon to be presented, the wind speed is assumed to be equal to zero for all problems and derivations in this book.

Air density is a function of both elevation and temperature, as indicated in Table 2.1. Equation 2.3 indicates that as the air becomes denser, total aerodynamic resistance increases. The drag coefficient (C_D) is a term that implicitly accounts for all three of the aerodynamic resistance sources discussed above. Drag coefficients are measured from empirical data either from wind tunnel experiments or actual field tests in which the vehicle is allowed to decelerate from a known speed with other sources of resistance (rolling and grade) taken into account. Table 2.2 provides a range of drag coefficients for different types of road vehicles, and Table 2.3 presents drag coefficients for specific automobiles starting with those in the mid-twentieth century.

Table 2.3 shows a wide range of drag coefficients over time. Vehicles with an emphasis on fuel efficiency tend to have lower drag coefficients. High-performance sports cars also seek low drag coefficients, but these cars can have bodies designed to generate downforce for performance braking and cornering, and this can adversely affect the drag coefficient. Table 2.3 also includes some larger personal vehicles, such as pickup trucks and sport utility vehicles, which generally represent the upper range of drag coefficients. Also, operating conditions can have a significant effect on drag coefficients. For example, even a small operational change, such as the opening of windows, can increase drag coefficients by 5% or more. More significant operational changes, such as having the top down on a convertible automobile, can increase the drag coefficient by more than 25%.

Finally, projected frontal area (approximated as the height of the vehicle multiplied by its width) typically ranges from 10 ft^2 to 30 ft^2 for passenger cars and frontal area is also a major factor in determining aerodynamic resistance.

Table 2.1 Typical Values of Air Density under Specified Atmospheric Conditions

Altitude (ft)	Temperature (°F)	Pressure (lb/in^2)	Air density (slugs/ft^3)
0	59.0	14.7	0.002378
5,000	41.2	12.2	0.002045
10,000	23.4	10.1	0.001755

Table 2.2 Ranges of Drag Coefficients for Typical Road Vehicles

Vehicle type	Drag coefficient (C_D)
Automobile	0.25–0.55
Bus	0.5–0.7
Tractor-Trailer	0.6–1.3
Motorcycle	0.27–1.8

Table 2.3 Drag Coefficients of Selected Automobiles

Vehicle	Drag coefficient (C_D)
1955 Mercedes Benz 300SL Gullwing	0.38
1967 Volkswagen Beetle	0.46
1977 Jaguar XJS	0.36
1987 Acura Integra	0.34
1987 Ford Taurus	0.32
1992 Mazda MX5 Miata	0.38
1993 Ford Ranger (truck)	0.45
1996 Dodge Viper RT/10	0.45
1997 Lexus LS400	0.29
2002 Acura NSX	0.30
2003 Dodge Caravan (minivan)	0.35
2003 Dodge Ram (truck)	0.53
2003 Hummer H2	0.57
2005 Chevrolet Corvette C5	0.29
2009 Nissan Cube	0.35
2012 Tesla Model S	0.24
2014 BMW I8	0.26
2016 Audi A5	0.31
2016 Dodge Charger	0.35
2016 Nissan Leaf	0.28
2016 Mercedes Benz GT AMG	0.36

Because aerodynamic resistance is proportional to the square of the vehicle's speed, it is clear that such resistance will increase rapidly at higher speeds. The magnitude of this increase can be underscored by considering an expression for the power (hp_{R_a}) required to overcome aerodynamic resistance. With power being the product of force and speed, the multiplication of Eq. 2.3 by speed gives

$$\text{hp}_{R_a} = \frac{\rho\, C_D A_f V^3}{1100} \tag{2.4}$$

where

hp_{R_a} = horsepower required to overcome aerodynamic resistance
(1 horsepower = 550 ft-lb/s),
other terms are as defined previously.

Thus, the amount of power required to overcome aerodynamic resistance increases with the cube of speed, indicating, for example, that eight times as much power is required to overcome aerodynamic resistance if the vehicle speed is doubled.

2.4   ROLLING RESISTANCE

Rolling resistance refers to the resistance generated from a vehicle's internal mechanical friction and from pneumatic tires and their interaction with the roadway surface. The primary source of this resistance is the deformation of the tire as it passes over the roadway surface. The force needed to overcome this deformation accounts for approximately 90% of the total rolling resistance. Depending on the vehicle's weight and the material composition of the roadway surface, the penetration of the tire into the surface and the corresponding surface compression can also be a significant source of rolling resistance. However, for typical vehicle weights and pavement types, penetration and compression constitute only around 4% of the total rolling resistance. Finally, frictional motion due to the slippage of the tire on the roadway surface and, to a lesser extent, air circulation around the tire and wheel (the fanning effect) are sources accounting for roughly 6% of the total rolling resistance [Taborek, 1957].

In considering the sources of rolling resistance, three factors are worthy of note. First, the rigidity of the tire and the roadway surface influence the degree of tire penetration, surface compression, and tire deformation. Hard, smooth, and dry roadway surfaces provide the lowest rolling resistance. Second, tire conditions, including inflation pressure and temperature, can have a substantial impact on rolling resistance. High tire inflation decreases rolling resistance on hard paved surfaces as a result of reduced friction but increases rolling resistance on soft unpaved surfaces due to additional surface penetration. Also, higher tire temperatures make the tire body more flexible, and thus less resistance is encountered during tire deformation. The third and final factor is the vehicle's operating speed, which affects tire deformation. Increasing speed results in additional tire flexing and vibration and thus a higher rolling resistance.

Due to the wide range of factors that determine rolling resistance, a simplifying approximation is used. Studies have shown that overall rolling resistance can be approximated as the product of a friction term (coefficient of rolling resistance) and the weight of the vehicle acting normal to the roadway surface. The coefficient of rolling resistance for road vehicles operating on paved surfaces is approximated as

$$f_{rl} = 0.01\left(1 + \frac{V}{147}\right) \tag{2.5}$$

where

f_{rl} = coefficient of rolling resistance (unitless), and
V = vehicle speed in ft/s.

By inspection of Fig. 2.1, the rolling resistance, in lb, will simply be the coefficient of rolling resistance multiplied by $W \cos \theta_g$, the vehicle weight acting normal to the roadway surface. For most highway applications, θ_g is quite small, so it can be assumed that $\cos \theta_g = 1$, giving the equation for rolling resistance (R_{rl}) as

$$R_{rl} = f_{rl} W \tag{2.6}$$

From this, the amount of power required to overcome rolling resistance is

$$hp_{R_{rl}} = \frac{f_{rl} W V}{550} \tag{2.7}$$

where

$hp_{R_{rl}}$ = horsepower required to overcome rolling resistance
(1 horsepower = 550 ft-lb/s),
W = total vehicle weight in lb.

EXAMPLE 2.1   AERODYNAMIC AND ROLLING RESISTANCE

A 2500-lb car is driven at sea level ($\rho = 0.002378$ slugs/ft^3) on a level paved surface. The car has $C_D = 0.38$ and 20 ft^2 of frontal area. It is known that at maximum speed, 50 hp is being expended to overcome rolling and aerodynamic resistance. Determine the car's maximum speed.

SOLUTION

It is known that at maximum speed (V_m),

$$\text{available horsepower} = R_a V_m + R_{rl} V_m$$

or

$$\text{available hp} = \frac{\frac{\rho}{2}C_D A_f V_m^3 + f_{rl} W V_m}{550}$$

Substituting, we have

$$50 = \frac{\frac{0.002378}{2}(0.38)(20)V_m^3 + 0.01\left(1 + \frac{V_m}{147}\right)(2500)V_m}{550}$$

or

$$27,500 = 0.00904 V_m^3 + 0.17 V_m^2 + 25 V_m$$

Solving for V_m gives

$$V_m = 133 \text{ ft/s or } 90 \text{ mi/h}$$

2.5 GRADE RESISTANCE

Grade resistance is simply the gravitational force (the component parallel to the roadway) acting on the vehicle. As suggested in Fig. 2.1, the expression for grade resistance (R_g) is

$$R_g = W \sin \theta_g \qquad (2.8)$$

As in the development of the rolling resistance formula (Eq. 2.6), highway grades are usually very small, so $\sin \theta_g \cong \tan \theta_g$. Rewriting Eq. 2.8, we get

$$R_g \cong W \tan \theta_g = WG \qquad (2.9)$$

where

G = grade, defined as the vertical rise per some specified horizontal distance (opposite side of the force triangle, Fig. 2.1, divided by the adjacent side) in ft/ft.

Grades are generally specified as percentages for ease of understanding. Thus a roadway that rises 5 ft vertically per 100 ft horizontally ($G = 0.05$ and $\theta_g = 2.86°$) is said to have a 5% grade.

EXAMPLE 2.2 **GRADE RESISTANCE**

A 2000-lb car has $C_D = 0.40$, $Af = 20$ ft^2, and an available tractive effort of 255 lb. If the car is traveling at an elevation of 5000 ft ($\rho = 0.002045$ slugs/ft^3) on a paved surface at a speed of 70 mi/h, what is the maximum grade that this car could ascend and still maintain the 70-mi/h speed?

SOLUTION

To maintain the speed, the available tractive effort will be exactly equal to the summation of resistances. Thus no tractive effort will remain for vehicle acceleration ($ma = 0$). Therefore, Eq. 2.2 can be written as

$$F = R_a + R_{rl} + R_g$$

For grade resistance (using Eq. 2.9),

$$R_g = WG = 2000G$$

for aerodynamic resistance (using Eq. 2.3),

$$R_a = \frac{\rho}{2} C_D A_f V^2$$
$$= \frac{0.002045}{2}(0.4)(20)(70 \times 5280 / 3600)^2$$
$$= 86.22 \text{ lb}$$

and for rolling resistance (using Eq. 2.6),

$$R_{rl} = f_{rl} W$$
$$= 0.01 \left(1 + \frac{70 \times 5280/3600}{147}\right) \times 2000$$
$$= 33.97 \text{ lb}$$

Therefore,

$$F = 255 = 86.22 + 33.97 + 2000G$$
$$G = \underline{0.0674} \quad \text{or a} \quad 6.74\% \text{ grade}$$

2.6   AVAILABLE TRACTIVE EFFORT

With the resistance terms in the basic equation of vehicle motion (Eq. 2.2) discussed, attention can now be directed toward available tractive effort (F) as used in Example 2.2. The tractive effort available to overcome resistance and/or to accelerate the vehicle is determined either by the force generated by the vehicle's engine or by some maximum value that will be a function of the vehicle's weight distribution and the characteristics of the roadway surface–tire interface. The basic concepts underlying these two determinants of available tractive effort are presented here.

2.6.1   Maximum Tractive Effort

No matter how much force a vehicle's engine makes available at the roadway surface, there is a point beyond which additional force merely results in the spinning of tires and does not overcome resistance or accelerate the vehicle. To explain what determines this point of maximum tractive effort (the limiting value beyond which tire spinning begins), a force and moment-generating diagram is provided in Fig. 2.2.

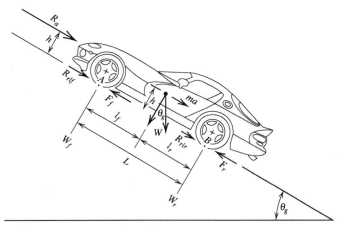

Figure 2.2 Vehicle forces and moment-generating distances.

R_a = aerodynamic resistance in lb,

R_{rlf} = rolling resistance of the front tires in lb,

R_{rlr} = rolling resistance of the rear tires in lb,

F_f = available tractive effort of the front tires in lb,

F_r = available tractive effort of the rear tires in lb,

W = total vehicle weight in lb,

W_f = weight of the vehicle on the front axle in lb,

W_r = weight of the vehicle on the rear axle in lb,

θ_g = angle of the grade in degrees,

m = vehicle mass in slugs,

a = acceleration in ft/s^2,

L = length of wheelbase,

h = height of the center of gravity above the roadway surface,

l_f = distance from the front axle to the center of gravity, and

l_r = distance from the rear axle to the center of gravity.

To determine the maximum tractive effort that the roadway surface-tire contact can support, it is necessary to examine the normal loads on the axles. The normal load on the rear axle (W_r) is given by summing the moments about point A (see Fig. 2.2):

$$W_r = \frac{R_a h + Wl_f \cos\theta_g + mah \pm Wh \sin\theta_g}{L} \qquad (2.10)$$

In this equation, the grade moment ($Wh \sin\theta_g$) is positive for an upward slope and negative for a downward slope. Rearranging terms (assuming $\cos\theta_g = 1$ for the small grades encountered in highway applications) and substituting into Eq. 2.2 gives

$$W_r = \frac{l_f}{L}W + \frac{h}{L}\left(F - R_{rl}\right) \qquad (2.11)$$

From basic physics, the maximum tractive effort as determined by the roadway surface–tire interaction will be the normal force multiplied by the coefficient of road adhesion (μ), so for a rear-wheel–drive car

$$F_{max} = \mu W_r \qquad (2.12)$$

and substituting Eq. 2.11 into Eq. 2.12,

$$F_{max} = \mu \left[\frac{l_f}{L} W + \frac{h}{L} (F_{max} - R_{rl}) \right] \tag{2.13}$$

$$F_{max} = \frac{\mu W \left(l_f - f_{rl} h \right) / L}{1 - \mu h / L} \tag{2.14}$$

Similarly, by summing moments about point B (see Fig. 2.2), it can be shown that for a front-wheel–drive vehicle

$$F_{max} = \frac{\mu W \left(l_r + f_{rl} h \right) / L}{1 + \mu h / L} \tag{2.15}$$

Note that in Eqs. 2.14 and 2.15, because of canceling of units, h, l_f, l_r, and L can be in any unit of length (feet, inches, etc.). However, all of these terms must be in the same chosen unit of measure. The units of F_{max} will be the same as the units for W (lb).

Typical values of the coefficient of road adhesion (μ) are provided in Table 2.4.

Table 2.4 Typical Coefficients of Road Adhesion (μ)

Pavement	Coefficient of road adhesion
Good, dry	1.00*
Good, wet	0.90
Poor, dry	0.80
Poor, wet	0.60
Packed snow or ice	0.25

*In some instances, the coefficient of road adhesion values can exceed 1.0. See discussion near the end of Section 2.6.1.

Source: S. G. Shadle, L. H. Emery, and H. K. Brewer, "Vehicle Braking, Stability, and Control," *SAE Transactions*, vol. 92, paper 830562, 1983.

EXAMPLE 2.3   **MAXIMUM TRACTIVE EFFORT**

A 2500-lb car is designed with a 120-inch wheelbase. The center of gravity is located 22 inches above the pavement and 40 inches behind the front axle. If the car is on a road with poor, wet pavement, what is the maximum tractive effort that can be developed if the car is (a) front-wheel drive and (b) rear-wheel drive?

SOLUTION

For the front-wheel–drive case Eq. 2.15 is used:

$$F_{max} = \frac{\mu W \left(l_r + f_{rl} h \right) / L}{1 + \mu h / L}$$

With poor, wet pavement, the coefficient of road adhesion $\mu = 0.60$ from Table 2.4, and from Eq. 2.5, $f_{rl} = 0.01$ because $V = 0$ ft/s, so

$$F_{max} = \frac{\left[0.60 \times 2500 \times \left(80 + 0.01(22)\right)\right]/120}{1 + \left(0.60 \times 22\right)/120}$$

$$= \underline{\underline{903.38 \text{ lb}}}$$

For the rear-wheel–drive case, Eq. 2.14 is used:

$$F_{max} = \frac{\left[0.60 \times 2500 \times \left(40 - 0.01(22)\right)\right]/120}{1 - \left(0.60 \times 22\right)/120}$$

$$= \underline{\underline{558.71 \text{ lb}}}$$

2.6.2   Engine-Generated Tractive Effort

The amount of tractive effort generated by the vehicle's engine is a function of a variety of engine and drivetrain design factors. For engine design, critical factors in determining output include the shape of the combustion chamber, the quantity of air drawn into the combustion chamber during the induction phase, the type of fuel used, fuel intake design, and so on. Although a complete description of engine design is beyond the scope of this book, an understanding of how engine output is measured and used is important in the study of vehicle performance. The two most commonly used measures of engine output are torque and power. Torque is the work generated by the engine (the twisting moment) and is expressed in foot-pounds (ft-lb). Power is the rate of engine work, expressed in horsepower (hp), and is related to the engine's torque by the following equation:

$$\text{hp}_e = \frac{2\pi M_e n_e}{550} \tag{2.16}$$

where

hp_e = engine-generated horsepower (1 horsepower equals 550 ft-lb/s),
M_e = engine torque in ft-lb, and
n_e = engine speed in crankshaft revolutions per second.

Figure 2.3 presents a torque-power diagram for a typical gasoline-powered engine.

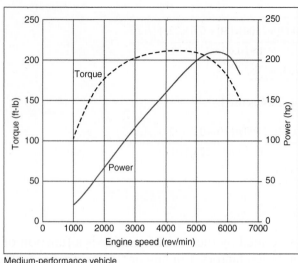
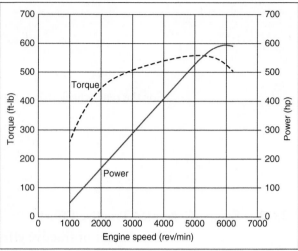

Medium-performance vehicle   High-performance vehicle

Figure 2.3 Example torque-power curves for a gasoline-powered automobile engine.

EXAMPLE 2.4   ENGINE TORQUE AND POWER

It is known that an experimental engine has a torque curve of the form $M_e = an_e - bn_e^2$ where M_e is engine torque in ft-lb, n_e is engine speed in revolutions per second, and a and b are unknown parameters. If the engine develops a maximum torque of 92 ft-lb at 3200 rev/min (revolutions per minute), what is the engine's maximum power?

SOLUTION

At maximum torque, n_e = 53.33 rev/s (3200/60) and

$$\frac{dM_e}{dn_e} = 0 = a - 2bn_e$$

$$a = 2(53.33)b = 106.67b$$

Also, at maximum torque,

$$M_e = an_e - bn_e^2$$

$$92 = a(53.33) - b(53.33)^2$$

Using these two equations to solve for the two unknowns (a and b), we find that b = 0.032 and a = 3.450. Using Eq. 2.16 and $M_e = an_e - bn_e^2$,

$$hp_e = \frac{2\pi(an_e - bn_e^2)n_e}{550}$$

$$= \frac{2\pi(3.450n_e - 0.032n_e^2)\,n_e}{550}$$

The first derivative of the power equation is used to solve for the engine speed at maximum power:

$$\frac{d\text{hp}_e}{dn_e} = 0 = (0.01142)\left(6.90n_e - 0.096n_e^2\right)$$

$$n_e = 71.88 \text{ rev/s}$$

so the engine's maximum power is

$$\text{hp}_e = \frac{2\pi(3.450n_e - 0.032n_e^2)n_e}{550}$$

$$= \frac{2\pi\left(3.450(71.88) - 0.032(71.88)^2\right)71.88}{550}$$

$$= \underline{\underline{67.87 \text{ hp}}}$$

Given the output measures of output of a vehicle's engine, focus can be directed toward the relationship between engine-generated torque and the tractive effort ultimately delivered to the driving wheels. The tractive effort needed for acceptable vehicle performance (to provide adequate acceleration characteristics) is greater at lower vehicle speeds, and because maximum engine torque is developed at fairly high engine speeds (crankshaft revolutions) for common engine types, such as gasoline- and diesel-powered engines, some form of gear reduction is necessary, as illustrated in Fig. 2.4 (in contrast, electric vehicles provide the same "maximum" torque at all engine speeds, so a gear reduction is not necessary). For gasoline and diesel engines, this gear reduction provides the mechanical advantage necessary for acceptable vehicle acceleration.

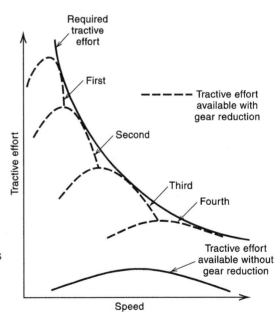

Figure 2.4 Tractive effort requirements and tractive effort generated by a typical gasoline-powered vehicle.

With gear reductions there are two factors that determine the amount of tractive effort reaching the drive wheels. First, the mechanical efficiency of the drivetrain (the engine and gear reduction devices, including the transmission and differential) must be considered. Typically 5% to 25% of the tractive effort generated by the engine is lost in gear reduction devices, which corresponds to a mechanical efficiency of the drivetrain (η_d) of 0.75 to 0.95. Second, the overall gear reduction ratio (ε_0), which includes the gear reductions of the transmission and differential, plays a key role in the determination of tractive effort.

By definition, the overall gear reduction ratio refers to the relationship between the revolutions of the engine's crankshaft and the revolutions of the drive wheels. For example, an overall gear reduction ratio of 4 to 1 ($\varepsilon_0 = 4$) means that the engine's crankshaft turns four revolutions for every one revolution of the drive wheels.

With these terms defined, the engine-generated tractive effort reaching the drive wheels is given as

$$F_e = \frac{M_e \varepsilon_0 \eta_d}{r}$$ (2.17)

where

F_e = engine-generated tractive effort reaching the drive wheels in lb,
M_e = engine torque in ft-lb,
ε_0 = overall gear reduction ratio,
η_d = mechanical efficiency of the drivetrain, and
r = radius of the drive wheels in ft.

It follows that the relationship between vehicle speed and engine speed is

$$V = \frac{2\pi r n_e (1 - i)}{\varepsilon_0}$$ (2.18)

where

V = vehicle speed in ft/s,
n_e = engine speed in crankshaft revolutions per second,
i = slippage of the drive axle, generally taken as 2 to 5% ($i = 0.02$ to 0.05) for passenger vehicles, and
other terms are as defined previously.

To summarize this section, the available tractive effort (F in Eq. 2.2) at any given speed is the lesser of the maximum tractive effort (F_{max}) and the engine-generated tractive effort (F_e).

2.7   VEHICLE ACCELERATION

As defined in the previous section, available tractive effort (F) can be used to determine a number of vehicle performance characteristics including vehicle acceleration and top speed. For determining vehicle acceleration, Eq. 2.2 can be applied with an additional term to account for the inertia of the vehicle's rotating parts that must be overcome during acceleration. This term is referred to as the mass factor (γ_m) and is introduced in Eq. 2.2 as

$$F - \sum R = \gamma_m m a \tag{2.19}$$

where the mass factor is approximated as

$$\gamma_m = 1.04 + 0.0025 \varepsilon_0^2 \tag{2.20}$$

Two measures of vehicle acceleration are worthy of note: the time to accelerate and the distance to accelerate. For both, the force available to accelerate is $F_{net} = F - \Sigma R$. The basic relationship between the force available to accelerate, F_{net}, the available tractive effort, F (which again is the lesser of F_{max} and F_e), and the summation of resistances is illustrated in Fig. 2.5. In this figure, F_{net} is the vertical distance between the lesser of the F_{max} and the F_e curves and the total resistance curve. So, referring to Fig. 2.5, at speed V', F_{net} will be $F_{max} - \Sigma R$, and at V'', F_{net} will be $F_e - \Sigma R$. It follows that when $F_{net} = 0$, the vehicle cannot accelerate and is at its maximum speed for specified conditions (grade, air density, engine torque, and so on). Such was the case for the vehicle described in Example 2.2. When F_{net} is greater than zero (the vehicle is traveling at a speed less than its maximum speed), Eq. 2.19 can be written in differential form as

$$F_{net} = \gamma_m m \frac{dV}{dt} \quad \text{or} \quad dt = \frac{\gamma_m m dV}{F_{net}}$$

and because F_{net} is itself a function of vehicle speed [$F_{net} = f(V)$], integration gives the time to accelerate as

$$t = \gamma_m m \int_{V_1}^{V_2} \frac{dV}{f(V)} \tag{2.21}$$

where V_1 is the initial vehicle speed and V_2 is the final vehicle speed.

Similarly, it can be shown that the distance to accelerate is

$$d_a = \gamma_m m \int_{V_1}^{V_2} \frac{V dV}{f(V)} \tag{2.22}$$

To solve Eqs. 2.21 and 2.22, numerical integration is necessary because the functional forms of these equations do not lend themselves to closed-form solutions. Such numerical integration is straightforward but requires a computer. Consequently, we do not provide an example of solving these equations.

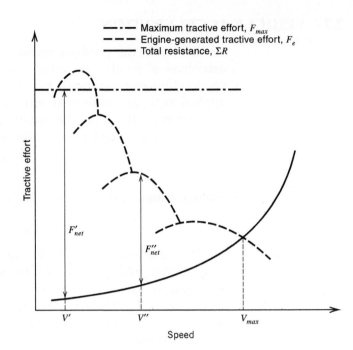

Figure 2.5 Relationship among the forces available to accelerate, available tractive effort, and total vehicle resistance.

EXAMPLE 2.5 VEHICLE ACCELERATION

A car is traveling at 10 mi/h on a roadway covered with hard-packed snow. The car has $C_D = 0.30$, $A_f = 20$ ft², and $W = 3000$ lb. The wheelbase is 120 inches, and the center of gravity is 20 inches above the roadway surface and 50 inches behind the front axle. The air density is 0.002045 slugs/ft³. The car's engine is producing 95 ft-lb of torque and is in a gear that gives an overall gear reduction ratio of 4.5 to 1, the wheel radius is 14 inches, and the mechanical efficiency of the drivetrain is 80%. If the driver needs to accelerate quickly to avoid an accident, what would the acceleration be if the car is (a) front-wheel drive and (b) rear-wheel drive?

SOLUTION

We begin by computing the resistances, tractive effort generated by the engine, and mass factor because all of these factors will be the same for both front- and rear-wheel drive.

The aerodynamic resistance is (from Eq. 2.3)

$$R_a = \frac{\rho}{2} C_D A_f V^2$$

$$= \frac{0.002045}{2} (0.3)(20)(10 \times 5280 / 3600)^2$$

$$= 1.32 \text{ lb}$$

The rolling resistance is (from Eq. 2.6)

$$R_{rl} = f_{rl} W$$

$$= 0.01 \left(1 + \frac{10 \times 5280/3600}{147} \right) \times 3000$$

$$= 32.99 \text{ lb}$$

The engine-generated tractive effort is (from Eq. 2.17)

$$F_e = \frac{M_e \varepsilon_0 \eta_d}{r}$$

$$= \frac{95(4.5)(0.8)}{14/12}$$

$$= 293.14 \text{ lb}$$

The mass factor is (from Eq. 2.20)

$$\gamma_m = 1.04 + 0.0025\varepsilon_0^2$$

$$= 1.04 + 0.0025(4.5)^2$$

$$= 1.091$$

Recall that, to determine acceleration, we need the resistances (already computed) and the available tractive effort, F, which is the lesser of F_e or F_{max}. For the case of the front-wheel–drive car, Eq. 2.15 can be applied to determine F_{max} (with the coefficient of road adhesion $\mu = 0.25$ for hard-packed snow from Table 2.4):

$$F_{\max} = \frac{\mu W(l_r + f_{rl}h)/L}{1 + \mu h/L}$$

$$= \frac{\left[0.25 \times 3000 \times (70 + 0.011(20))\right]/120}{1 + (0.25 \times 20)/120}$$

$$= \underline{\underline{421.32 \text{ lb}}}$$

Thus for a front-wheel–drive car $F = 293.14$ lb (the lesser of $F_e = 293.14$ and $F_{max} = 421.32$) and the acceleration is (from Eq. 2.19)

$$F - \sum R = \gamma_m ma$$

$$a = \frac{F - \sum R}{\gamma_m m} = \frac{293.14 - 34.31}{1.091(3000/32.2)} = \underline{\underline{2.546 \text{ ft/s}^2}}$$

For the case of the rear-wheel–drive car, Eq. 2.14 can be applied to determine F_{max} :

$$F_{max} = \frac{\left[0.2 \times 3000 \times (50 - 0.011(20))\right]/120}{1 - (0.2 \times 20)/120} = 257.48 \text{ lb}$$

Thus for a rear-wheel–drive car $F = 257.48$ lb (the lesser of 293.14 and 257.48) and the acceleration is (from Eq. 2.19)

$$a = \frac{F - \sum R}{\gamma_m m} = \frac{257.48 - 34.31}{1.091(3000/32.2)} = \underline{\underline{2.196 \text{ ft/s}^2}}$$

EXAMPLE 2.6   ENGINE-TORQUE AND VEHICLE ACCELERATION

A front-wheel–drive car is in gear with a gear-reduction ratio 8 to 1 and is traveling at 25 mi/h. The engine torque of the car is given by the equation $M_e = 10n_e - 0.06n_e^2$. The car has a frontal area of 20 ft², C_D of 0.30 and is traveling at sea level (59 degrees F) on a level road. The wheelbase is 120 inches and the center of gravity is 40 inches behind the front axle and 30 inches above the road surface and the car weighs 2800 pounds. If the car is on a road that is wet with poor pavement, what is the maximum acceleration from 25 mi/h (driveline efficiency is 90%, slippage of the drive axle is 2%, wheel radius is 15 inches)?

SOLUTION

To begin, the engine speed must be computed so that the torque equation provided in the problem statement can be applied to determine the engine-generated tractive effort. This is done by applying Eq. 2.18 with $i = 0.02$, $\varepsilon_0 = 8$, $r = 15/12$ ft, and $V = 20$ mi/h:

$$V = \frac{2\pi r n_e (1-i)}{\varepsilon_0} = 25 \times 5280/3600 = \frac{2(3.141)(15/12)n_e(1-0.02)}{8}$$

$$n_e = \frac{(25 \times 5280/3600)8}{(2)(3.141)(15/12)(1-0.02)} = 38.22 \text{ revolutions per second}$$

Substituting this engine speed into the torque equation provided in the problem statement gives

$$M_e = 10n_e - 0.06n_e^2 = 10(38.22) - 0.06(38.22)^2 = 294.55 \text{ ft-lb}$$

The engine-generated tractive effort is (from Eq. 2.17) with $\eta_d = 0.9$

$$F_e = \frac{M_e \varepsilon_0 \eta_d}{r} = \frac{294.55(8)(0.9)}{15/12} = 1696.63 \text{ lb}$$

As in Example 2.5, the available tractive effort for maximum acceleration, F, is the lesser of F_e or F_{max}. F_{max} for a front-wheel–drive car is computed using Eq. 2.15. With given values of $l_r = 80$ inches, $h = 30$ inches, $L = 120$ inches, $W = 2800$ lb, $\mu = 0.6$ (from Table 2.4 with poor, wet pavement), and f_{rl} from Eq. 2.5, we find

$$f_{rl} = 0.01\left(1+\frac{V}{147}\right) = 0.01\left(1+\frac{25 \times 5280/3600}{147}\right) = 0.0125$$

Application of Eq. 2.15 yields (with the coefficient of road adhesion $\mu = 0.60$ for poor, wet pavement from Table 2.4)

$$F_{max} = \frac{\mu W\left(l_r + f_{rl}h\right)/L}{1+\mu h/L}$$

$$= \frac{\left[0.60 \times 2800 \times \left(80 + 0.0125(30)\right)\right]/120}{1+(0.60 \times 30)/120}$$

$$= 978.48 \text{ lb}$$

Thus $F = 978.48$ lb (the lesser of 978.48 and 1696.63). With this, acceleration is determined by applying Eq. 2.19. For input into Eq. 2.19, the rolling resistance (from Eq. 2.6) is

$$R_{rl} = f_{rl}W$$
$$= 0.0125 \times 2800$$
$$= 35 \text{ lb}$$

The aerodynamic resistance with $\rho = 0.002378$ slugs/ft³ (from Table 2.1), $C_D = 0.30$, $A_f = 20$ ft² and $V = 25$ mi/h (from Eq. 2.3) is

$$R_a = \frac{\rho}{2}C_D A_f V^2$$
$$= \frac{0.002378}{2}(0.3)(20)(25 \times 5280 / 3600)^2$$
$$= 9.63 \text{ lb}$$

The mass factor (from Eq. 2.20) is

$$\gamma_m = 1.04 + 0.0025\varepsilon_0^2$$
$$= 1.04 + 0.0025(8)^2$$
$$= 1.2$$

Thus, application of Eq. 2.19 gives the maximum acceleration as

$$F - \sum R = \gamma_m ma$$

$$a = \frac{F - \sum R}{\gamma_m m} = \frac{978.48 - 35 - 9.63}{1.2(2800/32.2)} = \underline{\underline{8.95 \text{ ft/s}^2}}$$

The values given for the coefficient of road adhesion (μ) in the previous examples are drawn from Table 2.4. However, in determining acceleration (as well as braking and cornering, as will be shown later in this book) two points are worthy of note. First, it is possible for the coefficient of road adhesion to exceed 1.0. This is because a micro-interaction at the tire–pavement interface results in a "cog-type" effect that, for some high-performance tires that use softer compounds, can increase μ to above 1.0. This explains why many race cars, particularly drag-racing cars, have initial acceleration rates well in excess of 1 g (32.2 ft/s²). Second, at high speed, vehicle aerodynamics can create downward forces that effectively increase W in the preceding equations, and this facilitates greater acceleration. Drag-racing cars and some open-wheel race cars (such as Formula One–style cars) are examples of aerodynamic designs that use air deflectors designed to generate significant downward forces to enhance acceleration, braking, and cornering.

2.8 FUEL EFFICIENCY

Given the factors discussed in the preceding sections of this chapter, the elements that determine a vehicle's fuel efficiency are clear. One of the most critical determinants relates to engine design (how the engine-generated tractive effort is produced). Engine designs that increase the quantity of air entering the combustion chamber, improve fuel delivery to the combustion chamber, and decrease internal engine friction lead to improved fuel efficiency. Improvements in other mechanical components, such as decreasing slippage and improving the mechanical efficiency of the transmission and driveshaft, also increase the overall fuel efficiency.

In terms of resistance-reducing options, decreasing overall vehicle weight (W) will lower grade and rolling resistances, thus reducing fuel consumption (all other factors held constant). Similarly, aerodynamic improvements such as a lower drag coefficient (C_D) and a reduced frontal area (A_f) can produce significant fuel savings. Finally, improved tire designs with lower rolling resistance can improve overall fuel efficiency. However, there can sometimes be a trade-off between improved efficiency and safety. For example, low rolling resistance tires may not be able to the same high coefficient of road adhesion values that high-performance race tires can. Possible safety-efficiency trade-offs must be given careful consideration.

2.9   PRINCIPLES OF BRAKING

In highway design and traffic analysis, the braking characteristics of road vehicles are arguably the single most important aspect of vehicle performance. The braking behavior of road vehicles is critical in the determination of stopping sight distance, roadway surface design, and accident avoidance and mitigation systems. Moreover, ongoing advances in braking technology make it essential that transportation engineers have a basic comprehension of the underlying principles involved.

2.9.1   Braking Forces

To begin the discussion of braking principles, consider the force and moment-generating diagram in Fig. 2.6. During vehicle braking there is a load transfer from the rear to the front axle. To illustrate this, expressions for the normal loads on the front and rear axles can be written by summing the moments about roadway surface–tire contact points A and B (as was done in deriving Eqs. 2.14 and 2.15, with cos θ_g assumed equal to 1 because of the small grades encountered in highway applications):

$$W_f = \frac{1}{L}\left[Wl_r + h\left(ma - R_a \pm W\sin\theta_g\right)\right] \qquad (2.23)$$

and

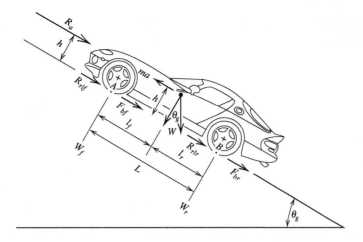

Figure 2.6 Forces acting on a vehicle during braking, with drivetrain resistance ignored.

R_a = aerodynamic resistance in lb,

R_{rlf} = rolling resistance of the front tires in lb,

R_{rlr} = rolling resistance of the rear tires in lb,

F_{bf} = braking force on the front tires in lb,

F_{br} = braking force on the rear tires in lb,

W = total vehicle weight in lb,

W_f = weight of the vehicle on the front axle in lb,

W_r = weight of the vehicle on the rear axle in lb,

θ_g = angle of the grade in degrees,

m = vehicle mass in slugs,

a = acceleration in ft/s^2,

L = length of wheelbase,

h = height of the center of gravity above the roadway surface,

l_f = distance from the front axle to the center of gravity, and

l_r = distance from the rear axle to the center of gravity.

$$W_r = \frac{1}{L}\left[Wl_f - h(ma - R_a \pm W\sin\theta_g) \right] \tag{2.24}$$

where, in this case, the contribution of grade resistance ($W\sin\theta_g$) is negative for uphill grades and positive for downhill grades.

From the summation of forces along the vehicle's longitudinal axis,

$$F_b + f_{rl}W = ma - R_a \pm W\sin\theta_g \tag{2.25}$$

with the rolling resistance equal to the coefficient of rolling resistance multiplied by the vehicle weight (from Eq. 2.6, $R_{rl} = f_{rl}W$) and $F_b = F_{bf} + F_{br}$. Substituting Eq. 2.25 into Eqs. 2.23 and 2.24 gives

$$W_f = \frac{1}{L}\left[Wl_r + h(F_b + f_{rl}W) \right] \tag{2.26}$$

and

$$W_r = \frac{1}{L}\left[Wl_f - h(F_b + f_{rl}W) \right] \tag{2.27}$$

Because the maximum vehicle braking force ($F_{b\ max}$) is equal to the coefficient of road adhesion (μ), multiplied by the vehicle weights normal to the roadway surface,

$$F_{bf\ max} = \mu W_f$$
$$= \frac{\mu W}{L} \left[l_r + h(\mu + f_{rl}) \right] \tag{2.28}$$

and

$$F_{br\ max} = \mu W_r$$
$$= \frac{\mu W}{L} \left[l_f - h(\mu + f_{rl}) \right] \tag{2.29}$$

To develop maximum braking forces, the tires should be at the point of an impending slide. If the tires begin to slide (the brakes lock), a significant reduction in road adhesion results. An indication of the extent of the reduction in the coefficient of road adhesion as the result of tire slide, under various pavement and weather conditions, is presented in Table 2.5. It is clear from this table that the braking forces decline dramatically when the wheels are locked (resulting in tire slide). Avoiding this locked condition is the function of antilock braking systems in cars. Such systems are discussed later in this chapter.

2.9.2   Braking Force Ratio and Efficiency

On a given roadway surface, the maximum attainable vehicle deceleration (using the vehicle's braking system) is equal to μg, where μ is the coefficient of road adhesion and g is the gravitational constant (32.2 ft/s^2). To achieve this maximum vehicle deceleration, vehicle braking systems must correctly distribute braking forces between the vehicle's front and rear brakes. This is typically done by allocation of hydraulic pressures within the braking system. This front-rear proportioning of braking forces (within the vehicle's braking system) will be optimal (achieving a deceleration rate equal to μg) when it is in exactly the same proportion as the ratio of the maximum braking forces on the front and rear axles ($F_{bf\ max}$ / $F_{br\ max}$). Thus maximum braking forces (with the tires at the point of impending slide) will be developed when the brake force ratio (front force over rear force) is

Table 2.5 Typical Values of Coefficients of Road Adhesion (Maximum and Slide)

Pavement	Coefficient of road adhesion	
	Maximum	Slide
Good, dry	1.00*	0.80
Good, wet	0.90	0.60
Poor, dry	0.80	0.55
Poor, wet	0.60	0.30
Packed snow or ice	0.25	0.10

*In some instances, the coefficient of road adhesion values can exceed 1.0. See discussion near the end of Section 2.6.1.

Source: S. G. Shadle, L. H. Emery, and H. K. Brewer, "Vehicle Braking, Stability, and Control," *SAE Transactions,* vol. 92, paper 830562, 1983.

$$BFR_{f/r\,max} = \frac{l_r + h(\mu + f_{rl})}{l_f - h(\mu + f_{rl})} \qquad (2.30)$$

where

$BFR_{f/r\,max}$ = the brake force ratio, allocated by the vehicle's braking system, that results in maximum (optimal) braking forces, and

other terms are as defined previously.

It follows that the percentage of braking force that the braking system should allocate to the front axle (PBF_f) for maximum braking is

$$PBF_f = 100 - \frac{100}{1 + BFR_{f/r\,max}} \qquad (2.31)$$

and the percentage of braking force that the braking system should allocate to the rear axle (PBF_r) for maximum braking is

$$PBF_r = \frac{100}{1 + BFR_{f/r\,max}} \qquad (2.32)$$

EXAMPLE 2.7 BRAKE-FORCE PROPORTIONING

A car has a wheelbase of 100 inches and a center of gravity that is 40 inches behind the front axle at a height of 24 inches. If the car is traveling at 80 mi/h on a road with poor pavement that is wet, determine the percentages of braking force that should be allocated to the front and rear brakes (by the vehicle's braking system) to ensure that maximum braking forces are developed.

SOLUTION

The coefficient of rolling resistance is

$$f_{rl} = 0.01\left(1 + \frac{80 \times 5280 / 3600}{147}\right) = 0.018$$

and $\mu = 0.6$ from Table 2.5 (maximum because we want the tires to be at the point of impending slide). Applying Eq. 2.30 gives

$$BFR_{f/r\,max} = \frac{l_r + h(\mu + f_{rl})}{l_f - h(\mu + f_{rl})}$$
$$= \frac{60 + 24(0.6 + 0.018)}{40 - 24(0.6 + 0.018)}$$
$$= 2.973$$

Using Eq. 2.31, the percentage of the force allocated to the front brakes should be

$$PBF_f = 100 - \frac{100}{1 + BFR_{f/r\,max}}$$

$$= 100 - \frac{100}{1 + 2.973}$$

$$= \underline{\underline{74.83\,\%}}$$

and using Eq. 2.32 (or simply $100 - PBF_f$), the percentage of the force allocated to the rear brakes should be

$$PBF_r = \frac{100}{1 + BFR_{f/r\,max}}$$

$$= \frac{100}{1 + 2.973}$$

$$= \underline{\underline{25.17\,\%}}$$

It is clear from Eq. 2.30 that the design of a vehicle's braking system is not an easy task because the optimal brake force proportioning changes with both vehicle and road conditions. For example, the addition of vehicle cargo and/or passengers will change not only the weight of the vehicle (which affects f_{rl} in Eq. 2.30) but also the distribution of the weight, shifting the height of the center of gravity and its location along the vehicle's longitudinal axis, and this will change the optimal brake force proportioning ($BFR_{fr\,max}$). This is particularly problematic for trucks because of the large weight and center of gravity differences between loaded and unloaded conditions. Similarly, changes in road conditions produce different coefficients of adhesion, again changing optimal brake force proportioning. As a result of the uncertainties in weight and road conditions, vehicle designers often choose a compromise value of brake force proportioning that, on average, provides good braking but is rarely, if ever, optimal.

It is important to note that studies have indicated that if wheel lockup is to occur, it is preferable to have the front wheels lock first because having the rear wheels lock first can result in uncontrollable vehicle spin. Front-wheel lockup results in loss of steering control, but the vehicle will at least continue to brake in a straight line. Technological advancements in braking systems since the late 1970s have resulted in vehicles that are increasingly capable of proportioning brake forces in a manner that is closer to optimal and avoids the dangerous rear-wheel–first lockup due to front-wheel under-braking.

Because true optimal brake force proportioning is seldom achieved in standard non-antilock braking systems, it is useful to define a braking-efficiency term that reflects the degree to which the braking system is operating below optimal. Simply stated, braking efficiency is defined as the ratio of the maximum rate of deceleration, expressed in g's (g_{max}), achievable prior to any wheel lockup to the coefficient of road adhesion:

$$\eta_b = \frac{g_{max}}{\mu} \tag{2.33}$$

where

η_b = braking efficiency,

g_{max} = maximum deceleration in g units (with the absolute maximum = μ), and

μ = coefficient of road adhesion.

2.9.3   Antilock Braking Systems

Many modern cars have braking systems designed to prevent the wheels from locking during braking applications (antilock braking systems). In theory, antilock braking systems serve two purposes. First, they prevent the coefficient of road adhesion from dropping to slide values (see Table 2.5). Second, they have the potential to raise the braking efficiency to 100%. In practice, designing an antilock braking system that avoids slide coefficients of adhesion and achieves 100% braking efficiency (η_b = 1.0) is a difficult task. This is because most antilock braking system technologies detect which wheels have locked and release them momentarily before reapplying the brake on the locking wheel. The wheel lock detection speed, speed of brake force reallocation, and braking system design (the amount of braking forces that can be accommodated by the vehicle's front and rear brake discs and calipers) all impact the overall effectiveness of the antilock braking system. Early antilock braking systems often fell short of achieving 100% braking efficiency, and in many cases, an expert driver operating a non-antilock braking car could modulate the brakes to achieve shorter stopping distances than cars equipped with antilock brakes. However, advances in antilock braking system technology continue to bring us closer to 100% braking efficiency.

2.9.4   Theoretical Stopping Distance

With a basic understanding of brake force proportioning and the resulting braking efficiency, attention can now be directed toward developing expressions for minimum stopping distances. By inspection of Fig. 2.6, it can be seen that the relationship among stopping distance, braking force, vehicle mass, and vehicle speed is

$$a \, ds = \left[\frac{F_b + \sum R}{\gamma_b m} \right] ds \tag{2.34}$$

$$= V \, dV$$

where

γ_b = mass factor accounting for moments of inertia during braking, which is given the value of 1.04 for automobiles [Wong, 2008], and

other terms are as defined in Fig. 2.6.

Integrating to determine stopping distance (S) gives

$$S = \int_{V_2}^{V_1} \gamma_b m \frac{V \, dV}{F_b + \sum R} \tag{2.35}$$

Substituting in the resistances (see Fig. 2.6), we obtain

$$S = \gamma_b m \int_{V_2}^{V_1} \frac{V \, dV}{F_b + R_a + f_{rl}W \pm W \sin\theta_g} \tag{2.36}$$

where

V_1 = initial vehicle speed in ft/s,

V_2 = final vehicle speed in ft/s,

$f_{rl}W$ = rolling resistance,

$W \sin\theta_g$ = grade resistance (positive for uphill slopes and negative for downhill slopes), and

other terms are as defined previously.

To simplify notation, let

$$K_a = \frac{\rho}{2} C_D A_f \tag{2.37}$$

so that Eq. 2.3 is

$$R_a = K_a V^2 \tag{2.38}$$

Continuing, assume that the effect of speed on the coefficient of rolling resistance, f_{rl}, is constant and can be approximated by using the average of initial (V_1) and final (V_2) speeds in Eq. 2.5 [$V = (V_1 + V_2) / 2$]. With this assumption (which introduces only a very small amount of error), and letting $m = W/g$ and $F_b = \mu W$, integration of Eq. 2.36 gives

$$S = \frac{\gamma_b W}{2gK_a} \ln\left[\frac{\mu W + K_a V_1^2 + f_{rl}W \pm W \sin\theta_g}{\mu W + K_a V_2^2 + f_{rl}W \pm W \sin\theta_g}\right] \tag{2.39}$$

If the vehicle is assumed to stop ($V_2 = 0$),

$$S = \frac{\gamma_b W}{2gK_a} \ln\left[1 + \frac{K_a V_1^2}{\mu W + f_{rl}W \pm W \sin\theta_g}\right] \tag{2.40}$$

With braking efficiency considered, the actual braking force is

$$F_b = \eta_b \mu W \tag{2.41}$$

Therefore, by substitution into Eq. 2.40, the theoretical stopping distance is

$$S = \frac{\gamma_b W}{2 g K_a} \ln\left[1 + \frac{K_a V_1^2}{\eta_b \mu W + f_{rl} W \pm W \sin\theta_g}\right] \tag{2.42}$$

Similarly, Eq. 2.39 can be written to include braking efficiency. Finally, if aerodynamic resistance is ignored (due to its comparatively small contribution to braking), integration of Eq. 2.35 gives the theoretical stopping distance as

$$S = \frac{\gamma_b \left(V_1^2 - V_2^2\right)}{2 g \left(\eta_b \mu + f_{rl} \pm \sin\theta_g\right)} \tag{2.43}$$

EXAMPLE 2.8   **THEORETICAL MINIMUM STOPPING DISTANCE**

A new experimental 2500-lb car, with $C_D = 0.25$ and $A_f = 18$ ft², is traveling at 90 mi/h down a 10% grade. The coefficient of road adhesion is 0.7 and the air density is 0.0024 slugs/ft³. The car has an advanced antilock braking system that gives it a braking efficiency of 100%. Determine the theoretical minimum stopping distance for the case where aerodynamic resistance is considered and the case where aerodynamic resistance is ignored.

SOLUTION

With aerodynamic resistance considered, Eq. 2.42 can be applied with $\gamma_b = 1.04$, $\theta_g = 5.71°$, and

$$f_{rl} = 0.01\left(1 + \frac{\left(\dfrac{90 \times 5280 / 3600 + 0}{2}\right)}{147}\right) = 0.0145$$

$$K_a = \frac{0.0024}{2}(0.25)(18) = 0.0054$$

Then

$$S = \frac{1.04(2500)}{2(32.2)(0.0054)} \ln\left[1 + \frac{0.0054(90 \times 5280 / 3600)^2}{(1.0)(0.7)(2500) + (0.0145)(2500) - 2500\sin(5.71°)}\right]$$

$$= \underline{\underline{444.07 \ \text{ft}}}$$

With aerodynamic resistance excluded, Eq. 2.43 is used:

$$S = \frac{1.04(90 \times 5280 / 3600)^2}{2(32.2)(0.7 + 0.0145 - \sin(5.71°))} = \underline{\underline{457.53 \ \text{ft}}}$$

EXAMPLE 2.9 **EFFECTS OF GRADE ON THEORETICAL MINIMUM STOPPING DISTANCE**

A car is traveling at 80 mi/h and has a braking efficiency of 80%. The brakes are applied to miss an object that is 150 ft from the point of brake application, and the coefficient of road adhesion is 0.85. Ignoring aerodynamic resistance and assuming the theoretical minimum stopping distance, estimate how fast the car will be going when it strikes the object if (a) the surface is level and (b) the surface is on a 5% upgrade.

SOLUTION

In both cases, rolling resistance is approximated as

$$f_{rl} = 0.01\left(1 + \frac{\left(\dfrac{80 \times 5280/3600 + V_2}{2}\right)}{147}\right) = 0.014 + 0.000034V_2$$

Applying Eq. 2.43 for the level grade with $\gamma_b = 1.04$, $\theta_g = 0°$,

$$S = \frac{\gamma_b\left(V_1^2 - V_2^2\right)}{2g\left(\eta_b\mu + f_{rl} \pm \sin\theta_g\right)}$$

$$150 = \frac{1.04\left((80 \times 5280/3600)^2 - V_2^2\right)}{2(32.2)\left[0.8(0.85) + (0.014 + 0.000034V_2) \pm 0\right]}$$

$$V_2 = \underline{\underline{85.40 \text{ ft/s}}} \quad \text{or} \quad \underline{\underline{58.23 \text{ mi/h}}}$$

On a 5% grade with $\theta_g = 2.86°$,

$$150 = \frac{1.04\left((80 \times 5280/3600)^2 - V_2^2\right)}{2(32.2)\left[0.8(0.85) + (0.014 + 0.000034V_2) + 0.05\right]}$$

$$V_2 = \underline{\underline{82.64 \text{ ft/s}}} \quad \text{or} \quad \underline{\underline{56.35 \text{ mi/h}}}$$

EXAMPLE 2.10 **THEORETICAL MINIMUM STOPPING DISTANCE WITH AND WITHOUT ANTILOCK BRAKES**

A car is traveling up a 3% grade on a road that has good, wet pavement. The engine is running at 2500 revolutions per minute. The radius of the wheels is 15 inches, the driveline slippage is 3%, and the overall gear reduction ratio is 2.5 to 1. A deer jumps out onto the road and the driver applies the brakes 291 ft from it. The driver hits the deer at a speed of 20mi/h. If the driver did not have antilock brakes, and the wheels were locked the entire distance, would a deer-impact speed of 20 mi/h be possible?

SOLUTION

The speed of the car must first be determined by applying Eq. 2.18 with $i = 0.03$, $\varepsilon_0 = 2.5$, $r = 15/12$ ft, and $n_e = 2500/60$ revolutions per second:

$$V = \frac{2\pi r n_e (1-i)}{\varepsilon_0}$$

$$= \frac{2(3.141)(15/12)(2500/60)(1-0.03)}{2.5}$$

$$= 126.92 \text{ ft/s (86.52 mi/h)}$$

Next, using Eq. 2.5 the coefficient of rolling resistance is computed using the average speed, $(V_1 + V_2)/2$ as an approximation of V (with a V_2 of 20 mi/h):

$$f_{rl} = 0.01\left(1 + \frac{V}{147}\right) = 0.01\left(1 + \frac{\frac{126.92 + 20(5280/3600)}{2}}{147}\right) = 0.01532$$

If the car did not have an antilock braking system and the brakes were locked, the slide value on good, wet pavement is a coefficient of road adhesion of 0.6 ($\mu = 0.6$) from Table 2.5. Applying Eq. 2.43 with $\mu = 0.6$, a 3% grade (so $\sin\theta_g \approx 0.03$), and $\gamma_b = 1.04$,

$$S = \frac{\gamma_b \left(V_1^2 - V_2^2\right)}{2g(\eta_b \mu + f_{rl} \pm \sin\theta_g)}$$

$$291 = \frac{1.04\left(126.92^2 - \left(20(5280/3600)\right)^2\right)}{2(32.2)\left[\eta_b(0.6) + 0.01532 + 0.03\right]}$$

or $\eta_b = \underline{\underline{1.33}}$

Because a braking efficiency of 1.33 is not possible, the driver would hit the deer at a higher speed if the wheels were locked the entire distance. Note that to achieve a deer-impact speed of 20 mi/h or less, $\eta_b \mu$ in Eq. 2.43 must be 0.8 (1.33×0.6) or greater. The maximum coefficient of road adhesion for good, wet pavements is 0.9 (from Table 2.5). So, to achieve a deer-impact speed of 20 mi/h or less (with $\mu = 0.9$),

$$\eta_b \mu = 0.8 \text{ so,}$$
$$\eta_b = 0.8/\mu$$
$$= 0.8/0.9$$
$$= 0.89$$

Thus a braking efficiency of at least 89% is needed (if the vehicle has a functional antilock braking system) in order to achieve a deer-impact speed of 20 mi/h or less.

2.9.5   Practical Stopping Distance

As mentioned earlier, one of the most critical concerns in the design of a highway is the provision of adequate driver sight distance to permit a safe stop. The theoretical assessment of vehicle stopping distance presented in the previous section provided the principles of braking for an individual vehicle under specified roadway surface conditions. However, highway engineers face a more complex problem because they must design for a variety of driver skill levels (which can affect whether or not the brakes lock and reduce the coefficient of road adhesion to slide values), vehicle types (with varying aerodynamics, weight distributions, and brake efficiencies), and weather conditions (which change the roadway's coefficient of adhesion). As a result of the wide variability inherent in the determination of braking distance, an equation is required that provides an estimate of typical observed braking distances and is more simplistic and usable than Eq. 2.42.

The basic physics equation on rectilinear motion, assuming constant deceleration, is chosen as the basis of a practical equation for stopping distance:

$$V_2^2 = V_1^2 + 2ad \tag{2.44}$$

where

V_2 = final vehicle speed in ft/s,
V_1 = initial vehicle speed in ft/s,
a = acceleration (negative for deceleration) in ft/s², and
d = deceleration distance (practical stopping distance) in ft.

Rearranging Eq. 2.44 and assuming a is negative for deceleration gives

$$d = \frac{V_1^2 - V_2^2}{2a} \tag{2.45}$$

If $V_2 = 0$ (the vehicle comes to a complete stop), the practical stopping distance equation is

$$d = \frac{V_1^2}{2a} \tag{2.46}$$

To make this equation generally applicable for design purposes, a deceleration rate, a, must be chosen that is representative of appropriately conservative braking behavior. AASHTO [2018] recommends a deceleration rate of 11.2 ft/s². Empirical studies [Fambro et al., 1997] have shown that approximately 90% of drivers decelerate at rates greater than this, and that this deceleration rate is well within a driver's capability to maintain steering control during a braking maneuver on wet surfaces. Additionally, empirical studies [Fambro et al., 1997] have confirmed that most vehicle braking systems and tire-pavement friction levels are capable of supporting this deceleration rate, even under wet conditions.

To account for the effect of grade, Eq. 2.46 is modified as follows:

$$d = \frac{V_1^2}{2g\left(\left(\dfrac{a}{g}\right) \pm G\right)} \tag{2.47}$$

where

g = gravitational constant, 32.2 ft/s^2,

G = roadway grade (+ for uphill, − for downhill) in percent/100, and

other terms are as defined previously.

It is important to note that Eq. 2.47 is consistent with Eq. 2.43 (the theoretical stopping distance ignoring aerodynamic resistance). Rewriting Eq. 2.43 with the assumption that the vehicle comes to a stop ($V_2 = 0$), that sin θ_g = tan θ_g = G (for small grades), and that γ_b and f_{rl} can be ignored due to their small and essentially offsetting effects, we have

$$S = \frac{V_1^2}{2g\left(\eta_b\mu \pm G\right)} \qquad (2.48)$$

Recall that $\eta_b\mu$ = g_{max} (Eq. 2.33). However, rather than determining the maximum deceleration rate (in g's) for a specific vehicle braking efficiency and specific coefficient of road adhesion, the AASHTO-recommended maximum deceleration rate (again, an appropriately conservative value for the overall driver and vehicle population) is used. Thus, a maximum deceleration of 0.35 g's (11.2/32.2) is used for Eq. 2.47.

   The recommended deceleration rate as determined empirically already accounts for the effects of aerodynamic resistance, braking efficiency, coefficient of road adhesion, and inertia during braking (the braking mass factor). This value reflects current vehicle technologies and driving behavior. It is important to recognize that as vehicle braking technology and other vehicle characteristics change, as well as possibly driver behavior, the recommended value of a should be reviewed to determine if it is still applicable for highway design purposes. The relationship between changing vehicle characteristics and changing highway design guidelines is one that must always be kept in the design engineer's mind.

EXAMPLE 2.11   BRAKING EFFICIENCY AND STOPPING DISTANCE

A car [W = 2200 lb, C_D = 0.25, A_f = 21.5 ft^2] has an antilock braking system that gives it a braking efficiency of 100%. The car's stopping distance is tested on a level roadway with poor, wet pavement (with tires at the point of impending skid), and ρ = 0.00238 slugs/ft^3. How inaccurate will the stopping distance predicted by the practical-stopping-distance equation be compared with the theoretical stopping distance, assuming the car is initially traveling at 60 mi/h? How inaccurate will the practical-stopping-distance equation be if the same car has a braking efficiency of 85%?

SOLUTION

First, to calculate the theoretical minimum stopping distance, Eq. 2.42 is applied with γ_b = 1.04, θ_g = 0°, μ = 0.60 (maximum for poor, wet pavement, from Table 2.5), and

$$f_{rl} = 0.01\left[1 + \frac{\left(\dfrac{60 \times 5280/3600 + 0}{2}\right)}{147}\right] = 0.013$$

$$K_a = \frac{0.00238}{2}(0.25)(21.5) = 0.0064$$

Thus, from Eq. 2.42,

$$S = \frac{1.04(2200)}{2(32.2)(0.0064)}\ln\left[1 + \frac{0.0064(60 \times 5280/3600)^2}{(1.0)(0.60)(2200) + (0.013)(2200) \pm 0}\right] = \underline{\underline{200.35 \text{ ft}}}$$

For the same conditions but with a vehicle braking efficiency of 85%, Eq. 2.42 gives

$$S = \frac{1.04(2200)}{2(32.2)(0.0064)}\ln\left[1 + \frac{0.0064(60 \times 5280/3600)^2}{(0.85)(0.60)(2200) + (0.013)(2200) \pm 0}\right]$$
$$= \underline{\underline{234.11 \text{ ft}}}$$

Now applying Eq. 2.46 (since $G = 0$) for the practical stopping distance, we find

$$d = \frac{(60 \times 5280/3600)^2}{2(11.2)} = \underline{\underline{345.71 \text{ ft}}}$$

In the first case, the error is 145.36 ft. In the case of 85% braking efficiency, the error is 111.60 ft. Rearranging Eq. 2.46 to solve for a, we find that stopping distances of 200.35 ft and 234.11 ft correspond to deceleration rates of 19.33 ft/s^2 and 16.54 ft/s^2, respectively. Studies [Fambro et al., 1997] have shown that most drivers decelerate at rates of 18.4 ft/s^2 or greater in emergency stopping situations. Thus, this range of theoretical values is consistent with observed distances for situations in which minimum stopping distances are being attempted. Comparing these theoretical values to the AASHTO-recommended deceleration rate of 11.2 ft/s^2, it is readily apparent that a considerable level of conservatism is built into the deceleration rate for practical stopping distance.

2.9.6   Distance Traveled during Driver Perception/Reaction

Until now the focus has been directed toward the distance required to stop the vehicle from the point of brake application. However, in providing sufficient sight distance for a driver to stop safely, it is also necessary to consider the distance traveled during the time the driver is perceiving and reacting to the need to stop. The distance traveled during perception/reaction (d_r) is given by

$$d_r = V_1 \times t_r \tag{2.49}$$

where

V_1 = initial vehicle speed in ft/s, and

t_r = time required to perceive and react to the need to stop, in s.

The perception/reaction time of a driver is a function of a number of factors, including the driver's age, physical condition, and emotional state, as well as the complexity of the situation and the strength of the stimuli requiring a stopping action. For highway design, a conservative perception/reaction time has been determined to be 2.5 seconds [AASHTO, 2018]. For comparison, average drivers have perception/reaction times of approximately 1.0 to 1.5 seconds.

Thus, the total required stopping distance is a combination of the braking distance, either theoretical (Eq. 2.42 or 2.43) or practical (Eq. 2.47), and the distance traveled during perception/reaction (Eq. 2.49), as shown in Eq. 2.50:

$$d_s = d + d_r \tag{2.50}$$

where

- d_s = total stopping distance (including perception/reaction) in ft,
- d = distance traveled during braking in ft, and
- d_r = distance traveled during perception/reaction in ft.

The combination of practical stopping distance and the distance traveled during perception/reaction is a primary consideration in highway design, as will be discussed in detail in Chapter 3.

EXAMPLE 2.12   **PRACTICAL STOPPING DISTANCE AND PERCEPTION/REACTION TIMES**

Two drivers each have a reaction time of 2.5 seconds. One is obeying a 55-mi/h speed limit and the other is traveling illegally at 70 mi/h. How much distance will each of the drivers cover while perceiving/reacting to the need to stop, and what will the total stopping distance be for each driver (using practical stopping distance and assuming $G = -2.5\%$)?

SOLUTION

The distances traveled by each driver during perception/reaction will be calculated first, using Eq. 2.49. For the driver traveling at 55 mi/h,

$$d_r = V_1 \times t_r = (55 \times 5280 / 3600)(2.5) = \underline{201.67\ \text{ft}}$$

For the driver traveling at 70 mi/h,

$$d_r = V_1 \times t_r = (70 \times 5280 / 3600)(2.5) = \underline{256.67\ \text{ft}}$$

Therefore, driving at 70 mi/h increases the distance traveled during perception/reaction by 55.0 ft.

Next, the distance traveled during braking will be calculated for each driver, using the equation for practical stopping distance (Eq. 2.47). For the driver traveling at 55 mi/h,

$$d = \frac{(55 \times 5280/3600)^2}{2(32.2)\left(\left(\dfrac{11.2}{32.2}\right) - 0.025\right)} = \frac{6507.11}{20.79} = 312.99\ \text{ft}$$

For the driver traveling at 70 mi/h,

$$d = \frac{(70 \times 5280/3600)^2}{2(32.2)\left(\left(\frac{11.2}{32.2}\right) - 0.025\right)} = \frac{10540.44}{20.79} = 507.00 \text{ ft}$$

The total stopping distance for each driver is now calculated with Eq. 2.50. For the driver traveling at 55 mi/h,

$$d_s = d + d_r = 312.99 + 201.67 = \underline{\underline{514.66 \text{ ft}}}$$

For the driver traveling at 70 mi/h,

$$d_s = d + d_r = 507.00 + 256.67 = \underline{\underline{763.67 \text{ ft}}}$$

Therefore, driving at 70 mi/h increases the total stopping distance by a very substantial 249.01 ft.

EXAMPLE 2.13   **PRACTICAL STOPPING DISTANCES FOR HUMAN-DRIVEN VEHICLE AND AUTONOMOUS VEHICLE**

Two vehicles, one operated by a human and the other computer controlled, are traveling on a level highway. If the autonomous vehicle is traveling at 70 mi/h, and its perception/reaction time is considered negligible, what is the maximum speed the human-controlled vehicle can travel and have a stopping distance that does not exceed that of the autonomous vehicle?

SOLUTION

While autonomous vehicle perception/reaction time will not be zero in practice, it will be very small. So for the purposes of this example, we will assume t_r to be negligible and thus set it to zero. Correspondingly, d_r for the autonomous vehicle will be zero and the total stopping distance will just be equal to the distance traveled during braking. Using the practical stopping distance equation (Eq. 2.47) to calculate the braking distance gives,

$$d = \frac{(70 \times 5280/3600)^2}{2(11.2)} = 470.56 \text{ ft}$$

Substituting Eqs. 2.47 and 2.49 into Eq. 2.50 yields (for level grade)

$$d_s = \frac{V^2}{2a} + V \times t_r$$

Solving the resulting quadratic for V, with coefficient $a = 1/(2 \times 11.2)$, coefficient $b = 2.5$, and constant $c = -470.56$ gives two solutions. One solution is negative, thus disregarded. The other solution is 78.4 ft/s (<u>53.5 mi/h</u>), which is reasonable. Thus, the human-controlled vehicle must travel 16.5 mi/h (70 − 53.5) slower than the autonomous vehicle to attain the same stopping distance.

This problem assumed the same currently recommended deceleration rate of 11.2 ft/s^2 (0.35g) for the autonomous vehicle. This value is rooted in empirical studies of human behavior, as well as roadway surface condition considerations (such as wet pavement), but corresponds to a conservative value to account for the variability in driver, vehicle, and roadway characteristics. By eliminating the human driver variability component with autonomous vehicles, it can be argued that higher deceleration rates could be applied, thus reducing the stopping distances used for roadway design purposes (which will be discussed in Chapter 3). While deceleration rates under good pavement and tire conditions for a modern automobile can approach 1g, the design value adopted in practice will still have to consider poor pavement conditions and human, tolerances for rapid deceleration. Vehicle types with lower braking efficiencies, such as commercial trucks, may impose additional practical constraints to the design deceleration rate used for roadway design.

2.10 PRACTICE PROBLEMS

PRACTICE PROBLEM 2.1

ENGINE-GENERATED TRACTIVE EFFORT, AERODYNAMIC AND ROLLING RESISTANCE, ACCELERATION

A car weighs 2200 lb and is traveling 100 mi/h on a race track that is on a 3% upgrade. The car is preparing to pass a slower car and its torque/engine speed curve is given by the equation,

$$M_e = 8n_e - 0.05n_e^2$$

with M_e in ft-lb and n_e in revolutions per second. Drivetrain efficiency is 90%, drive axle slippage is 2%, wheel radius is 15 inches, frontal area is 22 ft², drag coefficient is 0.35, and air density is 0.0022 slugs/ft³. If the car is in a gear that produces maximum torque, what would the car's maximum acceleration be?

SOLUTION

Note: Open boxes in equations "☐" are to be completed by the reader

This problem combines a number of concepts covered in this chapter. We want to determine the acceleration, so eventually we want to apply Eq. 2.19,

$$F - \sum R = \gamma_m ma,$$

and solve for a. In this equation, all necessary terms for determining a can be readily calculated with the information provided except for the available tractive effort F. It is known that the available tractive effort is the lesser of the engine-generated tractive effort F_e, and the maximum tractive effort F_{max}. Because the car is traveling at 100 mi/h, the engine-generated tractive effort will almost certainly be less than the maximum tractive effort since at this speed the amount of engine-generated force required to have F_e exceed F_{max} would require an unrealistically powerful engine. This can be seen in Fig. 2.5 where the F_{max} curve is well above the F_e curves at higher speeds.

Given this, we know from Eq. 2.17 that,

$$F = F_e = \frac{M_e \varepsilon_0 \eta_d}{r}$$

The problem provides all of the terms in this equation except the torque M_e and the gear reduction ratio ε_0. To determine the values of these terms, the following steps are taken:

1. Note that in the last sentence there is the wording "the car is in a gear that produces maximum torque". First consider the latter portion of this wording and find the maximum torque. As in Example 2.4, this is done by taking the first derivative

of $M_e = 8n_e - 0.05n_e^2$ with respect to engine speed n_e and setting it equal to zero (to get the maximum):

$$\frac{dM_e}{dn_e} = 8 - \boxed{} \times n_e = 0$$

$$n_e = \frac{8}{\boxed{}} = 80 \text{ revolutions per second}$$

Note that this gives an equation with engine speed n_e at maximum torque being the only unknown. Solving, it is found that $n_e = 80$ revolutions per second which is the engine speed at maximum torque. Next, substitute this value of n_e into the given torque equation:

$$M_e = 8 \times \boxed{} - 0.05 \times \boxed{}^2 = 320 \text{ ft-lb.}$$

2. The problem states that the car is in "a gear that produces maximum torque." We need to find the gear reduction ratio ε_0, and Eq. 2.18 $[V = [2\pi r n_e (1-i)]/\varepsilon_0]$ is the equation that relates gear reduction ratio to engine speed. This equation is rearranged to get the gear reduction ratio as,

$$\varepsilon_0 = \frac{2\pi \times \left[\boxed{}/12\right] \times \boxed{} \times \left[1 - \boxed{}\right]}{100 \times \boxed{}} = 4.22$$

With the above, and all elements of Eq. 2.17 known, so

$$F = F_e = \frac{M_e \varepsilon_0 \eta_d}{r} = \frac{\boxed{} \times \boxed{} \times \boxed{}}{\boxed{}/12} = 972.29 \text{ lb}$$

Next, the resistances are now computed as follows:

for grade resistance (using Eq. 2.9),

$$R_g = WG = \boxed{} \times \boxed{} = 66 \text{ lb}$$

for aerodynamic resistance (using Eq. 2.3),

$$R_a = \frac{\rho}{2} C_D A_f V^2$$

$$= \frac{\boxed{}}{2} \times \boxed{} \times \boxed{} \times \left[\boxed{} \times 1.467\right]^2$$

$$= 182.3 \text{ lb}$$

and for rolling resistance (using Eqs. 2.5 and 2.6),

$$R_{rl} = f_{rl} W$$

$$= \left[0.01\left(1 + \frac{\boxed{} \times 1.467}{147}\right)\right] \times \boxed{}$$

$$= 44 \text{ lb}$$

The last variable needed to compute the maximum acceleration is the mass factor. This is determined from Eq. 2.20 as,

$$\gamma_m = 1.04 + 0.0025\varepsilon_0^2$$
$$= 1.04 + 0.0025 \times \boxed{}^2$$
$$= 1.085$$

Rearranging Eq. 2.19 to solve for acceleration gives,

$$F - \sum R = \gamma_m ma$$

$$a = \frac{F - \sum R}{\gamma_m m} = \frac{\boxed{} - \boxed{} - \boxed{} - \boxed{}}{\boxed{} \times \left[\boxed{}/32.2\right]} = \underline{\underline{9.17 \text{ ft/s}^2}}.$$

PRACTICE PROBLEM 2.2

VEHICLE DESIGN, ENGINE-GENERATED TRACTIVE EFFORT, ROLLING RESISTANCE, ACCELERATION

A rear-wheel–drive car has an engine running at 3296 revolutions per minute. It is known that at this engine speed the engine produces 80 horsepower. The car has an overall gear reduction ratio of 10, a wheel radius of 16 inches, and a 95% drivetrain mechanical efficiency. The weight of the car is 2600 lb, the wheelbase is 95 inches, and the center of gravity is 22 inches above the roadway surface. What is the closest distance the center of gravity can be behind the front axle to have the vehicle achieve its maximum acceleration from rest on good, wet pavement?

SOLUTION

Note: Open boxes in equations "$\boxed{}$" are to be completed by the reader

This problem illustrates the relationship between engine-generated tractive effort F_e, and the maximum tractive effort F_{max}. To achieve maximum acceleration, we can see from Eq. 2.19 ($F - \sum R = \gamma_m ma$) that we want the available tractive effort F (which is the lesser of the engine-generated tractive effort F_e, and the maximum tractive effort F_{max}) to be as high as possible. The problem provides information that will allow us to calculate the engine-generated tractive effort F_e at the conditions specified. So the point of interest in this problem is to determine an F_{max} that will enable the vehicle to use all of the available engine-generated tractive effort under the conditions given. It is known from Eq. 2.14 that F_{max} is a function of the location of the center of gravity from the front axle. In this problem, if the center of gravity is too close to the front axle, F_{max} will be less than F_e and the vehicle will not be able to achieve it maximum acceleration based on it engine-generated tractive effort. This condition is illustrated in the lowest speed region of Fig. 2.5 where F_e curve exceeds the F_{max} line. If $F_{max} < F_e$ the vehicle will not be able to achieve maximum acceleration based on current engine operating conditions because F_{max} will limit the available tractive effort F in Eq. 2.19. For maximum acceleration to be achieved $F_{max} \geq F_e$. As shown in Eq. 2.14, as the center of gravity gets farther behind the front axle, F_{max} will become larger (see Eq. 2.14). Since we want the center of gravity as close to the front axle as possible, we need to solve for the limiting case where $F_{max} = F_e$.

To show this equality using Eq. 2.17 we will have,

$$F_{max} = F_e = \frac{M_e \varepsilon_0 \eta_d}{r}$$

All values in the right-side of the above equation are given in the problem except M_e. However, we are given horsepower and engine speed so Eq. 2.16 can be rearranged to solve for M_e as,

$$M_e = \frac{550 hp_e}{2\pi n_e} = \frac{550 \times \boxed{}}{6.28 \times (\boxed{}/\boxed{})} = 127.54 \text{ ft-lb}$$

Using this value of Me, Eq. 2.17 can now be applied to give,

$$F_{max} = F_e = \frac{M_e \varepsilon_0 \eta_d}{r} = \frac{\boxed{} \times \boxed{} \times \boxed{}}{\boxed{}/\boxed{}} = 908.27 \text{ lb}$$

With F_{max} known, Eq. 2.14 can be applied to solve for l_f which will be the closest distance the center of gravity can be behind the front axle (with the coefficient of road adhesion $\mu = 0.9$ for good, wet pavement from Table 2.4),

$$F_{max} = \frac{\mu W (l_f - f_{rl} h)/L}{1 - \mu h/L}$$

The coefficient of rolling resistance f_{rl}, which can be determined from Eq. 2.5 (with $V = \boxed{}$ since acceleration is from rest) as,

$$f_{rl} = 0.01\left(1 + \frac{V}{147}\right) = 0.01\left(1 + \frac{\boxed{}}{147}\right) = \boxed{}$$

Substituting given and calculated values into Eq. 2.14 gives,

$$908.72 = \frac{\boxed{} \times \boxed{} \times (l_f - \boxed{} \times \boxed{})/95}{1 - \boxed{} \times 22/\boxed{}}$$

Which gives $l_f = \underline{29.42 \text{ inches}}$.

PRACTICE PROBLEM 2.3

ENGINE-GENERATED TRACTIVE EFFORT AND THEORETICAL MINIMUM STOPPING DISTANCE

A car has 11-inch radius wheels and is traveling up a 3% grade with a gear reduction ratio of 2 to 1 and a driveline slippage of 2%. The car has a torque/engine speed curve that is given by the equation,

$$M_e = 7n_e - 0.08n_e^2$$

with M_e in ft-lb and n_e in revolutions per second. It is also known that the car is traveling at a speed where its engine is generating maximum torque. If the car is on good, wet pavement with antilock brakes and a 95% braking efficiency, ignoring air resistance, what would be the theoretical stopping distance (distance traveled until a stop after the brakes are applied)?

SOLUTION

Note: Open boxes in equations "$\boxed{}$" are to be completed by the reader

To solve this problem we must first determine the speed at which the vehicle was traveling before the brakes were applied. The problem states that the car was traveling is traveling at a speed that has its engine producing maximum torque. So we need to find the engine speed at this maximum torque, and then use this engine speed along with the given gear reduction ratio and driveline slippage to arrive at vehicle speed (by applying Eq. 2.18).

As in Practice Problem 2.1, finding the engine speed n_e at maximum torque is achieved by taking the first derivative of $M_e = 7n_e - 0.08n_e^2$ with respect to engine speed n_e and setting it equal to zero (to get the maximum):

$$\frac{dM_e}{dn_e} = 7 - \boxed{} \times n_e = 0$$

$$n_e = \frac{7}{\boxed{}} = 43.75 \text{ revolutions per second}$$

With the engine speed, Eq. 2.18 is applied as follows,

$$V = \frac{2\pi r n_e (1-i)}{\varepsilon_0}$$

$$V = \frac{\boxed{} \times \boxed{} / \boxed{} \times \boxed{} \left(1 - \boxed{}\right)}{\boxed{}} = 123.45 \text{ ft/s}$$

Eq. 2.43 can now be applied to determine the theoretical stopping distance,

$$S = \frac{\gamma_b \left(V_1^2 - V_2^2\right)}{2g\left(\eta_b \mu + f_{rl} \pm \sin\theta_g\right)}$$

In the above equation, η_b and $\sin\theta_g$ are given in the problem, γ_b is a known constant (see text), μ can be obtained from Table 2.5, and f_{rl} can be determined from Eq. 2.5 as,

$$f_{rl} = 0.01\left(1 + \frac{\left(\dfrac{\boxed{} - \boxed{}}{\boxed{}}\right)}{147}\right) = 0.0142$$

So Eq. 2.43 is solved as,

$$S = \frac{\boxed{} \times \left(\boxed{}^2 - \boxed{}^2\right)}{2 \times \boxed{}\left(\boxed{} \times \boxed{} + \boxed{} \pm \boxed{}\right)} = \underline{\underline{273.69 \text{ ft.}}}$$

PRACTICE PROBLEM 2.4

THEORETICAL MINIMUM STOPPING DISTANCE WITH INTERMITENT ANTILOCK BRAKE FAILURE

A car is traveling 50 mi/h when the driver applies the brakes. The car is traveling on good wet pavement and is going down a 3% grade. The antilock braking system works intermittently during the stop. When the antilock brakes system is functioning, the braking efficiency is 100%. When the antilock system is not functioning the wheels lock and the braking efficiency is 70%. The antilock braking system turns off/on every 60 ft while the brakes are being applied (it is on for the first 60 ft of the stop, off for the second 60 ft, on for the third 60 ft of the stop, and so on). Under these conditions, what is the total distance needed for the car to stop after the brakes are applied? (Assume theoretical stopping distance, ignore aerodynamic resistance, and let $f_{rl} = 0.011$.)

SOLUTION

Note: Open boxes in equations "☐" are to be completed by the reader

Eq. 2.43 can be applied to solve this problem,

$$S = \frac{\gamma_b \left(V_1^2 - V_2^2 \right)}{2g \left(\eta_b \mu + f_{rl} \pm \sin \theta_g \right)}$$

but it will have to be applied in 60-ft distance increments to account for the intermittent antilock brake failure. For the first 60 ft with the antilock brakes functioning, in Eq. 2.43 S will be 60 ft, f_{rl} and $\sin \theta_g$ are given in the problem, γ_b is a known constant (see text), μ can be obtained from Table 2.5 and will alternate between maximum and slide values depending on whether or not the antilock braking system is functioning, and η_b will alternate between 1.0 and 0.7 depending on whether the antilock braking system is functioning or not.

We wish to determine the vehicle speed at the end of the first 60-ft interval (an interval where the antilock brakes are working), so rearranging Eq. 2.43 we have,

$$V_2 = \sqrt{\left(\boxed{} \times 1.467 \right)^2 - \frac{60 \times 2 \times \boxed{} \left(\boxed{} \times 0.9 + \boxed{} - \boxed{} \right)}{\boxed{}}} = 45.91 \text{ ft/s}$$

The final vehicle speed of 45.91 ft/s after the first 60-ft interval will become the initial speed of the second 60-ft interval, an interval in which the antilock braking system does not function. For the second 60-ft interval, μ and η_b will change to their non-antilock brake values and Eq. 2.43 is applied again to give,

$$V_2 = \sqrt{\left(46.14 \right)^2 - \frac{60 \times 2 \times \boxed{} \left(\boxed{} \times \boxed{} + \boxed{} - \boxed{} \right)}{\boxed{}}} = 24.86 \text{ ft/s}$$

The final vehicle speed of 24.86 ft/s after the second 60-ft interval will become the initial speed of the third 60-ft interval, an interval in which the antilock braking system functions again. For the third 60-ft interval μ and η_b will change to their antilock brake values. The reader can show that the car will come to a stop at some time in this third 60-ft interval. So in the third distance interval Eq. 2.43 is applied to give,

$$S = \frac{\boxed{} \times \left(\boxed{}^2 - 0 \right)}{2 \times \boxed{} \times \left(\boxed{} \times \boxed{} + \boxed{} - \boxed{} \right)} = 11.3 \text{ ft}$$

Thus, the total stopping distance is,

$$S = \boxed{} + \boxed{} + \boxed{} = \underline{\underline{131.3 \text{ ft}}}$$

NOMENCLATURE FOR CHAPTER 2

a	acceleration (deceleration if negative)	L	vehicle wheelbase
A_f	frontal area of vehicle	l_f	distance from vehicle's center of gravity to front axle
$BFR_{flr\,max}$	brake force ratio (front over rear) for maximum braking force	l_r	distance from vehicle's center of gravity to rear axle
C_D	coefficient of aerodynamic drag	M_e	engine torque
d	practical stopping distance	m	mass
d_a	distance to accelerate	n_e	engine speed in crankshaft revolutions per second
d_r	distance traveled during driver perception/reaction	PBF_f	optimal percent of braking force on the front axle
d_s	total stopping distance (vehicle braking distance plus perception/reaction distance)	PBF_r	optimal percent of braking force on the rear axle
F	total available tractive effort	R_a	aerodynamic resistance
F_b	total braking force	R_g	grade resistance
F_{bf}	front-axle braking force	R_{rl}	rolling resistance
$F_{bf\,max}$	maximum front-axle braking force	r	radius of vehicle drive wheels
F_{br}	rear-axle braking force	S	minimum theoretical stopping distance
$F_{br\,max}$	maximum rear-axle braking force	t_r	driver perception/reaction time
F_e	engine-generated tractive effort	V	vehicle speed
F_f	available tractive effort at the front axle	W	total vehicle weight
F_{max}	maximum tractive effort	W_f	vehicle weight acting normal to the roadway surface on the front axle
F_r	available tractive effort at the rear axle	W_r	vehicle weight acting normal to the roadway surface on the rear axle
f_{rl}	coefficient of rolling resistance		
G	roadway grade in ft/ft (percent grade divided by 100; $G = 0.05$ is a 5% grade)	γ_b	braking mass factor
g	gravitational constant (32.2 ft/s^2)	γ_m	acceleration mass factor
g_{max}	maximum deceleration achieved before wheel lockup	ε_0	gear reduction ratio
		η_b	braking efficiency
h	height of vehicle's center of gravity above the roadway surface	η_d	drivetrain efficiency
hp_e	engine-generated power, measured in horsepower	θ_g	angle of grade
		μ	coefficient of road adhesion
i	drive axle slippage	ρ	air density
K_a	elements of aerodynamic resistance that are not a function of speed		

REFERENCES

AASHTO (American Association of State Highway and Transportation Officials). *A Policy on Geometric Design of Highways and Streets*, 7th ed. Washington, DC: AASHTO, 2018.

Fambro, D. B., K. Fitzpatrick, and R. J. Koppa. *Determination of Stopping Sight Distances*. NCHRP Report 400. Transportation Research Board, Washington, DC, 1997.

S. G. Shadle, L. H. Emery, and H. K. Brewer, "Vehicle Braking, Stability, and Control," *SAE Transactions*, vol. 92, paper 830562, 1983.

Taborek, J. J. "Mechanics of Vehicles." *Machine Design*, 1957.

Wong, J. Y. *Theory of Ground Vehicles*. Fourth Edition. New York: John Wiley & Sons, 2008.

Chapter 3

Geometric Design of Highways

3.1 INTRODUCTION

The information on vehicle performance provided in Chapter 2 provides the fundamentals needed for highway design. Highway design encompasses a variety of design elements including the alignment required to provide adequate stopping sight distances (SSD), adequate passing sight distances on two-lane roads, length of acceleration and deceleration lanes for on- and off-ramps, number of lanes required to provide adequate mobility, and identification of the need for truck climbing lanes on steep grades. The relationship with these design elements and vehicle performance is obvious. For example, vehicle acceleration and deceleration characteristics have a direct impact on the design of acceleration and deceleration lanes (the length needed to provide a safe and orderly flow of traffic) and the highway alignment needed to provide adequate passing and stopping sight distances. Furthermore, vehicle performance characteristics determine the need for truck climbing lanes on steep grades (where the poor performance of large trucks necessitates a separate lane) as well as the number of lanes required because the observed spacing between vehicles in traffic is directly related to vehicle performance characteristics (this will be discussed further in Chapter 5). In addition, the physical dimensions of vehicles affect a number of design elements, such as the curve radii required for low-speed turning, height of highway overpasses, and lane widths.

Considering the range of vehicle performance characteristics likely to be encountered in highway traffic (from high-performance sports cars to heavily-loaded trucks), as well as the variance in the physical dimensions of highway vehicles, and the interaction of these characteristics with the many elements constituting highway design, it is clear that highway geometric design is a complex procedure that requires numerous compromises. Moreover, it is important that highway design guidelines evolve over time in response to changes in vehicle performance and dimensions, and in response to evidence collected on the effectiveness of existing highway design practices, such as the relationship between crash rates and various roadway design characteristics. Current guidelines of highway design are presented in detail in *A Policy on Geometric Design of Highways and Streets*, 7th Edition, published by the American Association of State Highway and Transportation Officials [AASHTO, 2018].

Because of the sheer number of geometric elements involved in highway design, a detailed discussion of each design element is beyond the scope of this book (the reader is referred to [AASHTO, 2018] for a complete discussion of current design practices). Instead, this book focuses exclusively on the key elements of highway alignment, which are arguably the most important components of geometric design. As will be shown, the alignment topic is particularly well suited for demonstrating the effect of vehicle performance (specifically braking performance) and vehicle dimensions (which determine critical factors such as the driver's eye height, vehicle headlight height, and vehicle taillight height) on the design of highways. By concentrating on the specifics of the highway alignment problem, the reader will develop an understanding of the procedures and compromises inherent in the design of all highway-related geometric elements.

3.2 PRINCIPLES OF HIGHWAY ALIGNMENT

The alignment of a highway is a three-dimensional problem measured in x, y, and z coordinates. This is illustrated, from a driver's perspective, in Fig. 3.1. However, in highway design practice, three-dimensional design computations are cumbersome, and, what is perhaps more important, the actual implementation and construction of a design based on three-dimensional coordinates has historically been prohibitively difficult. As a consequence, the three-dimensional highway alignment problem is reduced to two two-dimensional alignment problems, as illustrated in Fig. 3.2. One of the alignment problems in this figure corresponds roughly to x and z coordinates and is referred to as horizontal alignment. The other corresponds to highway length (measured along some constant elevation) and y coordinates (elevation) and is referred to as vertical alignment. Referring to Fig. 3.2, note that the horizontal alignment of a highway is referred to as the plan view, which is roughly equivalent to the perspective of an aerial photo of the highway. The vertical alignment is represented in a profile view, which gives the elevation of all points measured along the length of the highway (again, with length measured along a constant elevation reference).

Aside from considering the alignment problem as two two-dimensional problems, one further simplification is made: instead of using x and z coordinates, highway positioning and length are defined as the distance along the highway (usually measured along the centerline of the highway, on a horizontal, constant-elevation plane) from a specified point. This distance is measured in terms of stations, with each station consisting of 100 ft of highway alignment distance.

The notation for stationing distance is such that a point on a highway 4250 ft from a specified point of origin is said to be at station $42 + 50$ ft, that is, 42 stations and 50 ft, with the point of origin being at station $0 + 00$. This stationing

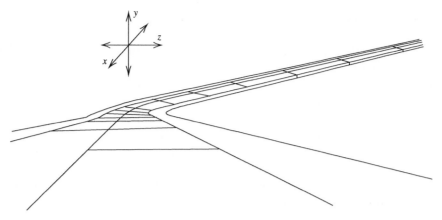

Figure 3.1 Highway alignment in three dimensions.

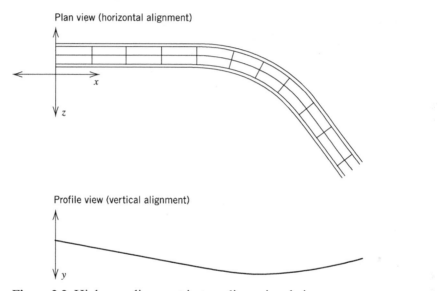

Figure 3.2 Highway alignment in two-dimensional views.

concept, combined with the highway's alignment direction given in the plan view (horizontal alignment) and the elevation corresponding to stations given in the profile view (vertical alignment), gives a unique identification of all highway points in a manner that is virtually equivalent to using true x, y, and z coordinates.

3.3 VERTICAL ALIGNMENT

Vertical alignment specifies the elevation of points along a roadway. The elevation of these roadway points is usually determined by the need to provide an acceptable level of driver safety, driver comfort, and proper drainage (from rainfall runoff). A primary concern in vertical alignment is establishing the transition of roadway elevations between two grades. This transition is achieved by means of a vertical curve.

Vertical curves can be broadly classified into crest vertical curves and sag vertical curves, as illustrated in Fig. 3.3. Note that in Fig. 3.3, the distance from the *PVC* to the *PVI* is *L*/2. This is used in this figure because in practice the vast majority of vertical curves are arranged such that half of the curve length is positioned before the *PVI* and half after. Curves that satisfy this criterion are called equal-tangent vertical curves.

For referencing points on a vertical curve, it is important to note that the profile views presented in Fig. 3.3 correspond to all highway points even if a horizontal curve occurs concurrently with a vertical curve (as in Figs. 3.1 and 3.2). Thus, each roadway point is uniquely defined by stationing (which is measured along a horizontal plane) and elevation. This will be made clearer through forthcoming examples.

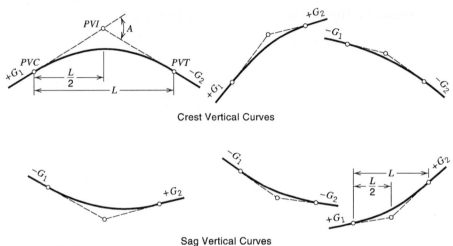

Figure 3.3 Types of vertical curves.

G_1 = initial roadway grade in percent or ft/ft (this grade is also referred to as the initial tangent grade, viewing Fig. 3.3 from left to right),

G_2 = final roadway (tangent) grade in percent or ft/ft,

A = absolute value of the difference in grades (initial minus final, usually expressed in percent),

L = length of the curve in stations or ft measured in a constant-elevation horizontal plane.

PVC = point of the vertical curve (the initial point of the curve),

PVI = point of vertical intersection (intersection of initial and final grades), and

PVT = point of vertical tangent, which is the final point of the vertical curve (the point where the curve returns to the final grade or, equivalently, the final tangent).

3.3.1   Vertical Curve Fundamentals

In connecting roadway grades (tangents) with an appropriate vertical curve, a mathematical relationship defining elevations at all points (or equivalently, stations) along the vertical curve is needed. A parabolic function has been found suitable in this regard because, among other things, it provides a constant rate of change of slope and implies equal curve tangents. The general form of the parabolic equation, as applied to vertical curves, is

$$y = ax^2 + bx + c \qquad (3.1)$$

where

y = roadway elevation at distance x from the beginning of the vertical curve (the PVC) in stations or ft,

x = distance from the beginning of the vertical curve in stations or ft,

a, b = coefficients defined below, and

c = elevation of the PVC (because $x = 0$ corresponds to the PVC) in ft.

In defining a and b, note that the first derivative of Eq. 3.1 gives the slope and is

$$\frac{dy}{dx} = 2ax + b \qquad (3.2)$$

At the PVC, $x = 0$, so, using Eq. 3.2,

$$b = \frac{dy}{dx} = G_1 \qquad (3.3)$$

where G_1 is the initial slope in ft/ft, as defined in Fig. 3.3. Also note that the second derivative of Eq. 3.1 is the rate of change of slope and is

$$\frac{d^2 y}{dx^2} = 2a \qquad (3.4)$$

However, the average rate of change of slope, by observation of Fig. 3.3, can also be written as

$$\frac{d^2 y}{dx^2} = \frac{G_2 - G_1}{L} \qquad (3.5)$$

Equating Eqs. 3.4 and 3.5 gives

$$a = \frac{G_2 - G_1}{2L} \qquad (3.6)$$

with all terms as defined previously (see Fig. 3.3). Please note that the units for coefficients a and b in Eqs. 3.3 and 3.6 must be such that they provide ft when multiplied by x^2 and x, respectively. The preceding equations define all the terms in the parabolic vertical curve equation (Eq. 3.1). The following example gives a typical application of this equation.

EXAMPLE 3.1   **VERTICAL CURVE STATIONS AND ELEVATIONS**

A 600-ft equal-tangent sag vertical curve has the *PVC* at station 170 + 00 and elevation 1000 ft. The initial grade is –3.5% and the final grade is +0.5%. Determine the stationing and elevation of the *PVI*, the *PVT*, and the lowest point on the curve.

SOLUTION

Since the curve is equal tangent, the *PVI* will be 300 ft or three stations (measured in a horizontal plane) from the *PVC*, and the *PVT* will be 600 ft or six stations from the *PVC*. Therefore, the stationing of the *PVI* and *PVT* is $\underline{173+00}$ and $\underline{176+00}$, respectively. For the elevations of the *PVI* and *PVT*, it is known that a –3.5% grade can be equivalently written as –3.5 ft/station (a 3.5 ft drop per 100 ft of horizontal distance). Since the *PVI* is three stations from the *PVC*, which is known to be at elevation 1000 ft, the elevation of the *PVI* is

$$1000 - 3.5 \text{ ft/station} \times (3 \text{ stations}) = \underline{989.5 \text{ ft}}$$

Similarly, with the *PVI* at elevation 989.5 ft, the elevation of the *PVT* is

$$989.5 + 0.5 \text{ ft/station} \times (3 \text{ stations}) = \underline{991.0 \text{ ft}}$$

It is clear from the values of the initial and final grades that the lowest point on the vertical curve will occur when the first derivative of the parabolic function (Eq. 3.1) is zero because the initial and final grades are opposite in sign. When initial and final grades are not opposite in sign, the low (or high) point on the curve will not be where the first derivative is zero because the slope along the curve will never be zero. For example, a sag curve with an initial grade of –2.0% and a final grade of –1.0% will have its lowest elevation at the *PVT*, and the first derivative of Eq. 3.1 will not be zero at any point along the curve. However, in our example problem the derivative will be equal to zero at some point, so the low point will occur when

$$\frac{dy}{dx} = 2ax + b = 0$$

From Eq. 3.3, we have

$$b = G_1 = -3.5$$

with G_1 in percent. From Eq. 3.6 (with L in stations and G_1 and G_2 in percent),

$$a = \frac{0.5 - (-3.5)}{2(6)} = 0.33333$$

Substituting for a and b gives

$$\frac{dy}{dx} = 2(0.33333)x + (-3.5) = 0$$

$$x = 5.25 \text{ stations}$$

This gives the stationing of the low point at $\underline{175+25}$ (5 + 25 stations from the *PVC*).

For the elevation of the lowest point on the vertical curve, the values of a, b, c (elevation of the *PVC*), and x are substituted into Eq. 3.1, giving

$$y = 0.33333(5.25)^2 + (-3.5)(5.25) + 1000$$
$$= 990.81 \text{ ft}$$

Note that the preceding equations can also be solved with grades expressed as the decimal equivalent of percent (for example, 0.02 ft/ft for 2%) if x is expressed in feet instead of stations. Care must be taken not to mix units. A dimensional analysis of Eq. 3.1 must ensure that each right-side element of the equation has resulting units of feet.

Another interesting vertical curve problem that is sometimes encountered is one in which the curve must be designed so that the elevation of a specific location is met. An example might be to have the roadway connect with another (at the same elevation) or to have the roadway at some specified elevation so as to pass under another roadway. This type of problem is referred to as a curve-through-a-point problem and is demonstrated by the following example.

EXAMPLE 3.2   ELEMENTS OF VERTICAL CURVE DESIGN

An equal-tangent vertical curve is to be constructed between grades of −2.0% (initial) and +1.0% (final). The *PVI* is at station 110 + 00 and at elevation 420 ft. Due to a street crossing the roadway, the elevation of the roadway at station 112 + 00 must be at 424.5 ft. Design the curve.

SOLUTION

The design problem is one of determining the length of the curve required to ensure that station 112 + 00 is at elevation 424.5 ft. To begin, we use Eq. 3.1:

$$y = ax^2 + bx + c$$

From Eq. 3.3,

$$b = G_1 = -2.0$$

and from Eq. 3.6,

$$a = \frac{G_2 - G_1}{2L}$$

Substituting $G_1 = -2.0$ and $G_2 = 1.0$, we have

$$a = \frac{G_2 - G_1}{2L} = \frac{1.0 - (-2.0)}{2L} = \frac{1.5}{L}$$

Now note that c (the elevation of the *PVC*) in Eq. 3.1 will be equal to the elevation of the *PVI* plus $G_1 \times 0.5L$ (this is simply using the slope of the initial grade to determine the elevation difference between the *PVI* and *PVC*). With G_1 in percent (which is ft/station) and the curve length L in stations, we have

$$c = 420 + 2.0(0.5L) = 420 + L$$

Finally, the value of x to be used in Eq. 3.1 will be $0.5L + 2$ because the point of interest (station 112 + 00) is two stations from the *PVI* (which is at station 110 + 00). Substituting $b = -2.0$, the expressions for a, c, and x, and $y = 424.5$ ft (the given elevation) into Eq. 3.1 give

$$424.5 = (1.5/L)(0.5L + 2)^2 + (-2.0)(0.5L + 2) + (420 + L)$$
$$4.5 = 0.375L + 3 + 6/L - 4$$
$$0 = -0.375 L^2 + 5.5L - 6$$

Solving this quadratic equation gives $L = 1.187$ stations (which is not feasible because we know that the point of interest is 2.00 stations beyond the *PVI*, so the curve must be longer than 1.187 stations) or $L = 13.466$ stations (which is the only feasible solution). This means that the curve must be 1346.6 ft long. Using this value of L,

elevation of $PVC = c = 420 + L = 420 + 13.466 = 433.47$ ft

station of $PVC = 110 + 00 - (13 + 46.6)/2 = 103 + 26.7$

elevation of PVT = elevation of $PVI + (0.5L)G_2 = 420 + \left[0.5(13.466) \right](1.0) = 426.73$ ft

station of $PVT = 110 + 00 + (13 + 46.6)/2 = 116 + 73.3$

and

$$x = 0.5L + 2.0 = 6.733 + 2.0 = 8.733 \text{ stations from the } PVC$$

To check the elevation of the curve at station 112 + 00, we apply Eq. 3.1 with $x = 8.733$:

$$y = ax^2 + bx + c$$

$$= \left(\frac{3}{2(13.466)} \right)(8.733)^2 + (-2.0)(8.733) + 433.47$$

$$= 424.5 \text{ ft}$$

Therefore, all calculations are correct.

Some additional properties of vertical curves can now be formalized. For example, offsets, which are vertical distances from the initial tangent to the curve, as illustrated in Fig. 3.4, are extremely important in vertical curve design and construction.

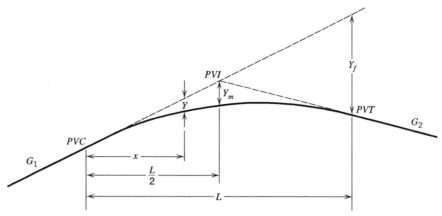

Figure 3.4 Offsets for equal-tangent vertical curves.

G_1 = initial roadway grade in percent or ft/ft (this grade is also referred to as the initial tangent grade, viewing Fig. 3.4 from left to right),

G_2 = final roadway (tangent) grade in percent or ft/ft,

Y = offset at any distance x from the PVC in ft,

Y_m = midcurve offset in ft,

Y_f = offset at the end of the vertical curve in ft,

x = distance from the PVC in ft,

L = length of the curve in stations or ft measured in a constant-elevation horizontal plane,

PVC = point of the vertical curve (the initial point of the curve),

PVI = point of vertical intersection (intersection of initial and final grades), and

PVT = point of vertical tangent, which is the final point of the vertical curve (the point where the curve returns to the final grade or, equivalently, the final tangent).

Referring to the elements shown in Fig. 3.4, the properties of an equal-tangent parabola can be used to give

$$Y = \frac{A}{200L}\, x^2 \qquad (3.7)$$

where

A = absolute value of the difference in grades ($|G_1 - G_2|$) expressed in percent, and other terms are as defined in Fig. 3.4.

Note that in this equation, 200 is used in the denominator instead of 2 because A is expressed in percent instead of ft/ft (this division by 100 also applies to Eqs. 3.8 and 3.9 below). It follows from Fig. 3.4 that

$$Y_m = \frac{AL}{800} \qquad (3.8)$$

and

$$Y_f = \frac{AL}{200} \tag{3.9}$$

Another useful vertical curve property is one that gives the length of curve required to effect a 1% change in slope. Because the parabolic equation used for roadway elevations (Eq. 3.1) gives a constant rate of change of slope, it can be shown that the horizontal distance required to change the slope by 1% is

$$K = \frac{L}{A} \tag{3.10}$$

where

K = value that is the horizontal distance, in ft, required to affect a 1% change in the slope of the vertical curve,

L = length of curve in ft, and

A = absolute value of the difference in grades ($|G_1 - G_2|$) expressed as a percentage.

This K-value can also be used to compute the high and low point locations of crest and sag vertical curves, respectively (provided the high or low point does not occur at the *PVC* or *PVT*). As shown in Example 3.1, setting $dy/dx = 0$ in Eq. 3.2 and solving for x gives the distance from the *PVC* to the high/low point. If Eq. 3.6 is used to substitute for a in Eq. 3.2 (with $L = KA$), it can be shown that setting $dy/dx = 0$ in Eq. 3.2 gives

$$x_{hl} = K \times |G_1| \tag{3.11}$$

where

x_{hl} = distance from the *PVC* to the high/low point in ft, and

other terms are as defined previously.

In addition to high/low point computations, K-values have an important application in the design of vertical curves, as will be demonstrated in Sections 3.3.3 and 3.3.4.

EXAMPLE 3.3   VERTICAL CURVE DESIGN WITH *K*-VALUES

A curve has initial and final grades of +3% and –4%, respectively, and is 700 ft long. The *PVC* is at elevation of 100 ft. Graph the vertical curve elevations and the slope of the curve against the length of curve. Compute the *K*-value and use it to locate the high point of the curve (distance from the *PVC*).

SOLUTION

Recall that to find the slope at any point on the curve, we take the derivative of Eq. 3.1, which gives Eq. 3.2. To apply this equation, a and b need to be determined. From Eq. 3.6,

$$a = \frac{-4.0 - 3.0}{2(7)} = -0.5$$

and from Eq. 3.3,

$$b = G_1 = 3$$

The results of applying Eq. 3.2 and solving for the slope at all points along the curve, as well as a profile view of the curve itself (by application of Eq. 3.1), are shown graphically in Fig. 3.5 (exaggerating the vertical scale). Figure 3.5 shows the constant rate of change of the slope along the length of the curve. The circular points on the slope-of-curve line correspond to changes in grade of 1%, and these points occur at equal intervals of 100 ft.

To show that this is consistent with the K-value, Eq. 3.10 gives

$$K = \frac{L}{A} = \frac{700}{|3 - (-4)|} = 100 \text{ ft}$$

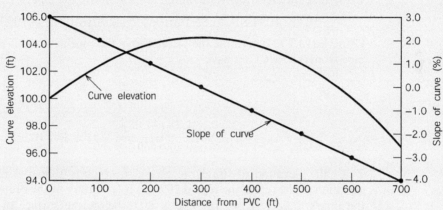

Figure 3.5 Profile view of vertical curve for Example 3.3 with the graph of the slope at all points along the curve overlaid.

This indicates that there should be a change in grade of 1% for every 100 ft of curve length (measured in the horizontal plane), and this is consistent with Fig. 3.5. Applying Eq. 3.11 with the K-value of 100 ft gives the high point at 300 ft from the beginning of the curve ($x_{hl} = 100 \times 3 = 300$ ft). This is shown in Fig. 3.5, where the slope of the curve at 300 ft is zero (the same result obtained by setting the derivative of Eq. 3.2 equal to zero and solving for x). This result can also be explained conceptually based on the definition of the K-value. The K-value gives the horizontal distance required to effect a 1% change in the slope of the curve, and for this curve that value is 100 ft. Thus, to go from an initial grade (G_1) of 3% to a grade of 0% (the high point) requires a horizontal distance equal to $K \times 3$, or 300 ft.

EXAMPLE 3.4   **VERTICAL CURVE DESIGN USING OFFSETS**

A vertical curve crosses a 4-ft diameter pipe at right angles. The pipe is located at station 110 + 85 and its centerline is at elevation 1091.60 ft. The *PVI* of the vertical curve is at station 110 + 00 and elevation 1098.4 ft. The vertical curve is equal tangent, 600 ft long, and connects an initial grade of +1.20% and a final grade of −1.08%. Using offsets, determine the depth, below the surface of the curve, of the top of the pipe and determine the station of the highest point on the curve.

SOLUTION

The *PVC* is at station 107 + 00 (110 + 00 minus 3 + 00, which is half of the curve length), so the pipe is 385 ft (110 + 85 minus 107 + 00) from the beginning of the curve (*PVC*). The elevation of the *PVC* will be the elevation of the *PVI* minus the drop in grade over one-half the curve length,

$$1098.4 - (3 \text{ stations} \times 1.2 \text{ ft/station}) = 1094.8 \text{ ft}$$

Using this, the elevation of the initial tangent above the pipe is

$$1094.8 + (3.85 \text{ stations} \times 1.2 \text{ ft/station}) = 1099.42 \text{ ft}$$

Using Eq. 3.7 to determine the offset above the pipe at $x = 385$ ft (the distance of the pipe from the *PVC*), we have

$$Y = \frac{A}{200L} x^2$$

$$Y = \frac{|1.2 - (-1.08)|}{200(600)} (385)^2 = 2.82 \text{ ft}$$

Thus, the elevation of the curve above the pipe is 1096.6 ft (1099.42 − 2.82). The elevation of the top of the pipe is 1093.60 ft (elevation of the centerline plus one-half of the pipe's diameter), so the pipe is <u>3.0 ft</u> below the surface of the curve (1096.6 − 1093.6).

To determine the location of the highest point on the curve, we find K from Eq. 3.10 as

$$K = \frac{600}{|1.2 - (-1.08)|} = 263.16$$

and the distance from the *PVC* to the highest point is (from Eq. 3.11)

$$x_{hl} = K \times |G_1| = 263.16 \times 1.2 = 315.79 \text{ ft}$$

This gives the station of the highest point at <u>110+15.79</u> (107 + 00 plus 3 + 15.79).

Note that this example could also be solved by applying Eq. 3.1, setting Eq. 3.2 equal to zero (for determining the location of the highest point on the curve), and following the procedure used in Example 3.1.

3.3.2   Stopping Sight Distance

Construction of a vertical curve is generally a costly operation requiring the movement of significant amounts of earthen material. Thus, one of the primary challenges facing highway designers is to minimize construction costs (usually by making the vertical curve as short as possible) while still providing an adequate level of safety. An appropriate level of safety is usually defined as that level of safety that gives drivers sufficient sight distance to allow them to safely stop their vehicles to avoid collisions with objects obstructing their forward motion. The provision of adequate roadway drainage is sometimes an important concern as well, but is not discussed in terms of vertical curves in this book (see [AASHTO, 2018]). Referring back to the vehicle braking performance concepts discussed in Chapter 2, we can compute this necessary SSD as the summation of vehicle practical stopping distance (Eq. 2.47) and the distance traveled during driver perception/reaction time (Eq. 2.49). That is,

$$\text{SSD} = \frac{V_1^2}{2g\left(\left(\dfrac{a}{g}\right) \pm G\right)} + V_1 \times t_r \tag{3.12}$$

where

$$
\begin{aligned}
\text{SSD} &= \text{stopping sight distance in ft,} \\
V_1 &= \text{initial vehicle speed in ft/s,} \\
g &= \text{gravitational constant, 32.2 ft/s}^2, \\
a &= \text{deceleration rate in ft/s}^2, \\
G &= \text{roadway grade (+ for uphill and – for downhill) in percent/100, and} \\
t_r &= \text{perception/reaction time in s.}
\end{aligned}
$$

Recall from Sections 2.9.5 and 2.9.6 that a value of 11.2 ft/s^2 for a and a value of 2.5 s for t_r were recommended for roadway design purposes. The design speed of the highway is defined as the maximum safe speed at which a highway can be negotiated assuming near-worst-case conditions (wet-weather conditions). The application of Eq. 3.12 (assuming $G = 0$) produces the SSDs presented in Table 3.1.

Table 3.1 Stopping Sight Distance

Design speed (mi/h)	Brake reaction distance (ft)	Braking distance on level (ft)	Stopping sight distance Calculated (ft)	Stopping sight distance Design (ft)
15	55.1	21.6	76.7	80
20	73.5	38.4	111.9	115
25	91.9	60.0	151.9	155
30	110.3	86.4	196.7	200
35	128.6	117.6	246.2	250
40	147.0	153.6	300.6	305
45	165.4	194.4	359.8	360
50	183.8	240.0	423.8	425
55	202.1	290.3	492.4	495
60	220.5	345.5	566.0	570
65	238.9	405.5	644.4	645
70	257.3	470.3	727.6	730
75	275.6	539.9	815.5	820
80	294.0	614.3	908.3	910

Note: Brake reaction distance is based on a time of 2.5 s; a deceleration rate of 11.2 ft/s^2 is used to determine calculated stopping sight distance.

Source: American Association of State Highway and Transportation Officials, *A Policy on Geometric Design of Highways and Streets*, 7th Edition, Washington, DC, 2018. Used by permission.

3.3.3   Stopping Sight Distance and Crest Vertical Curve Design

The length of curve (L in Fig. 3.3) is the critical element in providing sufficient SSD on a vertical curve. Longer curve lengths provide more SSD, all else being equal, but are more costly to construct. Shorter curve lengths are less expensive to construct but may not provide adequate SSD due to more rapid changes in slope. What is needed, then, is an expression for minimum curve length given a required SSD. In developing such an expression, crest and sag vertical curves are considered separately.

The case of designing a crest vertical curve for adequate SSD is illustrated in Fig. 3.6. To determine the minimum length of curve for a required sight distance, the properties of a parabola for an equal-tangent curve can be used to show that

For $S < L$

$$L_m = \frac{AS^2}{200\left(\sqrt{H_1} + \sqrt{H_2}\right)^2} \tag{3.13}$$

For $S > L$

$$L_m = 2S - \frac{200\left(\sqrt{H_1} + \sqrt{H_2}\right)^2}{A} \tag{3.14}$$

where

L_m = minimum length of vertical curve in ft,

A = absolute value of the difference in grades ($|G_1 - G_2|$), expressed as a percentage, and

other terms are as defined in Fig. 3.6.

For the sight distance required to provide adequate SSD, current AASHTO design guidelines [2018] use a driver eye height, H_1, of 3.5 ft and a roadway object height, H_2, of 2.0 ft (the height of an object to be avoided by stopping before a collision). In applying Eqs. 3.13 and 3.14 to determine the minimum length of curve required to provide adequate SSD, we set the sight distance, S, equal to the stopping sight distance, SSD (note that the relatively small distance from the driver's eye position to the front of the vehicle is ignored). Substituting AASHTO guidelines for H_1 and H_2 and letting S = SSD in Eqs. 3.13 and 3.14 gives

For SSD < L

$$L_m = \frac{A \times \text{SSD}^2}{2158} \qquad \textbf{(3.15)}$$

For SSD > L

$$L_m = 2 \times \text{SSD} - \frac{2158}{A} \qquad \textbf{(3.16)}$$

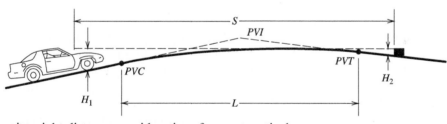

Figure 3.6 Stopping sight distance considerations for crest vertical curves.

S = sight distance in ft,

H_1 = height of driver's eye above roadway surface in ft,

H_2 = height of object above roadway surface in ft,

L = length of the curve in ft,

PVC = point of vertical curve (the initial point of the curve),

PVI = point of vertical intersection (intersection of initial and final grades), and

PVT = point of vertical tangent, which is the final point of the vertical curve (the point where the curve returns to the final grade or, equivalently, the final tangent).

EXAMPLE 3.5   **DESIGN SPEED AND CREST VERTICAL CURVE DESIGN**

A highway is being designed to AASHTO guidelines with a 70-mi/h design speed, and at one section, an equal-tangent vertical curve must be designed to connect grades of +1.0% and −2.0%. Determine the minimum length of curve necessary to meet SSD requirements.

SOLUTION

If we ignore the effect of grades ($Gs = 0$), the SSD can be read directly from Table 3.1. In this case, the SSD corresponding to a speed of 70 mi/h is 730 ft. If we assume that $L > $ SSD (an assumption that is typically made), Eq. 3.15 gives

$$L_m = \frac{A \times SSD^2}{2158} = \frac{3 \times 730^2}{2158} = \underline{\underline{740.82 \text{ ft}}}$$

Since 740.82 > 730, the assumption that $L > $ SSD was correct.

The assumption that $G = 0$, made at the beginning of Example 3.5, is not really correct. If $G \neq 0$, we cannot use the SSD values in Table 3.1 and instead must apply Eq. 3.12 with the appropriate G value. In this problem, if we use the initial grade in Eq. 3.12 (+1.0%), we will underestimate the SSD because the vertical curve has a slope as steeply positive as this only at the *PVC*. If we use the final grade in Eq. 3.12 (−2.0%), we will overestimate the SSD because the vertical curve has a slope as steeply negative as this only at the *PVT*. If we knew where the vehicle began to brake, we could use the first derivative of the parabolic curve function (from Eq. 3.2) to give G in Eq. 3.12 and set up the equation to solve for SSD exactly. In practice, policies vary as to how this grade issue is handled. Fortunately, because sight distance tends to be greater on downgrades (which require longer stopping distances) than on upgrades, a self-correction for the effect of grades is generally provided. As a consequence, some design agencies ignore the effect of grades completely, while others assume G is equal to zero for grades less than 3% and use simple adjustments to the SSD, depending on the initial and final grades, for grades of 3% or more. For the remainder of this chapter, we will ignore the effect of grades ($G = 0$ will be used in Eq. 3.12). However, it must be pointed out that the use of SSD grade corrections is very easy and straightforward, and all of the equations presented herein still apply.

The use of Eqs. 3.15 and 3.16 can be simplified if the initial assumption that $L > $ SSD is made, in which case Eq. 3.15 is always used. The advantage of this assumption is that the relationship between A and L_m is linear, and Eq. 3.10 can be used to give

$$L_m = KA \tag{3.17}$$

where $K = $ horizontal distance, in ft, required to effect a 1% change in the slope (as in Eq. 3.10), defined as

$$K = \frac{\text{SSD}^2}{2158} \qquad (3.18)$$

With known SSD for a given design speed (assuming $G = 0$), K-values can be computed for crest vertical curves as shown in Table 3.2. Thus, the minimum curve length can be obtained (as shown in Eq. 3.17) simply by multiplying A by the K-value read from Table 3.2.

Some discussion about the assumption that $L >$ SSD is warranted. This assumption is made because there are two complications that could arise when SSD $> L$. First, if SSD $> L$, the relationship between A and L_m is not linear, so K-values cannot be used in the $L = KA$ formula (Eq. 3.10). Second, at low values of A, it is possible to get negative minimum curve lengths (see Eq. 3.16). As a result of these complications, the assumption that $L >$ SSD is almost always made in practice, and Eqs. 3.17 and 3.18 and the K-values presented in Table 3.2 are used. It is important to note that the assumption that $L >$ SSD (upon which Eqs. 3.17 and 3.18 are based) is a good one because in many cases, L is greater than SSD, and when it is not (SSD $> L$), use of the $L >$ SSD formula (Eq. 3.15 instead of Eq. 3.16) gives longer curve lengths and thus the error is on the conservative, safe side.

A final point relates to the smallest allowable length of curve. Very short vertical curves can be difficult to construct and may not be warranted for safety purposes. As a result, it is common practice to set minimum curve length limits that range from 100 to 325 ft depending on individual jurisdictional guidelines. A common alternative to these limits is to set the minimum curve length limit at three times the design speed (with speed in mi/h and length in ft) [AASHTO, 2018].

Table 3.2 Design Controls for Crest Vertical Curves Based on Stopping Sight Distance

Design speed (mi/h)	Stopping sight distance (ft)	Rate of vertical curvature, K^*	
		Calculated	Design
15	80	3.0	3
20	115	6.1	7
25	155	11.1	12
30	200	18.5	19
35	250	29.0	29
40	305	43.1	44
45	360	60.1	61
50	425	83.7	84
55	495	113.5	114
60	570	150.6	151
65	645	192.8	193
70	730	246.9	247
75	820	311.6	312
80	910	383.7	384

*Rate of vertical curvature, K, is the length of curve per percent algebraic difference in intersecting grades (A): $K = L/A$.

Source: American Association of State Highway and Transportation Officials, *A Policy on Geometric Design of Highways and Streets*, 7th Edition, Washington, DC, 2018. Used by permission.

EXAMPLE 3.6 **DESIGN SPEED AND CREST VERTICAL CURVE DESIGN WITH K-VALUES**

Solve Example 3.5 using the K-values in Table 3.2.

SOLUTION

From Example 3.5, $A = 3$. For a 70-mi/h design speed, $K = 247$ (from Table 3.2). Therefore, application of Eq. 3.17 gives

$$L_m = KA = 247(3) = \underline{\underline{741.00 \text{ ft}}}$$

which is almost identical to the 740.82 ft obtained in Example 3.5. This difference is due to rounding. In this example, the rounded K of 247 was used as opposed to the calculated K of 246.9. The rounded values are typically used in design for computational convenience. Note, however, that fractional calculated values are always rounded up to the nearest integer value, to be conservative.

EXAMPLE 3.7 **STOPPING SIGHT DISTANCE AND CREST VERTICAL CURVE DESIGN**

If the grades in Example 3.5 intersect at station 100 + 00, determine the stationing of the *PVC*, *PVT*, and curve high point for the minimum curve length based on SSD requirements.

SOLUTION

Using the curve length from Example 3.6, $L = 741$ ft. Since the curve is equal tangent (as are virtually all curves used in practice), one-half of the curve will occur before the *PVI* and one-half after, so that

$$PVC \text{ is at } 100 + 00 - L/2 = 100 + 00 \text{ minus } 3 + 70.5 = \underline{\underline{96 + 29.5}}$$

$$PVT \text{ is at } 100 + 00 + L/2 = 100 + 00 \text{ plus } 3 + 70.5 = \underline{\underline{103 + 70.5}}$$

For the stationing of the high point, Eq. 3.11 is used:

$$x_{hl} = K \times |G_1| = 247(1) = 247 \text{ ft}$$

or

$$\text{station } 96 + 29.5 \text{ plus } 2 + 47 = \underline{\underline{98 + 76.5}}$$

3.3.4 Stopping Sight Distance and Sag Vertical Curve Design

Sag vertical curve design differs from crest vertical curve design in the sense that sight distance is governed by nighttime conditions because in daylight, sight distance on a sag vertical curve is unrestricted. Thus, the critical concern for sag vertical curve design is the length of roadway illuminated by the vehicle headlights, which is a function of the height of the headlight above the roadway and the inclined angle of the headlight beam, relative to the horizontal plane of the car. The sag vertical curve sight distance design problem is illustrated in Fig. 3.7.

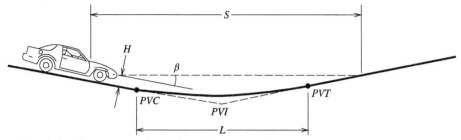

Figure 3.7 Stopping sight distance considerations for sag vertical curves.

S = sight distance in ft,

H = height of headlight in ft,

β = inclined angle of headlight beam in degrees,

L = length of the curve in ft,

PVC = point of the vertical curve (the initial point of the curve),

PVI = point of vertical intersection (intersection of initial and final grades), and

PVT = point of vertical tangent, which is the final point of the vertical curve (the point where the curve returns to the final grade or, equivalently, the final tangent).

To determine the minimum length of curve for a required sight distance, the properties of a parabola for an equal-tangent curve can be used to show that

For $S < L$

$$L_m = \frac{AS^2}{200\left(H + S \tan \beta\right)} \tag{3.19}$$

For $S > L$

$$L_m = 2S - \frac{200\left(H + S \tan \beta\right)}{A} \tag{3.20}$$

where

L_m = minimum length of vertical curve in ft,

A = absolute value of the difference in grades ($|G_1 - G_2|$), expressed as a percentage, and

other terms are as defined in Fig. 3.7.

For the sight distance required to provide adequate SSD, current AASHTO design guidelines [2018] use a headlight height of 2.0 ft and an upward angle of one degree. Substituting these design guidelines and S = SSD (as was done in the crest vertical curve case, and again ignoring the relatively small distance from the driver's eye position to the front of the vehicle) into Eqs. 3.19 and 3.20 gives

For SSD < L

$$L_m = \frac{A \times \text{SSD}^2}{400 + 3.5 \times \text{SSD}} \tag{3.21}$$

For SSD > L

$$L_m = 2 \times \text{SSD} - \frac{400 + 3.5 \times \text{SSD}}{A} \tag{3.22}$$

where

SSD = stopping sight distance in ft, and
other terms are as defined previously.

As was the case for crest vertical curves, K-values can be computed by assuming L > SSD, which gives us the linear relationship between L_m and A as shown in Eq. 3.21. Thus, for sag vertical curves (with $L_m = KA$),

$$K = \frac{\text{SSD}^2}{400 + 3.5 \, \text{SSD}} \tag{3.23}$$

where

K = horizontal distance, in ft, required to effect a 1% change in the slope (as in Eq. 3.10), and
other terms are as defined previously.

The K-values corresponding to design-speed–based SSDs are presented in Table 3.3. As was the case for crest vertical curves, some caution should be exercised in using this table because the assumption that $G = 0$ (for determining SSD) is used. Also, assume that L > SSD is a safe, conservative assumption (as was the case for crest vertical curves), and the smallest allowable curve lengths for sag curves are the same as those for crest curves (see discussion in Section 3.3.3).

Table 3.3 Design Controls for Sag Vertical Curves Based on Stopping Sight Distance

Design speed (mi/h)	Stopping sight distance (ft)	Rate of vertical curvature, K*	
		Calculated	Design
15	80	9.4	10
20	115	16.5	17
25	155	25.5	26
30	200	36.4	37
35	250	49.0	49
40	305	63.4	64
45	360	78.1	79
50	425	95.7	96
55	495	114.9	115
60	570	135.7	136
65	645	156.5	157
70	730	180.3	181
75	820	205.6	206
80	910	231.0	231

*Rate of vertical curvature, K, is the length of curve per percent algebraic difference in intersecting grades (A): $K = L/A$.

Source: American Association of State Highway and Transportation Officials, *A Policy on Geometric Design of Highways and Streets*, 7th Edition, Washington, DC, 2018. Used by permission.

EXAMPLE 3.8   SAG VERTICAL CURVE FUNDAMENTALS WITH DESIGN SPEED

An equal-tangent sag vertical curve has an initial grade of −2.5%. It is known that the final grade is positive and that the low point is at elevation 270 ft and station 141 + 00. The PVT of the curve is at elevation 274 ft and the design speed of the curve is 35 mi/h. Determine the station and elevation of the PVC and PVI.

SOLUTION

From Table 3.3, it can be seen the $K = 49$ for a design speed of 35 mi/h. With this, Eq. 3.11 is used to find the distance of the low point from the PVC:

$$x_{hl} = K \times |G_1| = 49(2.5) = 122.5 \text{ ft}$$

Knowing the elevation of the low point (270 ft) and the distance of the low point from the PVC (122.5 ft), Eq. 3.1 can be applied to determine the elevation of the PVC (c in the Eq. 3.1):

$$y = ax^2 + bx + c$$

From Eq. 3.3,

$$b = G_1 = -0.025$$

and from Eq. 3.6,

$$a = \frac{G_2 - G_1}{2L}$$

Because the final grade is known to be positive (and with G_2 being negative),

$$G_2 - G_1 = |G_1 - G_2| = A,$$

Using $L = KA$ from Eq. 3.17, Eq. 3.6 becomes

$$a = \frac{G_2 - G_1}{2L} = \frac{A/100}{2KA} = \frac{0.01}{2K} = \frac{0.01}{2(49)} = 0.000102$$

Note that A is divided by 100 to make certain that the units are consistent with the denominator since L is in feet from the $L = KA$ equation. At the low point, $y = 270$ ft so solving for c in Eq. 3.1 with $x = 122.5$, $a = 0.000102$ and $b = -0.025$ gives

$$270 = 0.000102(122.5)^2 + (-0.025)(122.5) + c$$
$$c = \underline{271.53} = \text{elevation of the } PVC$$

Knowing that the station of the low point is $141 + 00$ and the distance from the PVC to the low point is 122.5 ft,

$$\text{station of } PVC = 141 + 00 \text{ minus } 1 + 22.5 = \underline{139 + 77.5}$$

Next, the length of the curve is determined by applying Eq. 3.1. Because it is known that the elevation of the PVT is 274 ft, using $y = 274$ means that $x = L$ in Eq. 3.1, so (with $c = 271.53$)

$$y = aL^2 + bL + c$$
$$274 = 0.000102L^2 + (-0.025)L + 271.53$$
$$L = \underline{320.624 \text{ ft}}$$

The station of the PVI is

$$\text{station of } PVI = \text{station of } PVC + L/2$$
$$= 139 + 77.5 \text{ plus } (3 + 20.624/2) = \underline{141 + 37.812}$$

Finally, the elevation of the PVI is determined as

$$\text{elevation of } PVI = \text{elevation of } PVC + G_1(L/2)$$
$$= 271.53 - 0.025(320.624/2) = \underline{267.52}$$

EXAMPLE 3.9   COMBINED SAG AND CREST VERTICAL CURVES WITHOUT A CONSTANT-GRADE CONNECTION

An existing tunnel needs to be connected to a newly constructed bridge with sag and crest vertical curves. The profile view of the tunnel and bridge is shown in Fig. 3.8. Develop a vertical alignment to connect the tunnel and bridge by determining the highest possible common design speed for the sag and crest (equal-tangent) vertical curves needed. Compute the stationing and elevations of PVC, PVI, and PVT curve points.

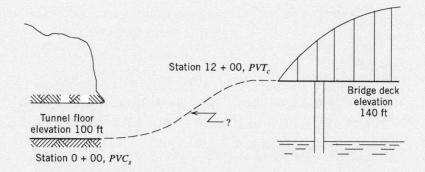

Figure 3.8 Profile view (vertical alignment diagram) for Example 3.9.

SOLUTION

From left to right (see Fig. 3.8), a sag vertical curve (with subscript s) and a crest vertical curve (with subscript c) are needed to connect the tunnel and bridge. From the given information, it is known that $G_{1s} = 0\%$ (the initial slope of the sag vertical curve) and $G_{2c} = 0\%$ (the final slope of the crest vertical curve). To obtain the highest possible design speed, we want to use all of the horizontal distance available. This means we want to connect the curve so that the PVT of the sag curve (PVT_s) will be the PVC of the crest curve (PVC_c). If this is the case, $G_{2s} = G_{1c}$ and since $G_{1s} = G_{2c} = 0$, $A_s = A_c = A$, the common algebraic difference in the grades.

Since 1200 ft separates the tunnel and bridge,

$$L_s + L_c = 1200$$

Also, the summation of the end-of-curve offset for the sag curve and the beginning-of-curve offset (relative to the final grade) for the crest curve must equal 40 ft. Using the equation for the final offset, Eq. 3.9, we have

$$\frac{AL_s}{200} + \frac{AL_c}{200} = 40$$

Rearranging,

$$\frac{A}{200}(L_s + L_c) = 40$$

and since $L_s + L_c = 1200$,

$$\frac{A}{200}(1200) = 40$$

Solving for A gives $A = 6.667\%$. The problem now becomes one of finding K-values that allow $L_s + L_c = 1200$. Since $L = KA$ (Eq. 3.17), we can write

$$K_s A + K_c A = 1200$$

Substituting $A = 6.667$,

$$K_s + K_c = 180$$

To find the highest possible design speed, Tables 3.2 and 3.3 are used to arrive at K-values to solve $K_s + K_c = 180$. From Tables 3.2 and 3.3, it is apparent that the highest possible design speed is 50 mi/h, at which speed $K_c = 84$ and $K_s = 96$ (the summation of K's is 180).

To arrive at the stationing of curve points, we first determine curve lengths as

$$L_s = K_s A = 96(6.667) = 640.0 \text{ ft}$$

$$L_c = K_c A = 84(6.667) = 560.0 \text{ ft}$$

Since the station of the PVC_s is $\underline{0+00}$ (given), it is clear that the $PVI_s = \underline{3+20.0}$, $PVT_s = PVC_c = \underline{6+40.0}$, $PVI_c = \underline{9+20.0}$, and $PVT_c = \underline{12+00.0}$. For elevations, $PVC_s = PVI_s = \underline{100 \text{ ft}}$ and $PVI_c = PVT_c = \underline{140 \text{ ft}}$. Finally, the elevation of PVT_s and PVC_c can be computed as

$$100 + \frac{AL_s}{200} = 100 + \frac{6.667(640.0)}{200} = 121.33 \text{ ft}$$

EXAMPLE 3.10   COMBINED SAG AND CREST VERTICAL CURVES WITH A CONSTANT-GRADE CONNECTION

Consider the conditions described in Example 3.9. Suppose that a design speed of only 35 mi/h is needed. Determine the lengths of curves required to connect the bridge and tunnel while keeping the connecting grade as small as possible.

SOLUTION

It is known that the 1200 ft separating the tunnel and bridge are more than enough to connect a 35-mi/h alignment because Example 3.9 showed that 50 mi/h is possible. Therefore, to connect the tunnel and bridge and keep the connecting grade as small as possible, we will place a constant-grade section between the sag and crest curves (as shown in Fig. 3.9).

The elevation change will be the final offsets of the sag and crest curves plus the change in elevation resulting from the constant-grade section connecting the two curves. Let G_{con} be the grade of the constant-grade section. This means that $G_{2s} = G_{1c} = G_{con}$, and since $G_{1s} = G_{2c} = 0$ (as in Example 3.9), $G_{con} = A_s = A_c = A$. The equation that will solve the vertical alignment for this problem is

$$\frac{AL_s}{200} + \frac{AL_c}{200} + \frac{A(1200 - L_s - L_c)}{100} = 40$$

where the third term accounts for the elevation difference attributable to the constant-grade section connecting the sag and crest curves (the 100 in the denominator of this term converts A from percent to ft/ft). Using $L = KA$, we have

$$\frac{A^2 K_s}{200} + \frac{A^2 K_c}{200} + \frac{A(1200 - K_s A - K_c A)}{100} = 40$$

From Table 3.2, $K_c = 29$, and from Table 3.3, $K_s = 49$. Putting these values in the above equation gives

$$0.39 A^2 + 12A - 0.78 A^2 = 40$$

$$-0.39 A^2 + 12A - 40 = 0$$

Solving this gives $A = 3.803$ and $A = 26.966$; $A = 3.803\%$ is chosen because we want to minimize the grade. For this value of A, the curve lengths are

$$L_s = K_s A = 49(3.803) = \underline{186.35 \text{ ft}}$$

$$L_c = K_c A = 29(3.803) = \underline{110.29 \text{ ft}}$$

and the length of the constant-grade section will be 903.36 ft. This means that about 34.35 ft of the elevation difference will occur in the constant-grade section, with the remainder of the elevation difference attributable to the final curve offsets.

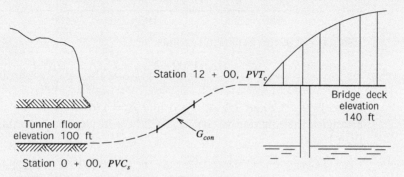

Figure 3.9 Profile view (vertical alignment diagram) for Example 3.10.

Another variation of this type of problem is the case when the initial and final grades are not equal to zero, as in the following example.

EXAMPLE 3.11   COMBINED SAG AND CREST VERTICAL CONNECTING HIGHWAY SEGMENTS WITH NONZERO GRADES

Two sections of highway are separated by 1800 ft, as shown in Fig. 3.10. Determine the curve lengths required for a 60-mi/h vertical alignment to connect these two highway segments while keeping the connecting grade as small as possible.

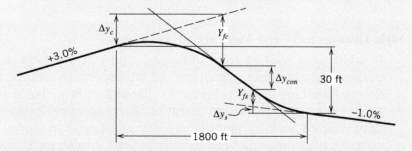

Figure 3.10 Profile view (vertical alignment diagram) for Example 3.11.

SOLUTION

Let Y_{fc} and Y_{fs} be the final offsets of the crest and sag curves, respectively. Let G_{con} be the slope of a constant-grade section connecting the crest and sag curves (we will assume that the horizontal distance is sufficient to connect the highway with a 60-mi/h alignment; if this assumption is incorrect, the following equations will produce an obviously erroneous answer and a lower design speed will have to be chosen). Finally, let Δy_{con} be the change in elevation over the constant-grade section, and let Δy_c and Δy_s be the changes in elevation due to the extended curve tangents. The elevation equation is then (see Fig. 3.10)

$$Y_{fc} + Y_{fs} + \Delta y_{con} + \Delta y_s = 30 + \Delta y_c$$

Substituting offset equations and equations for elevation changes (with subscripts c for crest and s for sag),

$$\frac{A_c L_c}{200} + \frac{A_s L_s}{200} + \frac{G_{con}(1800 - L_c - L_s)}{100} + \frac{1.0 L_s}{100} = 30 + \frac{3.0 L_c}{100}$$

Using $L = KA$, this equation becomes

$$\frac{A_c^2 K_c}{200} + \frac{A_s^2 K_s}{200} + \frac{G_{con}(1800 - K_c A_c - K_s A_s)}{100} + \frac{1.0 K_s A_s}{100} = 30 + \frac{3.0 K_c A_c}{100}$$

From Tables 3.2 and 3.3, $K_c = 151$ and $K_s = 136$. Substituting and defining A's (and arranging the equation so that G_{con} will be positive, and assuming G_{con} will be greater than 1%) gives

$$\frac{(3 + G_{con})^2\, 151}{200} + \frac{(G_{con} - 1)^2\, 136}{200} + \frac{G_{con}\left(1800 - 151(3 + G_{con}) - 136(G_{con} - 1)\right)}{100}$$
$$+ \frac{1.0\left(136(G_{con} - 1)\right)}{100} = 30 + \frac{3.0\left(151(3 + G_{con})\right)}{100}$$

or

$$-1.435 G_{con}^2 + 14.83 G_{con} - 37.475 = 0$$

which gives $G_{con} = 4.40$ (the other possible solution is 5.93, which is rejected because we want to minimize the grade). Using $L = KA$ gives $L_c = \underline{1117.40\ \text{ft}}$ (151×7.40) and $L_s = \underline{462.40\ \text{ft}}$ (136×3.40). Accordingly, the length of the constant-grade section is 220.20 ft ($1800 - L_c - L_s$). Elevations and the locations of curve points can be readily computed with this information.

3.3.5 Passing Sight Distance and Crest Vertical Curve Design

In addition to SSD, in some instances, it may be desirable to provide adequate passing sight distance, which can be an important issue in two-lane highway design (one lane in each direction). Passing sight distance is a factor only in crest vertical curve design because, for sag curves, the sight distance is unobstructed looking up or down the grade, and at night, the headlights of oncoming or opposing vehicles will be seen. In determining the sight distance required to pass on a crest vertical curve, Eqs. 3.13 and 3.14 will apply; however, whereas the driver's eye height, H_1, will remain at 3.5 ft, H_2 will now also be set to 3.5 ft. This value for H_2 is the assumed value for the portion of a vehicle's height necessary to be visible such that it can be recognized as an opposing vehicle to a driver performing a passing maneuver. Using the same height for both H_1 and H_2 provides a reciprocal design relationship; that is, if the driver of the passing vehicle can see the opposing vehicle, then the opposing vehicle driver can see the passing vehicle. Substituting these H values into Eqs. 3.13 and 3.14 and letting the sight distance S equal the passing sight distance, PSD, gives

For PSD < L

$$L_m = \frac{A \times \text{PSD}^2}{2800}$$ (3.24)

For PSD > L

$$L_m = 2 \times \text{PSD} - \frac{2800}{A}$$ (3.25)

where

   L_m = minimum length of vertical curve in ft,

   A = absolute value of the difference in grades ($|G_1-G_2|$), expressed as a percentage, and

   PSD = passing sight distance in ft.

As was the case for SSD, it is typically assumed that the length of curve is greater than the required sight distance (in this case, L > PSD), so

$$K = \frac{\text{PSD}^2}{2800}$$ (3.26)

where

   K = horizontal distance, in ft, required to effect a 1% change in the slope (as in Eq. 3.10), and

   PSD = passing sight distance in ft.

The passing sight distance (PSD) used for design is assumed to consist of four distances: (1) the initial maneuver distance (which includes the driver's perception/reaction time and the time it takes to bring the vehicle from its trailing speed to the point of encroachment on the left lane), (2) the distance that the passing vehicle traverses while occupying the left lane, (3) the clearance length between the passing and opposing vehicles at the end of the passing maneuver, and (4) the distance traversed by the opposing vehicle during two-thirds of the time the passing vehicle occupies the left lane. The determination of these distances is undertaken using assumptions regarding the time of the initial maneuver, average vehicle acceleration, and the speeds of passing, passed, and opposing vehicles. The sum of these four distances gives the required passing sight distance. The reader is referred to [AASHTO, 2018] for a complete description of the assumptions made in determining required passing sight distances.

The minimum distances needed to pass (PSDs) at various design speeds, along with the corresponding K-values as computed from Eq. 3.26, are presented in Table 3.4. Notice that the K-values in this table are much higher than those required for SSD (as given in Table 3.2). As a result, designing a crest curve to provide adequate passing sight distance is often an expensive proposition (due to the length of curve required).

Table 3.4 Design Controls for Crest Vertical Curves Based on Passing Sight Distance.

Design speed (mi/h)	Passing sight distance (ft)	Rate of vertical curvature, K*
20	400	57
25	450	72
30	500	89
35	550	108
40	600	129
45	700	175
50	800	229
55	900	289
60	1000	357
65	1100	432
70	1200	514
75	1300	604
80	1400	700

*Rate of vertical curvature, K, is the length of curve per percent algebraic difference in intersecting grades (A): $K = L/A$.

Source: American Association of State Highway and Transportation Officials, *A Policy on Geometric Design of Highways and Streets*, 7th Edition, Washington, DC, 2018. Used by permission.

EXAMPLE 3.12   VERTICAL CURVE DESIGN WITH PASSING SIGHT DISTANCE

An equal-tangent crest vertical curve is 1000 ft long and connects a +2.5% and a −1.5% grade. If the design speed of the roadway is 55 mi/h, does this curve have adequate passing sight distance?

SOLUTION

To determine the length of curve required to provide adequate passing sight distance at a design speed of 55 mi/h, we use $L = KA$ with $K = 289$ (as read from Table 3.4). This gives

$$L = 289(4.0) = 1156 \text{ ft}$$

Since the curve is only 1000 ft long, it is not long enough to provide adequate passing sight distance. Alternatively, the K-value for the existing design can be compared with that required for a PSD-based design. The K-value for the existing design is

$$K = \frac{1000}{4} = 250$$

Since the K-value of 250 for the existing curve design is less than 289, this curve does not provide adequate PSD for a 55-mi/h design speed.

3.3.6   Underpass Sight Distance and Sag Vertical Curve Design

As mentioned in Section 3.3.4, design for sag curves is based on nighttime conditions because during daytime conditions a driver can see the entire sag curve. However, in the case of a sag curve being built under an overhead structure (such as roadway or railroad crossing), a driver's line of sight may be restricted so that the entire curve length is not visible. An example of this situation is shown in Fig. 3.11.

In designing the sag curve, it is essential that the curve be long enough to provide a suitably gradual rate of curvature such that the overhead structure does not block the line of sight and allows the required SSD for the specified design speed to be maintained.

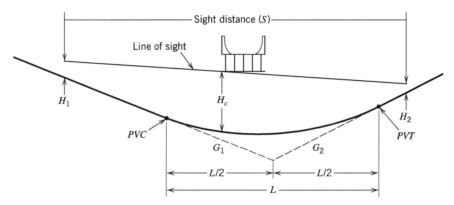

Figure 3.11 Stopping sight distance considerations for underpass sag curves.

Used by permission from American Association of State Highway and Transportation Officials, *A Policy on Geometric Design of Highways and Streets*, 7th Edition, Washington, DC, 2018.

S = sight distance in ft,

H_1 = height of driver's eye in ft,

H_2 = height of object in ft,

H_c = clearance height of overpass structure above roadway in ft,

L = length of the curve in ft

G_1 = initial roadway grade in percent or ft/ft,

G_2 = final roadway grade in percent or ft/ft,

PVC = point of the vertical curve (the initial point of the curve), and

PVT = point of vertical tangent, which is the final point of the vertical curve (the point where the curve returns to the final grade or, equivalently, the final tangent).

Again, by using the properties of a parabola for an equal-tangent vertical curve, it can be shown that the minimum length of sag curve for a required sight distance and clearance height is

For $S < L$

$$L_m = \frac{AS^2}{800\left(H_c - \left(\frac{H_1 + H_2}{2}\right)\right)} \tag{3.27}$$

For $S > L$

$$L_m = 2S - \frac{800\left(H_c - \left(\frac{H_1 + H_2}{2}\right)\right)}{A} \tag{3.28}$$

where

$L_m = $ minimum length of vertical curve in ft,

$A = $ absolute value of the difference in grades ($|G_1 - G_2|$), expressed as a percentage, and

other terms are as defined in Fig. 3.11.

Current AASHTO design guidelines [2018] use a driver eye height, H_1, of 8 ft for a truck driver, and an object height, H_2, of 2 ft for the taillights of a vehicle. Substituting these values and $S = $ SSD into Eqs. 3.27 and 3.28 gives

For SSD $< L$

$$L = \frac{A \times \text{SSD}^2}{800(H_c - 5)} \tag{3.29}$$

For SSD $> L$

$$L = 2 \times \text{SSD} - \frac{800(H_c - 5)}{A} \tag{3.30}$$

where

SSD $= $ stopping sight distance in ft, and

other terms are as defined previously.

In the case where there is an existing sag curve alignment and a new overpass structure is to be built over it, the above equations can be rearranged to solve for the necessary clearance height, H_c, of the overpass structure to provide for the required SSD. When the clearance height is determined in this manner, it is necessary to check this value against the minimum clearance heights based on maximum vehicle height regulations and AASHTO recommendations. Maximum vehicle heights as regulated by state laws range from 13.5 to 14.5 ft. AASHTO [2018] recommends a minimum structure clearance height of 14.5 ft and a desirable clearance height of 16.5 ft. AASHTO [2018] also recommends that clearance heights be no less than 1 ft greater than the maximum allowable vehicle height. This provides a margin for snow or ice accumulation, some over-height vehicles, and future roadway resurfacings. Thus, in building a new overpass structure over an existing sag curve alignment, the clearance height must be determined for both required SSD and maximum allowable vehicle height, and the greater of the two values should be used.

EXAMPLE 3.13   **UNDERPASS VERTICAL CURVE CLEARANCE**

An equal-tangent sag curve has an initial grade of −4.0%, a final grade of +3.0%, and a length of 1270 ft. An overpass is being placed directly over the *PVI* of this curve. At what height above the roadway should the bottom of this sign be placed?

SOLUTION

For this situation, Eq. 3.29 or 3.30 must be used to solve for the necessary clearance height based on SSD. Thus, the required SSD must be determined for the given sag curve specifications, based on the design speed. The design speed for the curve can be determined from the *K*-value by applying Eq. 3.10 as follows:

$$K = \frac{L}{A} = \frac{1270}{|-4-3|} = 181.4$$

From Table 3.3, this *K*-value corresponds approximately to a design speed of 70 mi/h ($K = 181$). For a 70-mi/h design speed, the required SSD is 730 ft. Since the curve length is greater than the required SSD (1270 > 730), Eq. 3.29 applies:

$$L = \frac{A \times \text{SSD}^2}{800(H_c - 5)}$$

Rearranging this equation to solve for the clearance height, H_c, and substituting $A = 7\%$, SSD = 730 ft, and $L = 1270$ ft gives

$$H_c = \frac{A \times \text{SSD}^2}{800L} + 5$$

$$= \frac{7 \times 730^2}{800(1270)} + 5$$

$$= 8.67 \text{ ft}$$

Although only 8.67 ft is needed for SSD requirements, AASHTO [2018] recommends a minimum clearance height of 14.5 ft to take maximum vehicle height into account. Thus, the bottom of the overpass should be placed at least 14.5 ft above the roadway surface (at the *PVI*), but desirably at a height of 16.5 ft according to AASHTO [2018].

3.4   HORIZONTAL ALIGNMENT

The critical aspect of horizontal alignment is the horizontal curve, with the focus on design of the directional transition of the roadway in a horizontal plane. Stated differently, a horizontal curve provides a transition between two straight (or tangent) sections of roadway. A key concern in this directional transition is the ability of the vehicle to negotiate a horizontal curve. (Provision of adequate drainage is also important, but is not discussed in this book; see [AASHTO, 2018].) As was the case with the straight-line vehicle performance characteristics discussed at length in Chapter 2, the highway engineer must design a horizontal alignment to accommodate the cornering capabilities of a variety of vehicles, ranging from nimble sports cars to ponderous trucks. A theoretical assessment of vehicle cornering at the level of detail given to straight-line performance in Chapter 2 is beyond the scope of this book (see [Campbell, 1978] and [Wong, 2008]). Instead, vehicle cornering performance is viewed only at the practical design-oriented level, with equations simplified in a manner similar to that used for the stopping-distance equation discussed in Section 2.9.5.

3.4.1   Vehicle Cornering

Figure 3.12 illustrates the forces acting on a vehicle during cornering.

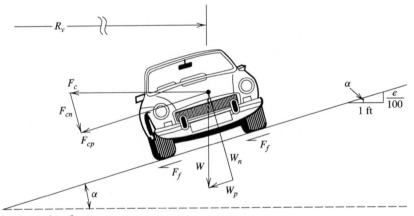

Figure 3.12 Vehicle cornering forces.

R_v = radius defined to the vehicle's traveled path in ft,

α = angle of incline in degrees,

e = number of vertical ft of rise per 100 ft of horizontal distance,

W = weight of the vehicle in lb,

W_n = vehicle weight normal to the roadway surface in lb,

W_p = vehicle weight parallel to the roadway surface in lb,

F_f = side frictional force (centripetal, in lb),

F_c = centripetal force (lateral acceleration × mass, in lb),

F_{cp} = centripetal force acting parallel to the roadway surface in lb, and

F_{cn} = centripetal force acting normal to the roadway surface in lb.

Some basic horizontal curve relationships can be derived by noting that

$$W_p + F_f = F_{cp} \tag{3.31}$$

From basic physics this equation can be written as [with $F_f = f_s(W_n + F_{cn})$]

$$W \sin \alpha + f_s \left(W \cos \alpha + \frac{WV^2}{gR_v} \sin \alpha \right) = \frac{WV^2}{gR_v} \cos \alpha \qquad (3.32)$$

where

    f_s = coefficient of side friction (unitless),
    V = vehicle speed in ft/s,
    g = gravitational constant, 32.2 ft/s², and
    other terms are as defined in Fig. 3.12.

Dividing both sides of Eq. 3.32 by $W \cos \alpha$ gives

$$\tan \alpha + f_s = \frac{V^2}{gR_v}\left(1 - f_s \tan \alpha\right) \qquad (3.33)$$

The term $\tan \alpha$ indicates the superelevation of the curve (banking) and can be expressed in percent; it is denoted e ($e = 100 \tan \alpha$). In words, the superelevation is the number of vertical feet of rise per 100 feet of horizontal distance (see Fig. 3.12). The term $f_s \tan \alpha$ in Eq. 3.33 is conservatively set equal to zero for practical applications due to the small values that f_s and α typically assume (this is equivalent to ignoring the normal component of centripetal force). With $e = 100 \tan \alpha$, Eq. 3.33 can be arranged as

$$R_v = \frac{V^2}{g\left(f_s + \dfrac{e}{100} \right)} \qquad (3.34)$$

EXAMPLE 3.14   SUPERELEVATION ON HORIZONTAL CURVES

A roadway is being designed for a speed of 70 mi/h. At one horizontal curve, it is known that the superelevation is 8.0% and the coefficient of side friction is 0.10. Determine the minimum radius of curve (measured to the traveled path) that will provide for safe vehicle operation.

SOLUTION

The application of Eq. 3.34 gives [with 1.467 (5280/3600) converting mi/h to ft/s]

$$R_v = \frac{V^2}{g\left(f_s + \dfrac{e}{100} \right)} = \frac{(70 \times 1.467)^2}{32.2(0.10 + 0.08)} = \underline{\underline{1819.40 \text{ ft}}}$$

This value is the minimum radius because radii smaller than 1819.40 ft will generate centripetal forces higher than those that can be safely supported by the superelevation and the side frictional force.

In the actual design of a horizontal curve, the engineer must select appropriate values of e and f_s. The value selected for superelevation, e, is critical because high rates of superelevation can cause vehicle steering problems on the horizontal curve, and in cold climates, ice on the roadway can reduce f_s such that vehicles traveling at less than the design speed on an excessively superelevated curve could slide inward off the curve due to gravitational forces. AASHTO provides general guidelines for the selection of e and f_s for horizontal curve design, as shown in Table 3.5. The values presented in this table are grouped by five values of maximum e. The selection of any one of these five maximum e values is dependent on the type of road (for example, higher maximum e's are permitted on freeways than on arterials and local roads) and local design practice. Limiting values of f_s are simply a function of design speed. Table 3.5 also presents calculated radii (given V, e, and f_s) by applying Eq. 3.34.

3.4.2 Horizontal Curve Fundamentals

In connecting straight (tangent) sections of roadway with a horizontal curve, several options are available. The most obvious of these is the simple circular curve, which is just a curve with a single, constant radius. Other options include reverse curves, compound curves, and spiral curves. Reverse curves generally consist of two consecutive curves that turn in opposite directions. They are used to shift the alignment of a highway laterally. The curves used are usually circular and have equal radii. Reverse curves, however, are not recommended because drivers may find it difficult to stay within their lane as a result of sudden changes to the alignment. Compound curves consist of two or more curves, usually circular, in succession. Compound curves are used to fit horizontal curves to very specific alignment needs, such as interchange ramps, intersection curves, or difficult topography. In designing compound curves, care must be taken not to have successive curves with widely different radii, as this will make it difficult for drivers to maintain their lane position as they transition from one curve to the next. Spiral curves are curves with a continuously changing radius. Spiral curves are sometimes used to transition a tangent section of roadway to a circular curve. In such a case, the radius of the spiral curve is equal to infinity where it connects to the tangent section and ends with the radius value of the connecting circular curve at the other end. Because motorists usually create their own transition paths between tangent sections and circular curves by utilizing the full lane width available, spiral curves are not often used. However, there are exceptions. Spiral curves are sometimes used on high-speed roadways with sharp horizontal curves and also to gradually introduce the superelevation of an upcoming horizontal curve. To illustrate the basic principles involved in horizontal curve design, this book will focus only on the single simple circular curve. For detailed information regarding these additional horizontal curve types, refer to standard route-surveying texts, such as Hickerson [1964]. Figure 3.13 shows the basic elements of a simple horizontal curve.

Table 3.5 Minimum Radius Using Limiting Values of e and f_s

Design speed (mi/h)	Maximum e (%)	Limiting values of f_s	Total ($e/100 + f_s$)	Calculated radius, R_v (ft)	Rounded radius, R_v (ft)	Design speed (mi/h)	Maximum e (%)	Limiting values of f_s	Total ($e/100 + f_s$)	Calculated radius, R_v (ft)	Rounded radius, R_v (ft)
10	4.0	0.38	0.42	15.9	16	10	10.0	0.38	0.48	13.9	14
15	4.0	0.32	0.36	41.7	42	15	10.0	0.32	0.42	35.7	36
20	4.0	0.27	0.32	86.0	86	20	10.0	0.27	0.37	72.1	72
25	4.0	0.23	0.27	154.3	154	25	10.0	0.23	0.33	126.3	126
30	4.0	0.20	0.24	250.0	250	30	10.0	0.20	0.30	200.0	200
35	4.0	0.18	0.22	371.2	371	35	10.0	0.18	0.28	291.7	292
40	4.0	0.16	0.20	533.3	533	40	10.0	0.16	0.26	410.3	410
45	4.0	0.15	0.19	710.5	711	45	10.0	0.15	0.25	540.0	540
50	4.0	0.14	0.18	925.9	926	50	10.0	0.14	0.24	694.4	694
55	4.0	0.13	0.17	1186.3	1190	55	10.0	0.13	0.23	876.8	877
60	4.0	0.12	0.16	1500.0	1500	60	10.0	0.12	0.22	1090.9	1090
						65	10.0	0.11	0.21	1341.3	1340
10	6.0	0.38	0.44	15.2	15	70	10.0	0.10	0.20	1633.3	1630
15	6.0	0.32	0.38	39.5	39	75	10.0	0.09	0.19	1973.7	1970
20	6.0	0.27	0.33	80.8	81	80	10.0	0.08	0.18	2370.4	2370
25	6.0	0.23	0.29	143.7	144						
30	6.0	0.20	0.26	230.8	231	10	12.0	0.38	0.50	13.3	13
35	6.0	0.18	0.24	340.3	340	15	12.0	0.32	0.44	34.1	34
40	6.0	0.16	0.22	484.8	485	20	12.0	0.27	0.39	68.4	68
45	6.0	0.15	0.21	642.9	643	25	12.0	0.23	0.35	119.0	119
50	6.0	0.14	0.20	833.3	833	30	12.0	0.20	0.32	187.5	188
55	6.0	0.13	0.19	1061.4	1060	35	12.0	0.18	0.30	272.2	272
60	6.0	0.12	0.18	1333.3	1330	40	12.0	0.16	0.28	381.0	381
65	6.0	0.11	0.17	1656.9	1660	45	12.0	0.15	0.27	500.0	500
70	6.0	0.10	0.16	2041.7	2040	50	12.0	0.14	0.26	641.0	641
75	6.0	0.09	0.15	2500.0	2500	55	12.0	0.13	0.25	806.7	807
80	6.0	0.08	0.14	3047.6	3050	60	12.0	0.12	0.24	1000.0	1000
						65	12.0	0.11	0.23	1224.6	1220
10	8.0	0.38	0.46	14.5	14	70	12.0	0.10	0.22	1484.8	1480
15	8.0	0.32	0.40	37.5	38	75	12.0	0.09	0.21	1785.7	1790
20	8.0	0.27	0.35	76.2	76	80	12.0	0.08	0.20	2133.3	2130
25	8.0	0.23	0.31	134.4	134						
30	8.0	0.20	0.28	214.3	214						
35	8.0	0.18	0.26	314.1	314						
40	8.0	0.16	0.24	444.4	444						
45	8.0	0.15	0.23	587.0	587						
50	8.0	0.14	0.22	757.6	758						
55	8.0	0.13	0.21	960.3	960						
60	8.0	0.12	0.20	1200.0	1200						
65	8.0	0.11	0.19	1482.5	1480						
70	8.0	0.10	0.18	1814.8	1810						
75	8.0	0.09	0.17	2205.9	2210						
80	8.0	0.08	0.16	2666.7	2670						

Note: In recognition of safety considerations, use of $e_{max} = 4.0\%$ should be limited to urban conditions.

Source: American Association of State Highway and Transportation Officials, *A Policy on Geometric Design of Highways and Streets*, 7th Edition, Washington, DC, 2018. Used by permission.

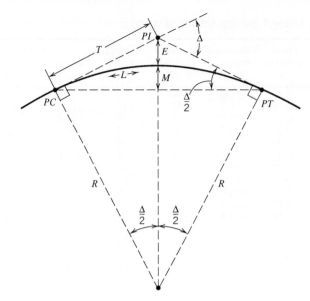

Figure 3.13 Elements of a simple circular horizontal curve.

R = radius, usually measured to the centerline of the road in ft,

Δ = central angle of the curve in degrees,

T = tangent length in ft,

E = external distance in ft,

M = middle ordinate in ft,

PC = point of curve (the beginning point of the horizontal curve),

PI = point of tangent intersection,

PT = point of tangent (the ending point of the horizontal curve), and

L = length of curve in ft.

Another important term is the degree of curve, which is defined as the angle subtended by a 100-ft arc along the horizontal curve. It is a measure of the sharpness of the curve and is frequently used instead of the radius in the construction of the curve. The degree of curve is directly related to the radius of the horizontal curve by

$$D = \frac{100\left(\dfrac{100}{\pi}\right)}{R} = \frac{18000}{\pi R} \qquad (3.35)$$

where

D = degree of curve [angle subtended by a 100-ft arc along the horizontal curve], and

other terms are as defined in Fig. 3.13.

Note that the quantity $180/\pi$ converts from radians to degrees.

Geometric and trigonometric analyses of Fig. 3.13 reveal the following relationships:

$$T = R \tan \frac{\Delta}{2} \qquad (3.36)$$

$$E = R\left(\frac{1}{\cos(\Delta/2)} - 1\right) \tag{3.37}$$

$$M = R\left(1 - \cos\frac{\Delta}{2}\right) \tag{3.38}$$

$$L = \frac{\pi}{180}R\Delta \tag{3.39}$$

where all terms are as defined in Fig. 3.13.

It is important to note that horizontal curve stationing, curve length, and curve radius (R) are usually measured to the centerline of the road. In contrast, the radius determined on the basis of vehicle forces (R_v in Eq. 3.34) is measured from the innermost vehicle path, which is assumed to be the midpoint of the innermost vehicle lane. Thus, a slight correction for lane width is required in equating the R_v of Eq. 3.34 with the R in Eqs. 3.35 through 3.39.

EXAMPLE 3.15   STATIONING ON HORIZONTAL CURVES

A horizontal curve is designed with a 2000-ft radius. The curve has a tangent length of 400 ft and the *PI* is at station 103 + 00. Determine the stationing of the *PT*.

SOLUTION

Equation 3.36 is applied to determine the central angle, Δ:

$$T = R\tan\frac{\Delta}{2}$$

$$400 = 2000\tan\frac{\Delta}{2}$$

$$\Delta = 22.62°$$

So, from Eq. 3.39, the length of the curve is

$$L = \frac{\pi}{180}R\Delta$$

$$L = \frac{3.1416}{180}2000(22.62) = 789.58 \text{ ft}$$

Given that the tangent length is 400 ft,

$$\text{station of } PC = 103 + 00 \text{ minus } 4 + 00 = 99 + 00$$

Since horizontal curve stationing is measured along the alignment of the road,

$$\text{station of } PT = \text{station of } PC + L$$
$$= 99 + 00 \text{ plus } 7 + 89.58 = \underline{\underline{106 + 89.58}}$$

3.4.3   Stopping Sight Distance and Horizontal Curve Design

As is the case for vertical curve design, adequate SSD must be provided in the design of horizontal curves. Sight distance restrictions on horizontal curves occur when obstructions are present, as shown in Fig. 3.14. Such obstructions are frequently encountered in highway design due to the cost of right-of-way acquisition or the cost of moving earthen materials, such as rock outcroppings. When such an obstruction exists, the SSD is measured along the horizontal curve from the center of the traveled lane (the assumed location of the driver's eyes). As shown in Fig. 3.14, for a specified stopping distance, some distance M_s (the middle ordinate of a curve that has an arc length equal to the SSD) must be visually cleared so that the line of sight is such that sufficient SSD is available.

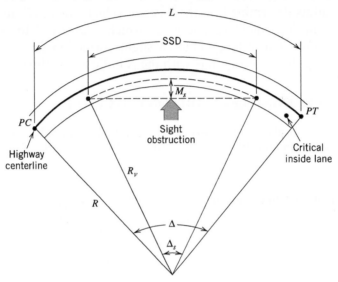

Figure 3.14 Stopping sight distance considerations for horizontal curves.

R = radius measured to the centerline of the road in ft,

R_v = radius to the vehicle's traveled path (usually measured to the center of the innermost lane of the road) in ft,

Δ = central angle of the curve in degrees,

Δ_s = angle (in degrees) subtended by an arc equal in length to the required stopping sight distance (SSD),

L = length of curve in ft,

M_s = middle ordinate necessary to provide adequate stopping sight distance (SSD) in ft.

SSD = stopping sight distance in ft,

PC = point of curve (the beginning point of the horizontal curve), and

PT = point of tangent (the ending point of the horizontal curve).

Equations for computing SSD relationships for horizontal curves can be derived by first determining the angle, Δ_s, for an arc length equal to the required SSD (see Fig. 3.14 and note that this is not the central angle, Δ of the horizontal curve whose arc length is equal to L). Assuming that the length of the horizontal curve exceeds the required SSD (as shown in Fig. 3.14), we have (as with Eq. 3.39)

$$SSD = \frac{\pi}{180} R_v \Delta_s \tag{3.40}$$

Rearranging terms,

$$\Delta_s = \frac{180 \, SSD}{\pi R_v} \tag{3.41}$$

Substituting this into the general equation for the middle ordinate of a simple horizontal curve (Eq. 3.38) to get an expression for M_s gives

$$M_s = R_v \left(1 - \cos \frac{90 \, SSD}{\pi R_v} \right) \tag{3.42}$$

Solving Eq. 3.42 for SSD gives

$$SSD = \frac{\pi R_v}{90} \left[\cos^{-1} \left(\frac{R_v - M_s}{R_v} \right) \right] \tag{3.43}$$

Note that Eqs. 3.40 to 3.43 can also be applied directly to determine sight distance requirements for passing. If these equations are to be used for passing, distance values given in Table 3.4 would apply and SSD in the equations would be replaced by PSD.

EXAMPLE 3.16   **HORIZONTAL CURVE STATIONING AND DESIGN SPEED**

A horizontal curve on a four-lane highway (two lanes each direction with no median) has a superelevation of 6% and a central angle of 40 degrees. The *PT* of the curve is at station 322 + 50 and the *PI* is at 320 + 08. The road has 10-ft lanes and 8-ft shoulders on both sides with high retaining walls going up immediately next to the shoulders. What is the highest safe speed of this curve (highest in 5 mi/h increments) and what is the station of the *PC*?

SOLUTION

The tangent will be equal to the *PI – PC* so $T = 320 + 08 - PC$. The length of the curve will be equal to the *PT – PC* so $L = 322 + 50 - PC$. With the equations for the tangent and length of curve both put in terms of the *PC*, Eqs. 3.36 and 3.39 can be rearranged, respectively, as,

$$R = \frac{T}{\tan \dfrac{\Delta}{2}} \quad \text{and} \quad R = \frac{L}{\dfrac{\pi}{180} \Delta} \quad \text{so that} \quad \frac{T}{\tan \dfrac{\Delta}{2}} = \frac{L}{\dfrac{\pi}{180} \Delta}$$

Substituting the previous tangent and length-of-curve equations ($T = 32008 - PC$ and $L = 32250 - PC$),

$$\frac{32008 - PC}{\tan\dfrac{40}{2}} = \frac{32250 - PC}{\dfrac{\pi}{180}40}$$

Which gives $PC = 317 + 44.25$. Using this value of PC, the tangent can be computed as $T = 32008 - 31744.25 = 263.75$ ft. This value of T can then be used to determine R (see equations above),

$$R = \frac{T}{\tan\dfrac{\Delta}{2}} = \frac{263.75}{\tan\dfrac{40}{2}} = 724.59 \text{ ft}$$

Because the curve radius is usually taken to the centerline of the roadway and there are two 10-ft lanes before the centerline (working from the inside of the curve to the outside), $R_v = R - 10 - 10/2 = 724.59 - 15 = 709.59$ ft. From Table 3.5 with a superelevation of 6%, at 45 mi/h a radius of 643 ft is needed; and at 50 mi/h a radius of 833 ft is needed. Therefore, the highest deign speed for centripetal force is 45 mi/h (since $709.59 > 643$, the design is acceptable for 45 mi/h because more than the needed radius is available, but with $833 > 709.59$ the design is not acceptable for 50 mi/h since insufficient radius is available).

To check for adequate sight distance, M_s is going to be the shoulder width plus half of the inside lane width or $8 + 10/2 = 13$ ft. Consider the SSD required at 40 mi/h. At 40 mi/h, the required SSD is 305 ft (from Table 3.1). Applying Eq. 3.42 gives,

$$M_s = R_v\left(1 - \cos\frac{90\,\text{SSD}}{\pi R_v}\right) = 709.59\left(1 - \cos\frac{90(305)}{\pi(709.59)}\right) = 16.34 \text{ ft}$$

Because 16.34 ft is greater than the 13 ft of available M_s, 40 mi/h is too fast. Consider a speed of 35 mi/h which gives SSD = 250 ft (from Table 3.1). The application of Eq. 3.42 then gives,

$$M_s = R_v\left(1 - \cos\frac{90\,\text{SSD}}{\pi R_v}\right) = 709.59\left(1 - \cos\frac{90(205)}{\pi(709.59)}\right) = 10.99 \text{ ft}$$

Because 10.99 ft is less than 13 ft, the highway is safe at 35 mi/h. Considering both the maximum safe speeds for centripetal force (45 mi/h) and sight distance (35 mi/h), the lower of the two speeds will govern. Thus, 35 mi/h (the highest safe speed for sight distance) is the lower of the two speeds and is the highest safe speed for this curve.

3.5 COMBINED VERTICAL AND HORIZONTAL ALIGNMENT

Thus far the discussion on highway alignment has treated vertical and horizontal curves independently. The combination of vertical and horizontal curves, however, is quite common in geometric design, and often necessary. Obvious examples are highways through mountainous terrain and freeway interchange ramp roadways, which typically have to make significant changes in direction and elevation over a relatively short distance.

As previously mentioned, the design of an alignment that consists of a vertical and horizontal curve in combination usually consists of 2 two-dimensional alignment problems. The following examples illustrate this process.

EXAMPLE 3.17   **COMBINED HORIZONTAL/VERTICAL ALIGNMENT—DESIGN ADEQUACY**

A two-lane highway (two 12-ft lanes) has a posted speed limit of 50 mi/h and, on one section, has both horizontal and vertical curves, as shown in Fig. 3.15. A recent daytime crash (driver traveling eastbound and striking a stationary roadway object) resulted in a fatality and a lawsuit alleging that the 50-mi/h posted speed limit is an unsafe speed for the curves in question and was a major cause of the crash. Evaluate and comment on the roadway design.

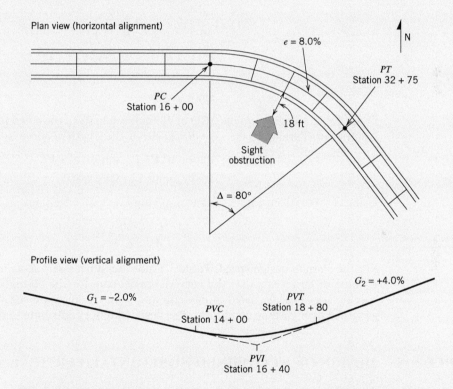

Figure 3.15 Horizontal and vertical alignment for Example 3.17.

SOLUTION

Begin with an assessment of the horizontal alignment. Two concerns must be considered: the adequacy of the curve radius and superelevation and the adequacy of the sight distance on the eastbound (inside) lane. For the curve radius, note from Fig. 3.15 that

$$L = \text{station of } PT - \text{station of } PC$$
$$= 32 + 75 \text{ minus } 16 + 00 = 1675 \text{ ft}$$

Rearranging Eq. 3.39, we get

$$R = \frac{180}{\pi \Delta} L = \frac{180}{\pi (80)} (1675) = 1199.63 \text{ ft}$$

Using the posted speed limit of 50 mi/h with $e = 8.0\%$, we find that Eq. 3.34 can be rearranged to give (with the vehicle traveling in the middle of the inside lane, $R_v = R -$ half the lane width, or $R_v = 1199.63 - 6 = 1193.63$ ft)

$$f_s = \frac{V^2}{gR_v} - e = \frac{(50 \times 1.467)^2}{32.2(1193.63)} - 0.08 = 0.060$$

From Table 3.5, the maximum f_s for 50 mi/h is 0.14. Since 0.060 does not exceed 0.14, the radius and superelevation are sufficient for the 50-mi/h design speed. For sight distance, the available M_s is 18 ft plus the 6-ft distance to the center of the eastbound (inside) lane, or 24 ft. Application of Eq. 3.43 gives

$$\begin{aligned} \text{SSD} &= \frac{\pi R_v}{90} \left[\cos^{-1} \left(\frac{R_v - M_s}{R_v} \right) \right] \\ &= \frac{\pi (1193.63)}{90} \left[\cos^{-1} \left(\frac{1193.63 - 24}{1193.63} \right) \right] \\ &= 479.5 \text{ ft} \end{aligned}$$

From Table 3.1, the required SSD at 50 mi/h is 425 ft, so the 479.5 ft of SSD provided is sufficient. Turning to the sag vertical curve, the length of curve is

$$\begin{aligned} L &= \text{ station of } PVT - \text{station of } PVC \\ &= 18 + 80 \text{ minus } 14 + 00 = 480 \text{ ft} \end{aligned}$$

Using $A = 6\%$ (from Fig. 3.15) and applying Eq. 3.10, we obtain

$$K = \frac{L}{A} = \frac{480}{6} = 80$$

For the 50-mi/h design speed, Table 3.3 indicates a necessary K-value of 96. Thus, the K-value of 80 reveals that the curve is inadequate for the 50-mi/h speed. However, because the crash occurred in daylight and sight distances on sag vertical curves are governed by nighttime conditions, this design did not contribute to the crash.

EXAMPLE 3.18   DESIGN OF A COMBINED HORIZONTAL/VERTICAL ALIGNMENT

A new highway is to be constructed over an existing highway. The two highways will intersect at right angles and are to be grade-separated. Both highways are level grade (constant elevation). The new highway will run east-west and the existing highway runs north-south at elevation of 565.5 ft. The proposed bridge structure for the new highway is such that the bridge girder thickness is 6 ft (measured from the road surface to the bottom of the girder). A single-lane ramp is to be constructed to allow eastbound traffic to go southbound. A single horizontal curve, with a central angle of 90 degrees, is to be used. With a design speed of 40 mi/h and a required superelevation of 4%, determine the following: the stationing of the *PC*, *PI*, and *PT*, assuming that the curve begins at station 40 + 00; the stationing and elevation of all key points along the vertical alignment; the distance that must be cleared from the inside of the horizontal curve so that the line of sight is sufficient to provide sufficient SSD. Fig. 3.16 displays the horizontal and vertical alignments.

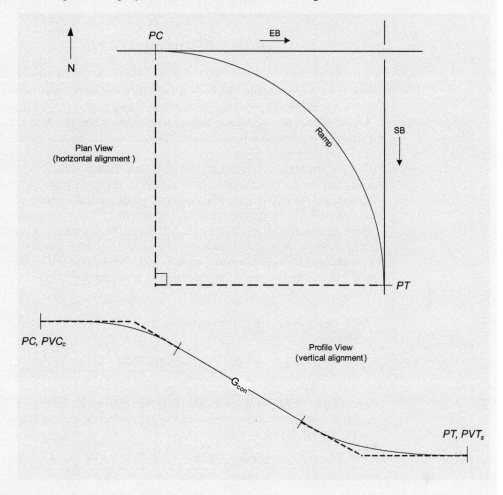

Figure 3.16 Horizontal and vertical alignment for Example 3.18.

SOLUTION

To begin, the required radius to the vehicle path (R_v) is determined to be 533 ft from Table 3.5 with a 40-mi/h design speed and 4% superelevation. Because the ramp is a single lane, the horizontal curve radius will be equal to the radius to the vehicle path ($R = R_v$). Applying Eq. 3.39 gives the length of the horizontal curve as (with $R = 533$ ft and $\Delta = 90$ degrees):

$$L = \frac{\pi}{180} R\Delta$$

$$= \frac{3.1416}{180} 533(90) = 837.24 \text{ ft}$$

Also, by inspection of Fig. 3.16 (or application of Eq. 3.36), the tangent length $T = R$ = 533 ft. The stationing for the horizontal curve is as follows:

$$\text{station of } PC = \underline{40+00}$$

$$\text{station of } PI = \text{station of } PC + T$$

$$= 40+00 \text{ plus } 5+33 = \underline{\underline{45+33}}$$

$$\text{station of } PT = \text{station of } PC + L$$

$$= 40+00 \text{ plus } 8+37.24 = \underline{\underline{48+37.24}}$$

For the vertical alignment, both a sag and crest curve will be needed, and adequate clearance must be provided over the existing highway. As shown in Section 3.3.6, AASHTO [2018] specifies a desirable clearance height of 16.5 ft. The bridge girder thickness is given as 6 ft so the total elevation difference between the two highways is 22.5 ft (16.5 + 6).

For the vertical alignment, the elevation change will be the final offsets of the sag and crest curves plus the change in elevation resulting from the constant-grade section connecting the two curves. The constant-grade section is included because the available distance of 837.24 ft (known from the length of the horizontal curve) is likely to be more than sufficient to affect a 22.5 ft elevation difference at a 40-mi/h design speed. Because both the sag and crest curves connect to a level grade at one end and have the constant grade in common at the other end, the A value will be the same for both curves and will also be the grade for the constant-grade section. That is,

$$A_s = A_c = A$$

With this information, the equation that will solve the vertical alignment for this problem is

$$\frac{AL_s}{200} + \frac{AL_c}{200} + \frac{A(837.24 - L_s - L_c)}{100} = 22.5$$

where the third term accounts for the elevation difference attributable to the constant-grade section connecting the sag and crest curves (see also Example 3.9 for comparison). Using $L = KA$, we have

$$\frac{A^2 K_s}{200} + \frac{A^2 K_c}{200} + \frac{A(837.24 - K_s A - K_c A)}{100} = 22.5$$

From Table 3.2 for 40 mi/h, $K_c = 44$, and from Table 3.3 for 40 mi/h, $K_s = 64$. Putting these values in the above equation gives

$$0.54 A^2 + 8.374 A - 1.08 A^2 = 22.5$$

$$-0.54 A^2 + 8.374 A - 22.5 = 0$$

Solving this gives $A = 3.458$ and $A = 12.049$; $A = 3.458\%$ is chosen because we want to minimize the grade. For this value of A, the curve lengths are

$$L_s = K_s A = 64(3.458) = \underline{\underline{221.31 \text{ ft}}}$$

$$L_c = K_c A = 44(3.458) = \underline{\underline{152.15 \text{ ft}}}$$

and the length of the constant-grade section (L_{con}) will be 463.78 ft (837.24 − 221.31 − 152.15). This means that about 16.04 ft of the elevation difference will occur in the constant-grade section, with the remainder of the elevation difference attributable to the final curve offsets.

The stationing and elevation of the key points along the vertical alignment can now be calculated:

$$\text{station of } PVC_c = \text{station of } PC$$
$$= \underline{\underline{40+00}}$$

$$\text{station of } PVI_c = 40+00 \text{ plus } L_c/2 = 40+00 \text{ plus } (1+52.15)/2 = \underline{\underline{40+76.075}}$$

$$\text{station of } PVT_c = \text{station of } PVC_c + L_c$$
$$= 40+00 \text{ plus } (1+52.15) = \underline{\underline{41+52.15}}$$

$$\text{station of } PVC_s = \text{station of } PVT_c + L_{con}$$
$$= 41+52.15 \text{ plus } 4+63.78 = \underline{\underline{46+15.93}}$$

$$\text{station of } PVI_s = \text{station of } PVC_s + L_s/2$$
$$= 46+15.93 \text{ plus } (1+10.66)/2 = \underline{\underline{47+26.59}}$$

$$\text{station of } PVT_s = \text{station of } PT$$
$$= \text{station of } PVC_s + L_s$$
$$= 46+15.93 \text{ plus } (2+21.31) = \underline{\underline{48+37.24}}$$

The elevation of the new east-west road will be 588 ft which is determined by adding 22.5 ft (the 16.5 ft clearance plus the 6 ft girder depth) above the north–south road elevation of 565.5 ft. The PC and PVC_c are both at station $40 + 00$ and elevation 588 ft (by inspection, the PVI_c will also be at this elevation)

$$\text{elevation } PVT_c = \text{elev } PVI_c - \frac{AL_c}{200}$$
$$= 588 - \frac{3.458(152.15)}{200} = \underline{\underline{585.37 \text{ ft}}}$$

$$\text{elevation } PVC_s = \text{elev } PVT_c - \frac{AL_{con}}{100}$$
$$= 585.37 - \frac{3.458(463.78)}{100} = \underline{\underline{569.33 \text{ ft}}}$$

$$\text{elevation } PVT_s = \text{elev } PVC_s - \frac{AL_s}{200}$$
$$= 569.33 - \frac{3.458(221.31)}{200} = \underline{\underline{565.00 \text{ ft}}}$$

The PT and PVT_s are both at station $48 + 37.24$ and will thus both be at 565 ft (by inspection, the PVI_s will also be at this elevation).

Finally, the distance that must be cleared from the inside of the horizontal curve to provide sufficient SSD is determined by applying Equation 3.42:

$$M_s = R_v \left(1 - \cos \frac{90\,\text{SSD}}{\pi R_v} \right)$$

With $R_v = 533$ and SSD = 305 ft (from Table 3.1 at 40 mi/h),

$$M_s = 533 \left(1 - \cos \frac{90(305)}{3.1416(533)} \right)$$
$$= \underline{\underline{21.67 \text{ ft}}}$$

Thus, a distance of at least 21.67 ft must be cleared from the center of the ramp's lane to the nearest sight obstruction on the inside of the curve.

3.6   PRACTICE PROBLEMS

PRACTICE PROBLEM 3.1   **DESIGN SPEED AND VERTICAL CURVE STATIONS AND ELEVATIONS**

An equal-tangent sag vertical curve (with a negative initial and a positive final grade) is designed for 55 mi/h. The *PVI* is at station 240 + 00 and elevation 122 ft. The *PVT* is at station 242 + 30 and elevation 127.75 ft. What is the station and elevation of the lowest point on the curve?

SOLUTION   **Note: Open boxes in equations "$\boxed{}$" are to be completed by the reader**

This problem shows the basic principles of vertical curve design as well as the determination of the high/low point on a curve. To begin, the length of the curve is determined as,

$$1/2\ L = \text{station of } PVT - \text{station of } PVI$$
$$L = 2\left(\boxed{} - \boxed{} \right) = \boxed{}.$$

The elevations of both *PVT* and *PVI* are given, so G_2 can be determined as,

$$G_2 = \frac{\text{elevation } PVT - \text{elevation } PVI}{1/2\,L} = \frac{127.75 - \boxed{}}{1/2\,\boxed{}} = 0.025 \text{ or } 2.5\%.$$

For a sag curve with a 55-mi/h design speed, Table 3.3 shows that the horizontal distance required to affect a 1% change in slope (the *K*-value) is equal to 115. With this, Eq. 3.10 can be rearranged to solve for *A* (the absolute value of the algebraic difference in the grades, in percent) such that,

$$A = \frac{L}{K} = \frac{\boxed{}}{115} = \boxed{},$$

which means, because $A = |G_1 - G_2|$ and the problem states that the initial grade is negative and the final grade is positive, that

$$-G_1 = A - G_2 = \boxed{} - 2.5, \text{ so } G_1 = -1.5.$$

With this initial grade, the elevation of the PVC is,

$$\text{elevation } PVC = \text{elevation } PVI + G_1(1/2\,L) = \boxed{} + \boxed{}(\boxed{}) = 125.45 \text{ ft.}$$

Also, the station of the PVC is determined as,

$$\text{station of } PVC = \text{station of } PVI - L/2 = \boxed{} \text{ minus } \boxed{} = 237 + 70.$$

With the station and elevation of the PVC known, we can now move forward to determine the station and elevation of the lowest point on the curve. For the station of the lowest point on the curve, we can first apply Eq. 3.11 to determine the distance of the low point from the PVC,

$$x_{hl} = K \times |G_1| = 115 \times \boxed{} = 172.5 \text{ ft.}$$

With this,

station of low point = station of PVC + distance from the PVC to the low point

station of low point = 237 + 70 plus 1 + $\boxed{}$ = <u>239 + 42.50</u>.

Solving for the elevation of the low point, Eq. 3.1 can be applied

$$y = ax^2 + bx + c$$

$$= \frac{\boxed{} - (-1.5)}{2(\boxed{})}(1.725)^2 + \boxed{}(1.725) + 125.45$$

$$= \underline{124.16 \text{ ft.}}$$

PRACTICE PROBLEM 3.2   **COMBINED SAG AND CREST VERTICAL CURVES**

West and east highway segments are separated by 1000 ft horizontally. The west segment has a 0% constant grade and the east segment has a –1% grade. The east segment has a higher elevation than the west segment, and the two segments are connected by a joining sag and crest curve combination (so $PVT_s = PVC_c$). If the road is designed for 60 mi/h, what is the elevation difference between the west and east highway segments?

SOLUTION   **Note: Open boxes in equations "$\boxed{}$" are to be completed by the reader**

The intent of this problem is to reinforce the concepts covered in Example 3.11, and the set up for this problem is similar to that shown in Example 3.11. In looking at Example 3.11, the current problem will be similar to what is shown in Fig. 3.10 except it will be reversed with the lower highway segment on the left of that figure (instead of on the right) and the higher roadway segment on the right instead of the left. Also, in comparison to Example 3.11, in the current problem the $\Delta y_{con} = 0$ (since the problem states that sag and crest curves are joined) and $\Delta y_s = 0$ since the initial grade of the sag curve (the grade of the west segment) is 0%.

To begin this problem, it is important to first get the K-values for both the sag and crest curves. With the 60-mi/h design speed given in the problem, from Table 3.2 we find the K-value for the crest curve is $K_c = 151$, and from Table 3.3 we find the sag curve's K-value is $K_s = 136$. As in Example 3.11, let G_{con} be the constant grade connecting the two curves so that $G_{2s} = G_{1c} = G_{con}$. To determine the absolute value of the algebraic difference in the grades (in percent) for the sag and crest curves, we have $A_s = G_{con}$ (since $G_{1s} = 0$ as given in the problem) and $A_c = G_{con} + 1$ (since $G_{1c} = G_{con}$ and $G_{2c} = -1$, note that the sign of G_{2c} becomes +1 since A's are based on the absolute value in the differences of the grades which means you add the absolute values of two grades if they are opposite in sign, as they are here, and you subtract the lower absolute value grade from the higher absolute value grade if the two grades are the same sign).

With these values of A_c and A_s, and with 1000 ft separating the two highway segments, we have,

$$1000 = L_s + L_c$$
$$= K_s A_s + K_c A_c$$
$$= 136 \times \boxed{} + \boxed{} \times (G_{con} + 1)$$

which gives $G_{con} = 2.958\%$.

To develop an equation for the elevation difference between the two highway segments, let Y_{fs} and Y_{fc} be the final offsets of the sag and vertical curves, respectively, and let Δy_c be the change in elevation due to the extended curve tangent for the crest curve, recalling that the extended curve tangent for the sag curve, Δy_s is now zero since the initial grade for the sag curve, $G_{1s} = 0$ with the west highway section a 0% grade (see these terms in Example 3.11 and Fig. 3.10 for comparison with this practice problem). With these terms and the elevation differences between the two highway segments (Δelevation), the elevation equality is (compare to Example 3.11),

$$\Delta\text{elevation} + \Delta y_c = Y_{fs} + Y_{fc}.$$

Rearranging terms and using Eq. 3.9 for final offsets and the 1% slope of the east highway segment (which produces a positive Δy_c), we have,

$$\Delta\text{elevation} = \frac{A_s L_s}{200} + \frac{A_c L_c}{200} - G_1 L_c.$$

With $L_s = K_s A_s$ and $L_c = K_c A_c$, substitution gives,

$$\Delta\text{elevation} = \frac{\boxed{}(\boxed{} \times \boxed{})}{200} + \frac{\boxed{}(\boxed{} \times \boxed{})}{200} - \boxed{}(\boxed{} \times \boxed{}),$$

and a final answer for the elevation difference between the west and east highway segments is Δelevation = 11.803 ft.

PRACTICE PROBLEM 3.3

HORIZONTAL CURVE STATIONING, DESIGN SPEED, AND SIGHT DISTANCE

A horizontal curve on a two-lane highway (with 11-ft lanes) has a central angle of 33 degrees and is designed for 60 mi/h with a 8% superelevation. First, if the *PI* is at station 300 + 00, what is the station of the *PT* and how many feet have to be cleared from the inside lane's lane-shoulder edge to provide adequate stopping sight distance? Second, if the normal component of centripetal force was considered, how much shorter would the radius of the curve be?

SOLUTION

Note: Open boxes in equations "$\boxed{}$" are to be completed by the reader

This problem illustrates a number of principles in horizontal curve design, including the magnitude of imprecision introduced by ignoring the normal component of centripetal force (which is standard practice in highway design). To begin, the radius to the vehicle (R_v) for a 60-mi/h design speed and a superelevation of 8% can be determined from Table 3.5. With this, to determine R (the radius to the center of the highway), half of the lane width is added to R_v obtained from Table 3.5 to get to the centerline of the roadway so,

$$R = \frac{1}{2}\text{lane-width} + R_v = \boxed{} \times \boxed{} = \boxed{}.$$

With R known and Δ given as 33 degrees, Eq. 3.39 can be applied to get length of the curve,

$$L = \frac{\pi}{180}R\Delta = \frac{\pi}{180}\boxed{} \times \boxed{} = 694.19 \text{ ft.}$$

Next, the tangent of the curve is determined by applying Eq. 3.36,

$$T = R\tan\frac{\Delta}{2} = \boxed{} \times \tan\frac{\boxed{}}{2} = \boxed{}.$$

To determine the station of the *PT*, we must first determine the station of the *PC* and then add the curve length *L*. The station of the *PC* is,

$$\text{station of } PC = \text{station of } PI - T$$

$$= \boxed{} - \boxed{} = 296 + 42.92$$

and the station *PT* is therefore,

$$\text{station of } PT = \text{station of } PC + L$$

$$= \boxed{} + \boxed{} = 303 + 37.01.$$

To determine how many feet have to be cleared from the inside lane's lane-shoulder edge to provide adequate stopping sight distance, Eq. 3.42 can be used to arrive at M_s,

$$M_s = R_v\left(1 - \cos\frac{90\,\text{SSD}}{\pi R_v}\right)$$

with SSD determined to be $\boxed{}$ from Table 3.1 (with a design speed of 60 mi/h), we have,

$$M_s = \boxed{}\left(1 - \cos\frac{90\times\boxed{}}{\pi\times\boxed{}}\right) = 33.70 \text{ ft.}$$

Because the problem asks for the distance from the inside lane's lane-shoulder edge and M_S is the distance to the path of the vehicle (which is assumed to be in the center of the lane), we must subtract half of the lane with from M_S to get the answer which is,

$$\text{required clearance} = M_s - \frac{1}{2}\text{lane width} = \boxed{} - \frac{1}{2}\boxed{} = 28.20 \text{ ft.}$$

Turning to the second part of the problem, to determine how much shorter the radius of the curve would be if the normal component of the centripetal force were considered, recall that highway design ignores the normal component of the centripetal force which gives a conservative answer (larger radius than actually needed). Thus, in highway design, Eq. 3.34 is used to arrive at the R_V values listed in Table 3.5. However, if the normal component of the centripetal force is considered, Eq. 3.34 becomes, with superscript N denoting the normal component of centripetal force (see also Eq. 3.33 and discussion below it),

$$R_v^N = \frac{V^2}{g\left(f_s + \dfrac{e}{100}\right)}\left(1 - f_s\tan\frac{e}{100}\right).$$

With $f_s = 0.12$ (from Table 3.5 with the given design speed of 60 mi/h), solving for the radius in the equation above gives,

$$R_v^N = \frac{\boxed{}^2}{32.2\left(\boxed{} + \dfrac{\boxed{}}{100}\right)}\left(1 - \boxed{}\tan\frac{\boxed{}}{100}\right) = \boxed{}.$$

With this, R_v minus R_v^N will tell us how much shorter the radius would be, so the answer is,

$$R_v - R_v^N = \boxed{} - \boxed{} = \underline{9.06 \text{ ft.}}$$

This rather modest difference shows that, with the small superelevations used in typical highway design, ignoring the normal component of centripetal force does not introduce great error.

PRACTICE PROBLEM 3.4

COMBINED VERTICAL AND HORIZONTAL ALIGNMENT

A highway segment has an equal-tangent sag curve and a horizontal curve with the same design speed. It is known that the sag curve has an initial grade of –1.8% and a positive final grade that is not known. The station of the *PVC* is 232 + 70, and the station of the low point of the sag vertical curve is 235 + 14.8. The horizontal curve has a superelevation of 8%, central angle of 30 degrees, and a *PT* at Station 302 + 20. The road has two 11-ft lanes (one in each direction). What is the station of the *PI*?

SOLUTION

Note: Open boxes in equations "$\boxed{}$" are to be completed by the reader

This problem shows how common design elements from horizontal and vertical alignments can be used to solve for unknown curve characteristics. The approach to this problem is to use the information from the equal-tangent vertical curve to determine the design speed, and then use this design speed to solve for the station of the horizontal curve's *PI*. By framing the problem in this way, the reader will gain additional insight into the various elements involved in the vertical and horizontal alignment of highways.

To begin, the design speed of the highway is determined from the vertical curve information provided. First, the distance from the *PVC* to the highpoint is determined as

$$x_{hl} = PVC - \text{station of low point} = \boxed{} - \boxed{} = 244.80 \text{ ft.}$$

With this, Eq. 3.11 can be rearranged to solve for the *K*-value such that,

$$K = \frac{x_{hl}}{|G_1|} = \frac{\boxed{}}{\boxed{}} = 136.$$

Table 3.3 for a sag vertical shows that a K-value of 136 corresponds to a design speed of 60 mi/h. With this design speed, the radius to the curve of the vehicle's traveled path R_v can be obtained from Table 3.5. With an 8% superelevation given and a 60-mi/h design speed, Table 3.5 gives,

$$R_v = \boxed{}$$

and, with 11-ft lanes (on a two-lane road), the radius of the curve measured to the roadway centerline is,

$$R = R_v + \frac{1}{2}\text{lane width} = \boxed{} + \frac{1}{2}\boxed{} = \boxed{}.$$

Eq. 3.39 can now be applied to arrive at the horizontal curve length L as,

$$L = \frac{\pi}{180}R\Delta = \frac{\pi}{180}\boxed{} \times \boxed{} = 631.08 \text{ ft.}$$

With this length of curve,

$$\text{station of } PC = \text{station of } PT - L$$
$$= \boxed{} - \boxed{} = \underline{295 + 88.92}.$$

To arrive at the station of the PI, we must add the tangent length to the station of the PC (note that this is not the same value as subtracting the tangent length from the PT since the tangents are not on the highway alignment). The tangent length is determined from Eq. 3.36 to be,

$$T = R\tan\frac{\Delta}{2} = \boxed{} \times \tan\frac{\boxed{}}{2} = \boxed{}$$

So,

$$\text{station of } PI = \text{station of } PC - T$$
$$= \boxed{} - \boxed{} = \underline{299 + 11.93}.$$

NOMENCLATURE FOR CHAPTER 3

A — absolute value of the algebraic difference in grades (in percent)

a — coefficient in the parabolic curve equation or the deceleration in the stopping distance equation

b — coefficient in the parabolic curve equation

c — elevation of the PVC

D — degree of curvature

e — rate of superelevation

F_f — frictional side force

F_c — centripetal force

F_{cn} — centripetal force normal to the roadway surface

F_{cp} — centripetal force parallel to the roadway surface

f_s — coefficient of side friction

G — grade

G_1 — initial roadway grade

G_2 — final roadway grade

g — gravitational constant

H — height of vehicle headlights

H_c — clearance height of structure above sag curve

H_1 — height of driver's eye

H_2 — height of roadway object for stopping, height of oncoming car for passing

K — horizontal distance required to effect a 1% change in slope

L — length of curve

L_m — minimum length of curve

M — middle ordinate

M_s — middle ordinate for stopping sight distance

PC — initial point of horizontal curve

PI — point of tangent intersection (horizontal curve)

PSD — passing sight distance

PT — final point of horizontal curve

PVC — initial point of vertical curve

PVI — point of tangent intersection (vertical curve)

PVT — final point of vertical curve

R — radius of curve measured to the roadway centerline

R_v — radius of curve to the vehicle's traveled path

S — sight distance

SSD — stopping sight distance

T — tangent length

V — vehicle speed

V_1 — initial vehicle speed

W — vehicle weight

W_n — vehicle weight normal to the roadway surface

W_p — vehicle weight parallel to the roadway surface

x — distance from the beginning of the vertical curve to specified point

x_{hl} — distance from the beginning of the vertical curve to high or low point

Y — vertical curve offset

Y_f — end-of-curve offset (vertical curve)

Y_m — midcurve offset (vertical curve)

α — angle of superelevation

β — upward angle of headlight beam

Δ — central angle

Δ_s — angle subtended by the stopping sight distance (SSD) arc

REFERENCES

AASHTO (American Association of State Highway and Transportation Officials). *A Policy on Geometric Design of Highways and Streets*, 7th ed. Washington, DC, 2018.

Campbell, C. *The Sports Car: Its Design and Performance*. Cambridge, MA: Robert Bently, 1978.

Hickerson, T. F. *Route Location and Design*, 5th ed. New York: McGraw-Hill, 1964.

Wong, J. Y. *Theory of Ground Vehicles*. New York: John Wiley & Sons, 2008.

Chapter 4

Pavement Design

4.1 INTRODUCTION

Pavements are among the most financially important highway-system assets due to their high construction and maintenance costs. Their costs are largely responsible for making the U.S. highway system the most expensive public works project undertaken by any society. Because pavements are so costly to construct and maintain, it is important for highway engineers to have an understanding of basic pavement design principles.

In the United States, there are over 4 million miles of highways. Of these, roughly 45% are lower-volume roads that are not paved (these roads generally have a gravel surface, or are composed of a stabilized material consisting of an aggregate bound together with a cementing agent such as Portland cement, lime fly ash, or asphaltic cement). However, environmental conditions such as moisture can seriously compromise the structural integrity of unpaved highway surfaces. As a result, highways that carry higher volumes of traffic, typically with heavy axle loads, require surfaces with asphalt concrete or Portland cement concrete (PCC) to provide for all-weather operations. Paved highways can cost several million dollars per mile to construct (based on the number of lanes, shoulders, soil conditions, etc.) and many thousands of dollars annually (per mile) to maintain. In the United States, roughly $40 billion dollars are spent annually on the construction and maintenance of pavements. Given the magnitude of this pavement-asset investment, it is easy to understand why the construction, maintenance, and rehabilitation of the pavement infrastructure must be done in a cost-effective manner.

While paved highway surfaces (and their associated lane markings) provide visual guidance to help vehicle operators traverse the highway alignment, the primary functions of pavement are to provide adequate friction, particularly to ensure safe stopping distances and acceptable vehicle cornering (the emphasis of Chapters 2 and 3), and to support vehicle loads while maintaining a safe and comfortable ride surface. This latter function (supporting vehicle loads and the quality of the riding surface) is the emphasis of the current chapter.

4.2 PAVEMENT TYPES

Pavement structures can be broadly categorized into two types: flexible pavements and rigid pavements. There are, however, many variations of these pavement types including composite pavements (which are made of both rigid and flexible pavement layers), continuously reinforced pavements, and post-tensioned pavements (these variations require specialized design procedures that not covered in this chapter).

As with most conventional civil-engineering structures, the underlying soil must ultimately carry the design-load forces. The intent of a pavement structure is to distribute traffic loads to the soil (subgrade) at a magnitude that will not shear or distort the soil. Typical soil-bearing capacities can be less than 50 lb/in^2, and in some cases can be as low as 2 to 3 lb/in^2. When soil is saturated with water, the load-bearing capacity can be very low, and in these cases it is very important for pavement to distribute tire loads to the supporting soil in such a way as to prevent permanent deformation of the underlying soil, which would classify as a failure of the pavement structure.

A typical automobile weighs approximately 4000 lb, with tire pressures around 35 lb/in^2. These loads are small compared with a typical tractor–semi-trailer truck, which can weigh up to 80,000 lb (the legal limit, in many states), on five axles with tire pressures of 100 lb/in^2 or higher. Truck loads such as these represent the standard loading used in pavement design. In this chapter, attention is directed toward an accepted procedure that can be used to design pavement structures for high–traffic-volume highways subjected to heavy truck traffic. The design of lower-volume facilities, which may have stabilized-soil and gravel-surfaced structures, is discussed elsewhere [Yoder and Witczak, 1991].

4.2.1 Flexible Pavements

A flexible pavement is constructed with asphaltic cement and aggregates and usually consists of several layers, as shown in Fig. 4.1. The lowest layer is referred to as the subgrade (the soil itself). The upper 6 to 8 inches of the subgrade is usually scarified and blended to provide a uniform material before it is compacted to a high density to assist with the overall stability and uniformity of the pavement structure. The next layer up in the structure, the subbase, usually consists of crushed aggregate (rock). This material has better engineering properties (higher modulus values) and more uniformity than the subgrade material in terms of its bearing capacity. The next layer up in the structure is the base layer and it is also often made of crushed aggregates (of a higher strength than those used in the subbase), which can in some cases be stabilized with a cementing material such as Portland cement, lime fly ash, or asphaltic cement.

The top layer of the flexible pavement structure is referred to as the wearing surface and usually consists of asphaltic concrete, which is a mixture of asphalt cement and aggregates. The purpose of the wearing layer is to protect the base layer from wheel abrasion and to waterproof the entire pavement structure. It also provides a skid-resistant surface that is important for safe vehicle stops and cornering. Typical thicknesses of the individual layers are shown in Fig. 4.1. These thicknesses vary with the types and magnitudes of axle loadings, available materials, and expected pavement design life (which is defined as the number of years the pavement is expected to provide adequate service before it must undergo major rehabilitation).

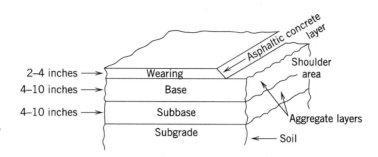

Figure 4.1 Typical flexible-pavement cross section.

4.2.2   Rigid Pavements

Rigid pavements are constructed with PCC and aggregates, as shown in Fig. 4.2. As with flexible pavements, the subgrade (the lower layer) is often scarified, blended, and compacted to high density before additional layers are added. In rigid pavements, the base layer (see Fig. 4.2) is optional, depending on the engineering properties of the subgrade. If the subgrade soil is poor and erodable, then it is advisable to use a base layer. However, if the soil has good engineering properties and drains well, a base layer need not be used. The top layer (wearing surface) is the PCC slab. Slab length varies from a spacing of 10 to 13 ft to a spacing of 40 ft or more.

Transverse contraction joints are built into the pavement to control cracking due to shrinkage of the concrete during the curing process. Load-transfer devices, such as dowel bars, are placed in the joints to minimize deflections between adjacent slabs and reduce stresses near the edges of the slabs. Slab thicknesses for PCC highway pavements usually vary from 8 to 12 inches, as shown in Fig. 4.2.

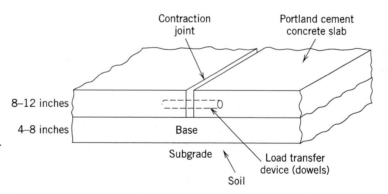

Figure 4.2 Typical rigid-pavement cross section.

4.3   PAVEMENT SYSTEM DESIGN: PRINCIPLES FOR FLEXIBLE PAVEMENTS

In supporting vehicle loads, the primary function of the pavement structure is to distribute the surface stresses (contact tire pressure) to an acceptable magnitude at the subgrade (to a magnitude that prevents permanent deformation). A flexible pavement reduces the stresses by distributing the traffic wheel loads over greater and greater areas, down through the individual layers, until the stress at the subgrade is at an acceptably low level. The traffic loads are transmitted to the subgrade by aggregate-to-aggregate particle contact. Confining pressures (lateral forces due to material weight) in the base and subbase layers increase the bearing strength of the materials in these layers. A cone of distributed loads spreads the stresses to the subgrade, as shown in Fig. 4.3.

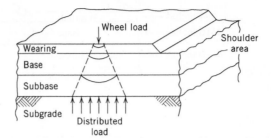

Figure 4.3 Distribution of load on a flexible pavement.

4.4   TRADITIONAL AASHTO FLEXIBLE-PAVEMENT DESIGN PROCEDURE

There are several accepted flexible-pavement design procedures available, including the Asphalt Institute method, the National Stone Association procedure, the Shell procedure, and the Mechanistic-Empirical Pavement Design Guide (which is briefly discussed later in this chapter). Most of the procedures have been field verified and used by highway agencies for many years and the selection of one procedure over another is usually based on highway-agency experience and satisfaction with previous design results.

A traditional and widely accepted flexible-pavement design procedure is presented in the *AASHTO Guide for Design of Pavement Structures*, which is published by the American Association of State Highway and Transportation Officials (AASHTO). The procedure was first published in 1972, with the latest revisions in 1993. Test data, used for the development of the design procedure, were collected at the AASHO Road Test in Illinois from 1958 to 1960 (AASHO, which stands for American Association of State Highway Officials, was the prior name of AASHTO).

Pavement designs must account for various failure possibilities as well as uncertainties in environmental conditions. For example, pavements can be subjected to a number of detrimental effects from traffic loads, including fatigue failures (cracking), which are the result of the accumulating effects of traffic passing over the pavement surface. The pavement is also placed in an uncontrolled environment that produces temperature extremes and moisture variations. In addition, pavement structures can have considerable variability in material quality (materials are often drawn from local sources due to the expense of transportation) and construction quality. The combination of the environment, traffic loads, material variations, and construction variations requires a robust design procedure. The AASHTO pavement design procedure accounts for most of the critical elements needed for an effective pavement design. It considers the environment, traffic loads, and material strengths in a methodology that is relatively easy to use. The AASHTO pavement design procedure has been widely accepted throughout the United States and around the world. Details of this procedure are presented in the following sections.

4.4.1   Serviceability Concept

Prior to the AASHO Road Test, there was no real consensus on the definition of pavement failure. In the eyes of an engineer, pavement failure occurred whenever cracking, rutting, or other surface distresses became visible. In contrast, the motoring public usually associated pavement failure with poor ride quality. Pavement engineers conducting the AASHO Road Tests of the late 1950s were faced with the task of accounting for these two failure definitions (engineer and public) in a single design procedure. The Pavement Serviceability-Performance Concept was developed by Carey and Irick [1962] to address the definition of pavement failure. Carey and Irick considered extensive pavement performance histories and noted that pavements usually begin their service life in excellent condition and deteriorate as traffic loads are applied in conjunction with prevailing environmental conditions. This gives rise to a pavement performance curve, which charts typical pavement conditions over time. Pavement performance, at any point in time, is known as the present serviceability index, or PSI. Examples of pavement performance trends (or PSI trends) are shown in Fig. 4.4.

The present serviceability index (PSI) of a pavement can be measured at any point in time. The PSI was originally based on the present serviceability rating (PSR) which was measured by a panel of raters who drove over the pavement section and rated the pavement performance on a scale of 0 to 5, with 5 being the smoothest ride and 0 being virtually impassible. Because having a panel of raters evaluate all pavements is not practical, the PSI was developed by correlating panel-rated PSRs with actual pavement measurements (roughness, cracking, etc.). Thus PSI values are based on a variety of pavement measurements and does not require a panel of raters. As shown in Fig. 4.4, the accumulation of traffic loads causes the pavement to deteriorate, and, as expected, the serviceability rating (PSI) drops. At some point, a terminal serviceability index (TSI) is reached and the pavement is in need of rehabilitation or replacement.

It has been found that new pavements usually have an initial PSI rating of approximately 4.2 to 4.5. The point at which pavements are considered to have failed (the TSI) varies by type of highway. Highway facilities such as interstate highways or principal arterials usually have TSIs of 2.5 or 3.0, whereas local roads can have TSIs as low as 2.0.

4.4.2   Flexible-Pavement Design Equation

The basic equation for flexible-pavement design given in the 1993 AASHTO design guide enables engineers to determine a structural number (SN) necessary to carry a designated traffic loading. The AASHTO equation is

$$
\log_{10} W_{18} = Z_R S_o + 9.36 \left[\log_{10} (SN+1) \right] - 0.20
$$
$$
+ \frac{\log_{10} \left[\Delta PSI / 2.7 \right]}{0.40 + \left[1094/(SN+1)^{5.19} \right]} \tag{4.1}
$$
$$
+ 2.32 \log_{10} M_R - 8.07
$$

where

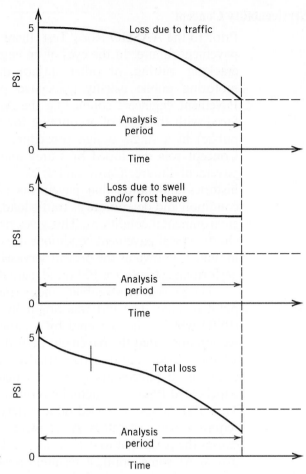

Figure 4.4 Pavement performance trends.

Redrawn from *AASHTO Guide for Design of Pavement Structures*, Washington, DC, The American Association of State Highway and Transportation Officials, 1993. Used by permission.

W_{18} = 18-kip–equivalent single-axle load (1 kip = 1000 lb),

Z_R = reliability (z-statistic from the standard normal curve),

S_o = overall standard deviation of traffic,

SN = structural number,

ΔPSI = loss in serviceability from the time the pavement is new until it reaches its TSI, and

M_R = soil resilient modulus of the subgrade in lb/in^2.

A graphical solution to Eq. 4.1 is shown in Fig. 4.5. Details on the variables that serve as inputs to Eq. 4.1 and Fig. 4.5 are as follows:

W_{18} Automobiles and truck traffic provide a wide range of vehicle axle types and axle loads. If one were to attempt to account for individual-vehicle traffic loads on a pavement, this input variable would require an incomprehensible amount of data collection and design evaluation. Instead, the problem of handling mixed traffic loadings is solved by axle classifications (by weight and axle type) and the adoption of a standard 18-thousand-pound-equivalent single-axle load or (with 1 kip = 1000 lb) an 18-kip–equivalent single-axle load (ESAL).

The idea is to determine the impact of any individual axle load on the pavement in terms of the equivalent amount of pavement impact that an 18-kip single-axle load would have. For example, if a 44-kip tandem-axle (double-axle) load has 2.88 times the impact on pavement structure as an 18-kip single-axle load, 2.88 would be the W_{18} value assigned to this tandem-axle load. The AASHO Road Test also found that the 18-kip–equivalent axle load is a function of the TSI of the pavement structure. The axle-load equivalency factors for flexible-pavement design, with a TSI of 2.5, are presented in Tables 4.1 (for single axles), 4.2 (for tandem axles), and 4.3 (for triple axles).

Z_R Represents the probability that serviceability will be maintained at adequate levels from a user's point of view throughout the design life of the facility. This factor estimates the likelihood that the pavement will perform at or above the TSI level during the design period and takes into account the inherent uncertainty in design. Equation 4.1 uses the z-statistic, which is obtained from the cumulative probabilities of the standard normal distribution (a normal distribution with mean equal to 0 and variance equal to 1). The z-statistics corresponding to various probability levels are given in Table 4.4. In the flexible-pavement–design nomograph (Fig. 4.5), the probabilities (in percent) are used directly (instead of Z_R as in the case of Eq. 4.1), and these percent probabilities are denoted R, the reliability (see Table 4.4).

Highways such as interstates and major arterials, which are costly to reconstruct (have their pavements rehabilitated) because of resulting traffic delay and disruption, require a high reliability level, whereas local roads, which will have lower impacts on users in the event of pavement rehabilitation, do not. Typical reliability values for interstate highways are 90% or higher, whereas local roads can have a reliability as low as 50%.

S_o The overall standard deviation, S_o, takes into account the designers' inability to accurately estimate the variation in future 18-kip–equivalent axle loads, and the statistical error in the equations resulting from variability in materials and construction practices. Typical values of S_o are on the order of 0.30 to 0.50.

SN   The structural number, SN, represents the overall structural requirement needed to sustain the design's traffic loadings. The SN is discussed further in Section 4.4.3.

ΔPSI   The amount of serviceability loss over the life of the pavement, ΔPSI, is determined during the pavement design process. The engineer must decide on the final PSI level for a particular pavement. Loss of serviceability is caused by pavement roughness, cracking, patching, and rutting. As pavement distress increases, serviceability decreases. If the design is for a pavement with heavy traffic loads, such as an interstate highway, then the serviceability loss may only be 1.2 (an initial PSI of 4.2 and a TSI of 3.0), whereas a low-volume road can be allowed to deteriorate further, with a possible total serviceability loss of 2.7 or more.

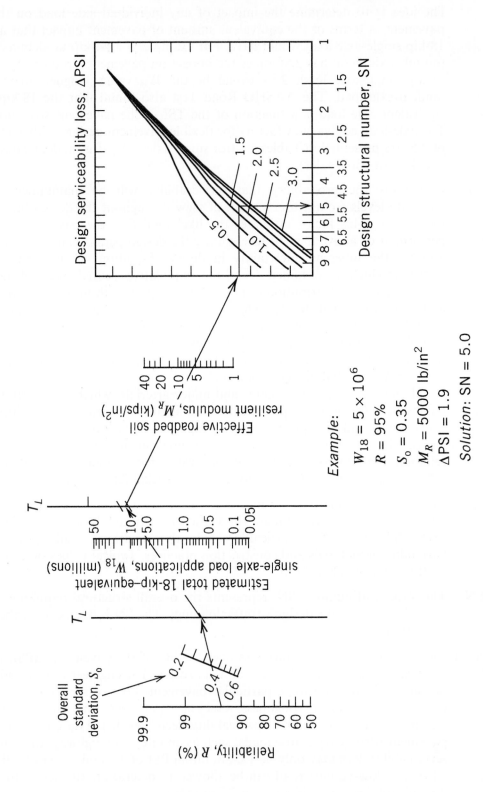

Figure 4.5 Design chart for flexible pavements based on the use of mean values for each input.
Redrawn from *AASHTO Guide for Design of Pavement Structures*, Washington, DC, The American Association of State Highway and Transportation Officials, 1993. Used by permission.

Example:

$W_{18} = 5 \times 10^6$
$R = 95\%$
$S_o = 0.35$
$M_R = 5000 \ lb/in^2$
$\Delta PSI = 1.9$

Solution: SN = 5.0

114

Table 4.1 Axle-Load Equivalency Factors for Flexible Pavements, Single Axles, and TSI = 2.5

Axle load (kips)	Pavement structural number (SN)					
	1	2	3	4	5	6
2	0.0004	0.0004	0.0003	0.0002	0.0002	0.0002
4	0.003	0.004	0.004	0.003	0.002	0.002
6	0.011	0.017	0.017	0.013	0.010	0.009
8	0.032	0.047	0.051	0.041	0.034	0.031
10	0.078	0.102	0.118	0.102	0.088	0.080
12	0.168	0.198	0.229	0.213	0.189	0.176
14	0.328	0.358	0.399	0.388	0.360	0.342
16	0.591	0.613	0.646	0.645	0.623	0.606
18	1.00	1.00	1.00	1.00	1.00	1.00
20	1.61	1.57	1.49	1.47	1.51	1.55
22	2.48	2.38	2.17	2.09	2.18	2.30
24	3.69	3.49	3.09	2.89	3.03	3.27
26	5.33	4.99	4.31	3.91	4.09	4.48
28	7.49	6.98	5.90	5.21	5.39	5.98
30	10.3	9.5	7.9	6.8	7.0	7.8
32	13.9	12.8	10.5	8.8	8.9	10.0
34	18.4	16.9	13.7	11.3	11.2	12.5
36	24.0	22.0	17.7	14.4	13.9	15.5
38	30.9	28.3	22.6	18.1	17.2	19.0
40	39.3	35.9	28.5	22.5	21.1	23.0
42	49.3	45.0	35.6	27.8	25.6	27.7
44	61.3	55.9	44.0	34.0	31.0	33.1
46	75.5	68.8	54.0	41.4	37.2	39.3
48	92.2	83.9	65.7	50.1	44.5	46.5
50	112.0	102.0	79.0	60.0	53.0	55.0

Source: *AASHTO Guide for Design of Pavement Structures*, The American Association of State Highway and Transportation Officials, Washington, DC, 1993. Used by permission.

Table 4.2 Axle-Load Equivalency Factors for Flexible Pavements, Tandem Axles, and TSI = 2.5

Axle load (kips)	Pavement structural number (SN)					
	1	2	3	4	5	6
2	0.0001	0.0001	0.0001	0.0000	0.0000	0.0000
4	0.0005	0.0005	0.0004	0.0003	0.0003	0.0002
6	0.002	0.002	0.002	0.001	0.001	0.001
8	0.004	0.006	0.005	0.004	0.003	0.003
10	0.008	0.013	0.011	0.009	0.007	0.006
12	0.015	0.024	0.023	0.018	0.014	0.013
14	0.026	0.041	0.042	0.033	0.027	0.024
16	0.044	0.065	0.070	0.057	0.047	0.043
18	0.070	0.097	0.109	0.092	0.077	0.070
20	0.107	0.141	0.162	0.141	0.121	0.110
22	0.160	0.198	0.229	0.207	0.180	0.166
24	0.231	0.273	0.315	0.292	0.260	0.242
26	0.327	0.370	0.420	0.401	0.364	0.342
28	0.451	0.493	0.548	0.534	0.495	0.470
30	0.611	0.648	0.703	0.695	0.658	0.633
32	0.813	0.843	0.889	0.887	0.857	0.834
34	1.06	1.08	1.11	1.11	1.09	1.08
36	1.38	1.38	1.38	1.38	1.38	1.38
38	1.75	1.73	1.69	1.68	1.70	1.73
40	2.21	2.16	2.06	2.03	2.08	2.14
42	2.76	2.67	2.49	2.43	2.51	2.61
44	3.41	3.27	2.99	2.88	3.00	3.16
46	4.18	3.98	3.58	3.40	3.55	3.79
48	5.08	4.80	4.25	3.98	4.17	4.49
50	6.12	5.76	5.03	4.64	4.86	5.28
52	7.33	6.87	5.93	5.38	5.63	6.17
54	8.72	8.14	6.95	6.22	6.47	7.15
56	10.3	9.6	8.1	7.2	7.4	8.2
58	12.1	11.3	9.4	8.2	8.4	9.4
60	14.2	13.1	10.9	9.4	9.6	10.7
62	16.5	15.3	12.6	10.7	10.8	12.1
64	19.1	17.6	14.5	12.2	12.2	13.7
66	22.1	20.3	16.6	13.8	13.7	15.4
68	25.3	23.3	18.9	15.6	15.4	17.2
70	29.0	26.6	21.5	17.6	17.2	19.2
72	33.0	30.3	24.4	19.8	19.2	21.3
74	37.5	34.4	27.6	22.2	21.3	23.6
76	42.5	38.9	31.1	24.8	23.7	26.1
78	48.0	43.9	35.0	27.8	26.2	28.8
80	54.0	49.4	39.2	30.9	29.0	31.7
82	60.6	55.4	43.9	34.4	32.0	34.8
84	67.8	61.9	49.0	38.2	35.3	38.1
86	75.7	69.1	54.5	42.3	38.8	41.7
88	84.3	76.9	60.6	46.8	42.6	45.6
90	93.7	85.4	67.1	51.7	46.8	49.7

Source: *AASHTO Guide for Design of Pavement Structures*, The American Association of State Highway and Transportation Officials, Washington, DC, 1993. Used by permission.

Table 4.3 Axle-Load Equivalency Factors for Flexible Pavements, Triple Axles, and TSI = 2.5

Axle load (kips)	Pavement structural number (SN)					
	1	2	3	4	5	6
2	0.0000	0.0000	0.0000	0.0000	0.0000	0.0000
4	0.0002	0.0002	0.0002	0.0001	0.0001	0.0001
6	0.0006	0.0007	0.0005	0.0004	0.0003	0.0003
8	0.001	0.002	0.001	0.001	0.001	0.001
10	0.003	0.004	0.003	0.002	0.002	0.002
12	0.005	0.007	0.006	0.004	0.003	0.003
14	0.008	0.012	0.010	0.008	0.006	0.006
16	0.012	0.019	0.018	0.013	0.011	0.010
18	0.018	0.029	0.028	0.021	0.017	0.016
20	0.027	0.042	0.042	0.032	0.027	0.024
22	0.038	0.058	0.060	0.048	0.040	0.036
24	0.053	0.078	0.084	0.068	0.057	0.051
26	0.072	0.103	0.114	0.095	0.080	0.072
28	0.098	0.133	0.151	0.128	0.109	0.099
30	0.129	0.169	0.195	0.170	0.145	0.133
32	0.169	0.213	0.247	0.220	0.191	0.175
34	0.219	0.266	0.308	0.281	0.246	0.228
36	0.279	0.329	0.379	0.352	0.313	0.292
38	0.352	0.403	0.461	0.436	0.393	0.368
40	0.439	0.491	0.554	0.533	0.487	0.459
42	0.543	0.594	0.661	0.644	0.597	0.567
44	0.666	0.714	0.781	0.769	0.723	0.692
46	0.811	0.854	0.918	0.911	0.868	0.838
48	0.979	1.015	1.072	1.069	1.033	1.005
50	1.17	1.20	1.24	1.25	1.22	1.20
52	1.40	1.41	1.44	1.44	1.43	1.41
54	1.66	1.66	1.66	1.66	1.66	1.66
56	1.95	1.93	1.90	1.90	1.91	1.93
58	2.29	2.25	2.17	2.16	2.20	2.24
60	2.67	2.60	2.48	2.44	2.51	2.58
62	3.09	3.00	2.82	2.76	2.85	2.95
64	3.57	3.44	3.19	3.10	3.22	3.36
66	4.11	3.94	3.61	3.47	3.62	3.81
68	4.71	4.49	4.06	3.88	4.05	4.30
70	5.38	5.11	4.57	4.32	4.52	4.84
72	6.12	5.79	5.13	4.80	5.03	5.41
74	6.93	6.54	5.74	5.32	5.57	6.04
76	7.84	7.37	6.41	5.88	6.15	6.71
78	8.83	8.28	7.14	6.49	6.78	7.43
80	9.92	9.28	7.95	7.15	7.45	8.21
82	11.1	10.4	8.8	7.9	8.2	9.0
84	12.4	11.6	9.8	8.6	8.9	9.9
86	13.8	12.9	10.8	9.5	9.8	10.9
88	15.4	14.3	11.9	10.4	10.6	11.9
90	17.1	15.8	13.2	11.3	11.6	12.9

Source: *AASHTO Guide for Design of Pavement Structures*, The American Association of State Highway and Transportation Officials, Washington, DC, 1993. Used by permission.

Table 4.4 Cumulative Percent Probabilities of Reliability, R, of the Standard Normal Distribution, and Corresponding Z_R

R	0	1	2	3	4	5	6	7	8	9	9.5	9.9
90	−1.282	−1.341	−1.405	−1.476	−1.555	−1.645	−1.751	−1.881	−2.054	−2.326	−2.576	−3.080
80	−0.842	−0.878	−0.915	−0.954	−0.994	−1.036	−1.080	−1.126	−1.175	−1.227	−1.253	−1.272
70	−0.524	−0.553	−0.583	−0.613	−0.643	−0.675	−0.706	−0.739	−0.772	−0.806	−0.824	−0.838
60	−0.253	−0.279	−0.305	−0.332	−0.358	−0.385	−0.412	−0.440	−0.468	−0.496	−0.510	−0.522
50	0	−0.025	−0.050	−0.075	−0.100	−0.125	−0.151	−0.176	−0.202	−0.228	−0.241	−0.251

Example: To be 95% confident that the pavement will remain at or above its TSI ($R = 95$ for use in Fig. 4.7), a Z_R value of −1.645 would be used in Eq. 4.1 (and in Eq. 4.4).

M_R The soil resilient modulus, M_R, is used to reflect the engineering properties of the subgrade (the soil). Each time a vehicle passes over pavement, stresses are developed in the subgrade. After the load passes, the subgrade soil relaxes and the stress is relieved. The resilient modulus test is used to determine the properties of the soil under this repeated load. The resilient modulus can be determined by AASHTO test method T274. Measurement of the resilient modulus is not performed by all transportation agencies; therefore, a relationship between M_R and the California bearing ratio (CBR) has been determined. The CBR has been widely used to determine the supporting characteristics of soils since the mid-1930s, and a significant amount of historical information is available. The CBR is the ratio of the load-bearing capacity of the soil to the load-bearing capacity of a high-quality aggregate, multiplied by 100. The relationship, used to provide a very basic approximation of M_R (in lb/in²) from a known CBR, is

$$M_R = 1500 \times \text{CBR} \tag{4.2}$$

The coefficient of 1500 in Eq. 4.2 is used for CBR values less than 10. Caution must be exercised in applying this equation to higher CBRs because the coefficient (the value 1500 shown in Eq. 4.2) has a range of 750 to 3000.

4.4.3 Structural Number

The objective of Eq. 4.1 and the nomograph in Fig. 4.5 is to determine a required SN for given axle loadings, reliability, overall standard deviation, change in PSI, and soil resilient modulus. As previously mentioned, there are many pavement material combinations and thicknesses that will provide satisfactory pavement service life. The following equation can be used to relate individual material types and thicknesses to the SN:

$$\text{SN} = a_1 D_1 + a_2 D_2 M_2 + a_3 D_3 M_3 \tag{4.3}$$

where

a_1, a_2, a_3 = structural-layer coefficients of the wearing surface, base, and subbase layers, respectively,

D_1, D_2, D_3 = thickness of the wearing surface, base, and subbase layers in inches, respectively, and

M_2, M_3 = drainage coefficients for the base and subbase, respectively.

Values for the structural-layer coefficients for various types of material are presented in Table 4.5. Drainage coefficients are used to modify the thickness of the lower pavement layers (base and subbase) to take into account a material's drainage characteristics. A value of 1.0 for a drainage coefficient represents a material with good drainage characteristics (a sandy material). A soil such as clay does not drain very well and, consequently, will have a lower drainage coefficient (less than 1.0) than a sandy material. The reader is referred to [AASHTO, 1993] for further information on drainage coefficients.

Because there are many combinations of structural-layer coefficients and thicknesses that solve Eq. 4.3, some guidelines are used to narrow the number of solutions. Experience has shown that wearing layers are typically 2 to 4 inches thick, whereas subbases and bases range from 4 to 10 inches thick. Knowing which of the materials is the most costly per inch of depth will assist in the determination of an initial layer thickness.

Table 4.5 Structural-Layer Coefficients

Pavement component	Coefficient
Wearing surface	
Sand-mix asphaltic concrete	0.35
Hot-mix asphaltic (HMA) concrete	0.44
Base	
Crushed stone	0.14
Dense-graded crushed stone	0.18
Soil cement	0.20
Emulsion/aggregate-bituminous	0.30
Portland cement/aggregate	0.40
Lime-pozzolan/aggregate	0.40
Hot-mix asphaltic (HMA) concrete	0.40
Subbase	
Crushed stone	0.11

EXAMPLE 4.1   FLEXIBLE-PAVEMENT DESIGN—STRUCTURAL NUMBER DETERMINATION

A pavement is to be designed to last 10 years. The initial PSI is 4.2, and the TSI (the final PSI) is determined to be 2.5. The subgrade has a soil resilient modulus of 15,000 lb/in^2. Reliability is 95% with an overall standard deviation of 0.4. For design, the daily car, pickup truck, and light van traffic is 30,000, and the daily truck traffic consists of 1000 passes of single-unit trucks with two single axles and 350 passes of tractor semi-trailer trucks with single, tandem, and triple axles. The axle weights are

$$\text{cars, pickups, light vans} = \text{two 2000-lb single axles}$$
$$\text{single-unit truck} = \text{8000-lb steering, single axle}$$
$$= \text{22,000-lb drive, single axle}$$
$$\text{tractor semi-trailer truck} = \text{10,000-lb steering, single axle}$$
$$= \text{16,000-lb drive, tandem axle}$$
$$= \text{44,000-lb trailer, triple axle}$$

M_2 and M_3 are equal to 1.0 for the materials in the pavement structure. Four inches of hot-mix asphalt (HMA) is to be used as the wearing surface and 10 inches of crushed stone as the subbase. Determine the thickness required for the base if soil cement is the material to be used.

SOLUTION

Because the axle-load equivalency factors presented in Tables 4.1, 4.2, and 4.3 are a function of the SN, we have to assume an SN to start the problem (later we will arrive at a SN and check to make sure that it is consistent with our assumed value). A typical assumption is to let SN = 4. Given this, the 18-kip–equivalent single-axle load for cars, pickups, and light vans is

$$\text{2-kip single-axle equivalent} = 0.0002 \text{ (Table 4.1)}$$

This gives an 18-kip ESAL total of 0.0004 for each vehicle. For single-unit trucks,

$$\text{8-kip single-axle equivalent} = 0.041 \text{ (Table 4.1)}$$
$$\text{22-kip single-axle equivalent} = 2.090 \text{ (Table 4.1)}$$

This gives an 18-kip ESAL total of 2.131 for single-unit trucks. For tractor semi-trailer trucks,

$$\text{10-kip single-axle equivalent} = 0.102 \text{ (Table 4.1)}$$
$$\text{16-kip tandem-axle equivalent} = 0.057 \text{ (Table 4.2)}$$
$$\text{44-kip triple-axle equivalent} = 0.769 \text{ (Table 4.3)}$$

This gives an 18-kip ESAL total of 0.928 for tractor semi-trailer trucks. Note the comparatively small effect of cars and other light vehicles in terms of the 18-kip ESAL. This small effect underscores the nonlinear relationship between axle loads and pavement damage. For example, from Table 4.2 with SN = 4, a 36-kip single-axle load has 14.4 times the impact on pavement as an 18-kip single-axle load (twice the weight has 14.4 times the impact).

Given the computed 18-kip ESAL, the daily traffic on this highway produces an 18-kip ESAL total of 2467.8 ($0.0004 \times 30{,}000 + 2.131 \times 1000 + 0.928 \times 350$). Traffic (total axle accumulations) over the 10-year design period will be

$$2467.8 \times 365 \times 10 = 9{,}007{,}470 \text{ 18-kip ESAL}$$

With an initial PSI of 4.2 and a TSI of 2.5, $\Delta\text{PSI} = 1.7$. Solving Eq. 4.1 for SN (using an equation solver on a calculator or computer) with $Z_R = -1.645$ (which corresponds to $R = 95\%$, as shown in Table 4.4) gives SN = 3.94 (Fig. 4.5 can also be used to arrive at an approximate solution for SN). Note that this is very close to the value that was assumed (SN = 4.0) to get the load equivalency factors from Tables 4.1, 4.2, and 4.3. If Eq. 4.1 gave SN = 5, we would go back and recompute total axle accumulations using the SN of 5 to read the axle-load equivalency factors in Tables 4.1, 4.2, and 4.3. Usually one iteration of this type is all that is needed. Later, Examples 4.3 and 4.5 will demonstrate this type of iteration.

Given that SN = 3.94, Eq. 4.3 can be applied with $a_1 = 0.44$ (surface course, hot-mix asphalt, Table 4.5), $a_2 = 0.20$ (base course, soil cement, Table 4.5), and $a_3 = 0.11$ (subbase, crushed stone, Table 4.5), $M_2 = 1.0$ (given), $M_3 = 1.0$ (given), $D_1 = 4.0$ inches (given), and $D_3 = 10.0$ inches (given). We have

$$\text{SN} = a_1 D_1 + a_2 D_2 M_2 + a_3 D_3 M_3$$

$$3.94 = 0.44(4) + 0.20 D_2(1.0) + 0.11(10.0)(1.0)$$

Solving for D_2 gives $D_2 = 5.4$ inches. Using $D_2 = \underline{5.5 \text{ inches}}$ would be a conservative estimate and allow for variations in construction. Rounding up to the nearest 0.5 inch is a safe practice.

EXAMPLE 4.2 FLEXIBLE-PAVEMENT DESIGN—RELIABILITY ASSESSMENT

A flexible pavement is constructed with 4 inches of HMA wearing surface, 8 inches of emulsion/aggregate-bituminous base, and 8 inches of crushed stone subbase. The subgrade has a soil resilient modulus of 10,000 lb/in², and M_2 and M_3 are equal to 1.0 for the materials in the pavement structure. The overall standard deviation is 0.5, the initial PSI is 4.5, and the TSI is 2.5. The daily traffic has 1080 20-kip single axles, 400 24-kip single axles, and 680 40-kip tandem axles. How many years would you estimate this pavement would last (how long before its PSI drops below a TSI of 2.5) if you wanted to be 90% confident that your estimate was not too high, and if you wanted to be 99% confident that your estimate was not too high?

SOLUTION

The pavement's SN is determined from Eq. 4.3, using Table 4.5 to find $a_1 = 0.44$, $a_2 = 0.30$ and $a_3 = 0.11$, and with $D_1 = 4$, $D_2 = 8$, $D_3 = 8$, $M_2 = M_3 = 1.0$ (all given) as

$$\text{SN} = a_1 D_1 + a_2 D_2 M_2 + a_3 D_3 M_3$$

$$\text{SN} = 0.44(4) + 0.30(8)(1.0) + 0.11(8.0)(1.0) = 5.04$$

For the daily axle loads, the equivalency factors (reading axle equivalents from Tables 4.1 and 4.2 while using SN = 5, which is very close to the 5.04 computed above) are

20-kip single-axle equivalent = 1.51 (Table 4.1)
24-kip single-axle equivalent = 3.03 (Table 4.1)
40-kip tandem-axle equivalent = 2.08 (Table 4.2)

Thus the total daily 18-kip ESAL is

Daily W_{18} = 1.51(1080) + 3.03(400) + 2.08(680) = 4257.2 18-kip ESAL

Applying Eq. 4.1, with S_o = 0.5, SN = 5.04, ΔPSI = 2.0 (4.5 – 2.5), and M_R = 10,000 lb/in², we find that at R = 90% (Z_R = –1.282 for purposes of Eq. 4.1, as shown in Table 4.4), W_{18} is 26,128,077. Therefore, the number of years is

$$\text{years} = \frac{26,128,077}{365 \times 4257.2}$$
$$= 16.82 \text{ years}$$

Similarly, with R = 99% (Z_R = –2.326 for purposes of Eq. 4.1, as shown in Table 4.4), W_{18} is 7,854,299, so the number of years is

$$\text{years} = \frac{7,854,299}{365 \times 4257.2}$$
$$= 5.05 \text{ years}$$

These results show that one can be 99% confident that the pavement will last (have a PSI above 2.5) at least 5.05 years, and one can be 90% confident that it will have a PSI above 2.5 for 16.82 years. This example demonstrates the large impact that the chosen reliability value can have on pavement design.

4.5   PAVEMENT SYSTEM DESIGN: PRINCIPLES FOR RIGID PAVEMENTS

Rigid pavements distribute wheel loads by the beam action of the PCC slab, which is made of a material that has a high modulus of elasticity, on the order of 4 to 5 million lb/in². This beam action (see Fig. 4.6) distributes the wheel loads over a large area of the pavement, thus reducing the high stresses experienced at the surface of the pavement to a level that is acceptable to the subgrade soil.

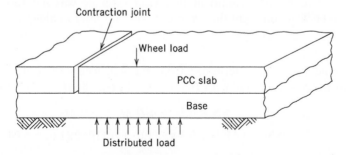

Figure 4.6 Beam action of a rigid pavement.

4.6   TRADITIONAL AASHTO RIGID-PAVEMENT DESIGN PROCEDURE

The design procedure for rigid pavements presented in the AASHTO design guide is also based on the field results of the AASHO Road Test. The AASHTO design procedure is applicable to jointed-plain concrete pavements, jointed-reinforced concrete pavements, and continuously-reinforced concrete pavements. Jointed-plain concrete pavements (JPCP) do not have slab-reinforcing material and can have doweled joints (steel bars to transfer loads between slabs as shown in Fig. 4.2) or undoweled joints. The traverse joints between slabs are spaced at about 10 to 13 ft. Jointed reinforced concrete pavements (JRCP) have steel reinforced slabs with joints that are 40 ft or more apart. Finally, continuously-reinforced concrete pavements (CRCP) do not have traverse expansion/contraction joints, necessitating the use of extensive steel-bar reinforcement in the slab. The idea with both jointed-reinforced and continuously-reinforced pavements is to permit slab cracking but to provide sufficient slab reinforcement to hold the cracks tightly together to ensure load transfer. It is important to note that faulting, which is a distress characterized by different slab elevations, was not a failure consideration in the AASHO Road Test, and thus the design of undoweled joints must be checked with a procedure other than that presented here (more information on faulting is provided in Section 4.7.5).

The design procedure for rigid pavements is based on a selected reduction in serviceability and is similar to the procedure for flexible pavements. However, instead of measuring pavement strength by using a SN, the thickness of the PCC slab is the measure of strength. The regression equation that is used to determine the thickness of a rigid-pavement PCC slab is

$$\log_{10}W_{18} = Z_R S_o + 7.35\left[\log_{10}(D+1)\right] - 0.06$$
$$+ \frac{\log_{10}\left[\Delta\text{PSI}/3.0\right]}{1 + \left[1.624\times10^7 / (D+1)^{8.46}\right]} \qquad \textbf{(4.4)}$$
$$+ (4.22 - 0.32\text{TSI})\log_{10}\left(\frac{S'_c C_d\left[D^{0.75} - 1.132\right]}{215.63J\left\{D^{0.75} - \left[18.42/(E_c/k)^{0.25}\right]\right\}}\right)$$

where

W_{18} = 18-kip–equivalent single-axle loads,

Z_R = Reliability (z-statistic from the standard normal curve),

S_o = Overall standard deviation of traffic,

D = PCC slab thickness in inches,

ΔPSI = Loss in serviceability from the time when the pavement is new until it reaches its TSI,

TSI = Pavement's terminal serviceability index,

S'_c = Concrete modulus of rupture in lb/in²,

C_d = Drainage coefficient,

J = Load transfer coefficient,

E_c = Concrete modulus of elasticity in lb/in², and

k = Modulus of subgrade reaction in lb/in³.

A graphic solution to Eq. 4.4 is shown in Figs. 4.7 and 4.8. The terms used in Eq. 4.4 and Figs. 4.8 and 4.9 are defined as follows:

W_{18} The 18-kip–equivalent single-axle load is the same concept as discussed for the flexible-pavement design procedure. However, instead of being a function of the SN, this value is a function of slab thickness. The axle-load equivalency factors used in rigid-pavement design are presented in Tables 4.6 (for single axles), 4.7 (for tandem axles), and 4.8 (for triple axles).

Z_R As in flexible-pavement design, the reliability, Z_R, is defined as the probability that serviceability will be maintained at adequate levels from a user's point of view throughout the design life of the facility (the PSI will stay above the TSI). In the rigid-pavement–design nomograph (Figs. 4.7 and 4.8), the probabilities (in percent) are used directly (instead of Z_R as in Eq. 4.4), and these percent probabilities are denoted R (see Table 4.4, which still applies).

S_o As in flexible-pavement design, the overall standard deviation, S_o, takes into account designers' inability to accurately estimate future 18-kip–equivalent axle loads and the statistical error in the equations resulting from variability in materials and construction practices.

TSI The pavement's terminal serviceability index, TSI, is the point at which the pavement can no longer perform in a serviceable manner, as discussed previously for the flexible-pavement design procedure.

ΔPSI The amount of serviceability loss, ΔPSI, over the life of the pavement is the difference between the initial PSI and the TSI, as discussed for the flexible-pavement design procedure.

S'_c The concrete modulus of rupture, S'_c, is a measure of the tensile strength of the concrete and is determined by loading a beam specimen, at the third points, to failure. The test method is ASTM C78, Flexural Strength of Concrete. Because concrete gains strength with age, the average 28-day strength is used for design purposes. Typical values are 500 to 1200 lb/in^2.

C_d The drainage coefficient, C_d, is slightly different from the value used in flexible-pavement design. In rigid-pavement design, it accounts for the drainage characteristics of the subgrade. A value of 1.0 for the drainage coefficient represents a material with good drainage characteristics (such as a sandy material). Soils with less-than-ideal drainage characteristics will have drainage coefficients less than 1.0.

J The load transfer coefficient, J, is a factor that is used to account for the ability of pavement to transfer a load from one PCC slab to another across the slab joints. Many rigid pavements have dowel bars across the joints to transfer loads between slabs. Pavements with dowel bars at the joints are typically designed with a J value of 3.2.

Table 4.6 Axle-Load Equivalency Factors for Rigid Pavements, Single Axles, and TSI = 2.5

Axle load (kips)	Slab thickness, D (inches)								
	6	7	8	9	10	11	12	13	14
2	0.0002	0.0002	0.0002	0.0002	0.0002	0.0002	0.0002	0.0002	0.0002
4	0.003	0.002	0.002	0.002	0.002	0.002	0.002	0.002	0.002
6	0.012	0.011	0.010	0.010	0.010	0.010	0.010	0.010	0.010
8	0.039	0.035	0.033	0.032	0.032	0.032	0.032	0.032	0.032
10	0.097	0.089	0.084	0.082	0.081	0.080	0.080	0.080	0.080
12	0.203	0.189	0.181	0.176	0.175	0.174	0.174	0.174	0.173
14	0.376	0.360	0.347	0.341	0.338	0.337	0.336	0.336	0.336
16	0.634	0.623	0.610	0.604	0.601	0.599	0.599	0.599	0.598
18	1.00	1.00	1.00	1.00	1.00	1.00	1.00	1.00	1.00
20	1.51	1.52	1.55	1.57	1.58	1.58	1.59	1.59	1.59
22	2.21	2.20	2.28	2.34	2.38	2.40	2.41	2.41	2.41
24	3.16	3.10	3.22	3.36	3.45	3.50	3.53	3.54	3.55
26	4.41	4.26	4.42	4.67	4.85	4.95	5.01	5.04	5.05
28	6.05	5.76	5.92	6.29	6.61	6.81	6.92	6.98	7.01
30	8.16	7.67	7.79	8.28	8.79	9.14	9.35	9.46	9.52
32	10.8	10.1	10.1	10.7	11.4	12.0	12.3	12.6	12.7
34	14.1	13.0	12.9	13.6	14.6	15.4	16.0	16.4	16.5
36	18.2	16.7	16.4	17.1	18.3	19.5	20.4	21.0	21.3
38	23.1	21.1	20.6	21.3	22.7	24.3	25.6	26.4	27.0
40	29.1	26.5	25.7	26.3	27.9	29.9	31.6	32.9	33.7
42	36.2	32.9	31.7	32.2	34.0	36.3	38.7	40.4	41.6
44	44.6	40.4	38.8	39.2	41.0	43.8	46.7	49.1	50.8
46	54.5	49.3	47.1	47.3	49.2	52.3	55.9	59.0	61.4
48	66.1	59.7	56.9	56.8	58.7	62.1	66.3	70.3	73.4
50	79.4	71.7	68.2	67.8	69.6	73.3	78.1	83.0	87.1

Source: *AASHTO Guide for Design of Pavement Structures*, The American Association of State Highway and Transportation Officials, Washington, DC, 1993. Used by permission.

Table 4.7 Axle-Load Equivalency Factors for Rigid Pavements, Tandem Axles, and TSI = 2.5

Axle load (kips)	Slab thickness, D (inches)								
	6	7	8	9	10	11	12	13	14
2	0.0001	0.0001	0.0001	0.0001	0.0001	0.0001	0.0001	0.0001	0.0001
4	0.0006	0.0006	0.0005	0.0005	0.0005	0.0005	0.0005	0.0005	0.0005
6	0.002	0.002	0.002	0.002	0.002	0.002	0.002	0.002	0.002
8	0.007	0.006	0.006	0.005	0.005	0.005	0.005	0.005	0.005
10	0.015	0.014	0.013	0.013	0.012	0.012	0.012	0.012	0.012
12	0.031	0.028	0.026	0.026	0.025	0.025	0.025	0.025	0.025
14	0.057	0.052	0.049	0.048	0.047	0.047	0.047	0.047	0.047
16	0.097	0.089	0.084	0.082	0.081	0.081	0.080	0.080	0.080
18	0.155	0.143	0.136	0.133	0.132	0.131	0.131	0.131	0.131
20	0.234	0.220	0.211	0.206	0.204	0.203	0.203	0.203	0.203
22	0.340	0.325	0.313	0.308	0.305	0.304	0.303	0.303	0.303
24	0.475	0.462	0.450	0.444	0.441	0.440	0.439	0.439	0.439
26	0.644	0.637	0.627	0.622	0.620	0.619	0.618	0.618	0.618
28	0.855	0.854	0.852	0.850	0.850	0.850	0.849	0.849	0.849
30	1.11	1.12	1.13	1.14	1.14	1.14	1.14	1.14	1.14
32	1.43	1.44	1.47	1.49	1.50	1.51	1.51	1.51	1.51
34	1.82	1.82	1.87	1.92	1.95	1.96	1.97	1.97	1.97
36	2.29	2.27	2.35	2.43	2.48	2.51	2.52	2.52	2.53
38	2.85	2.80	2.91	3.03	3.12	3.16	3.18	3.20	3.20
40	3.52	3.42	3.55	3.74	3.87	3.94	3.98	4.00	4.01
42	4.32	4.16	4.30	4.55	4.74	4.86	4.91	4.95	4.96
44	5.26	5.01	5.16	5.48	5.75	5.92	6.01	6.06	6.09
46	6.36	6.01	6.14	6.53	6.90	7.14	7.28	7.36	7.40
48	7.64	7.16	7.27	7.73	8.21	8.55	8.75	8.86	8.92
50	9.11	8.50	8.55	9.07	9.68	10.14	10.42	10.58	10.66
52	10.8	10.0	10.0	10.6	11.3	11.9	12.3	12.5	12.7
54	12.8	11.8	11.7	12.3	13.2	13.9	14.5	14.8	14.9
56	15.0	13.8	13.6	14.2	15.2	16.2	16.8	17.3	17.5
58	17.5	16.0	15.7	16.3	17.5	18.6	19.5	20.1	20.4
60	20.3	18.5	18.1	18.7	20.0	21.4	22.5	23.2	23.6
62	23.5	21.4	20.8	21.4	22.8	24.4	25.7	26.7	27.3
64	27.0	24.6	23.8	24.4	25.8	27.7	29.3	30.5	31.3
66	31.0	28.1	27.1	27.6	29.2	31.3	33.2	34.7	35.7
68	35.4	32.1	30.9	31.3	32.9	35.2	37.5	39.3	40.5
70	40.3	36.5	35.0	35.3	37.0	39.5	42.1	44.3	45.9
72	45.7	41.4	39.6	39.8	41.5	44.2	47.2	49.8	51.7
74	51.7	46.7	44.6	44.7	46.4	49.3	52.7	55.7	58.0
76	58.3	52.6	50.2	50.1	51.8	54.9	58.6	62.1	64.8
78	65.5	59.1	56.3	56.1	57.7	60.9	65.0	69.0	72.3
80	73.4	66.2	62.9	62.5	64.2	67.5	71.9	76.4	80.2
82	82.0	73.9	70.2	69.6	71.2	74.7	79.4	84.4	88.8
84	91.4	82.4	78.1	77.3	78.9	82.4	87.4	93.0	98.1
86	102.0	92.0	87.0	86.0	87.0	91.0	96.0	102.0	108.0
88	113.0	102.0	96.0	95.0	96.0	100.0	105.0	112.0	119.0
90	125.0	112.0	106.0	105.0	106.0	110.0	115.0	123.0	130.0

Source: *AASHTO Guide for Design of Pavement Structures*, The American Association of State Highway and Transportation Officials, Washington, DC, 1993. Used by permission.

Table 4.8 Axle-Load Equivalency Factors for Rigid Pavements, Triple Axles, and TSI = 2.5

Axle load (kips)	Slab thickness, D (inches)								
	6	7	8	9	10	11	12	13	14
2	0.0001	0.0001	0.0001	0.0001	0.0001	0.0001	0.0001	0.0001	0.0001
4	0.0003	0.0003	0.0003	0.0003	0.0003	0.0003	0.0003	0.0003	0.0003
6	0.001	0.001	0.001	0.001	0.001	0.001	0.001	0.001	0.001
8	0.003	0.002	0.002	0.002	0.002	0.002	0.002	0.002	0.002
10	0.006	0.005	0.005	0.005	0.005	0.005	0.005	0.005	0.005
12	0.011	0.010	0.010	0.009	0.009	0.009	0.009	0.009	0.009
14	0.020	0.018	0.017	0.017	0.016	0.016	0.016	0.016	0.016
16	0.033	0.030	0.029	0.028	0.027	0.027	0.027	0.027	0.027
18	0.053	0.048	0.045	0.044	0.044	0.043	0.043	0.043	0.043
20	0.080	0.073	0.069	0.067	0.066	0.066	0.066	0.066	0.066
22	0.116	0.107	0.101	0.099	0.098	0.097	0.097	0.097	0.097
24	0.163	0.151	0.144	0.141	0.139	0.139	0.138	0.138	0.138
26	0.222	0.209	0.200	0.195	0.194	0.193	0.192	0.192	0.192
28	0.295	0.281	0.271	0.265	0.263	0.262	0.262	0.262	0.262
30	0.384	0.371	0.359	0.354	0.351	0.350	0.349	0.349	0.349
32	0.490	0.480	0.468	0.463	0.460	0.459	0.458	0.458	0.458
34	0.616	0.609	0.601	0.596	0.594	0.593	0.592	0.592	0.592
36	0.765	0.762	0.759	0.757	0.756	0.755	0.755	0.755	0.755
38	0.939	0.941	0.946	0.948	0.950	0.951	0.951	0.951	0.951
40	1.14	1.15	1.16	1.17	1.18	1.18	1.18	1.18	1.18
42	1.38	1.38	1.41	1.44	1.45	1.46	1.46	1.46	1.46
44	1.65	1.65	1.70	1.74	1.77	1.78	1.78	1.78	1.78
46	1.97	1.96	2.03	2.09	2.13	2.15	2.16	2.16	2.16
48	2.34	2.31	2.40	2.49	2.55	2.58	2.59	2.60	2.60
50	2.76	2.71	2.81	2.94	3.02	3.07	3.09	3.10	3.11
52	3.24	3.15	3.27	3.44	3.56	3.62	3.66	3.68	3.68
54	3.79	3.66	3.79	4.00	4.16	4.26	4.30	4.33	4.34
56	4.41	4.23	4.37	4.63	4.84	4.97	5.03	5.07	5.09
58	5.12	4.87	5.00	5.32	5.59	5.76	5.85	5.90	5.93
60	5.91	5.59	5.71	6.08	6.42	6.64	6.77	6.84	6.87
62	6.80	6.39	6.50	6.91	7.33	7.62	7.79	7.88	7.93
64	7.79	7.29	7.37	7.82	8.33	8.70	8.92	9.04	9.11
66	8.90	8.28	8.33	8.83	9.42	9.88	10.17	10.33	10.42
68	10.1	9.4	9.4	9.9	10.6	11.2	11.5	11.7	11.9
70	11.5	10.6	10.6	11.1	11.9	12.6	13.0	13.3	13.5
72	13.0	12.0	11.8	12.4	13.3	14.1	14.7	15.0	15.2
74	14.6	13.5	13.2	13.8	14.8	15.8	16.5	16.9	17.1
76	16.5	15.1	14.8	15.4	16.5	17.6	18.4	18.9	19.2
78	18.5	16.9	16.5	17.1	18.2	19.5	20.5	21.1	21.5
80	20.6	18.8	18.3	18.9	20.2	21.6	22.7	23.5	24.0
82	23.0	21.0	20.3	20.9	22.2	23.8	25.2	26.1	26.7
84	25.6	23.3	22.5	23.1	24.5	26.2	27.8	28.9	29.6
86	28.4	25.8	24.9	25.4	26.9	28.8	30.5	31.9	32.8
88	31.5	28.6	27.5	27.9	29.4	31.5	33.5	35.1	36.1
90	34.8	31.5	30.3	30.7	32.2	34.4	36.7	38.5	39.8

Source: *AASHTO Guide for Design of Pavement Structures*, The American Association of State Highway and Transportation Officials, Washington, DC, 1993. Used by permission.

E_c The concrete modulus of elasticity, E_c, is derived from the stress-strain curve as taken in the elastic region. The modulus of elasticity is also known as Young's modulus. Typical values of E_c for PCC are between 3 and 7 million lb/in^2.

k The modulus of subgrade reaction, k, depends upon several different factors, including the moisture content and density of the soil. It should be noted that most highway agencies do not perform testing to measure the k value of the soil. At best, the agency will have a CBR value for the subgrade. Typical values for k range from 100 to 800 lb/in^3. Table 4.9 indicates the relationship between CBR and k values.

Table 4.9 Relationship between California Bearing Ratio (CBR) and Modulus of Subgrade Reaction, k

CBR	k, lb/in^3
2	100
10	200
20	250
25	290
40	420
50	500
75	680
100	800

EXAMPLE 4.3   RIGID-PAVEMENT DESIGN—SLAB THICKNESS DETERMINATION

A rigid pavement is to be designed to provide a service life of 20 years and has an initial PSI of 4.4 and a TSI of 2.5. The modulus of subgrade reaction is determined to be 300 lb/in^3. For design, the daily car, pickup truck, and light van traffic is 20,000; and the daily truck traffic consists of 200 passes of single-unit trucks with single and tandem axles, and 410 passes of tractor semi-trailer trucks with single, tandem, and triple axles. The axle weights are

cars, pickups, light vans   = two 2000-lb single axles

single-unit trucks   = 10,000-lb steering, single axle

                             = 22,000-lb drive, tandem axle

tractor semi-trailer trucks   = 12,000-lb steering, single axle

                             = 18,000-lb drive, tandem axle

                             = 50,000-lb trailer, triple axle

Reliability is 95%, the overall standard deviation is 0.45, the concrete's modulus of elasticity is 4.5 million lb/in^2, the concrete's modulus of rupture is 900 lb/in^2, the load transfer coefficient is 3.2, and the drainage coefficient is 1.0. Determine the required slab thickness.

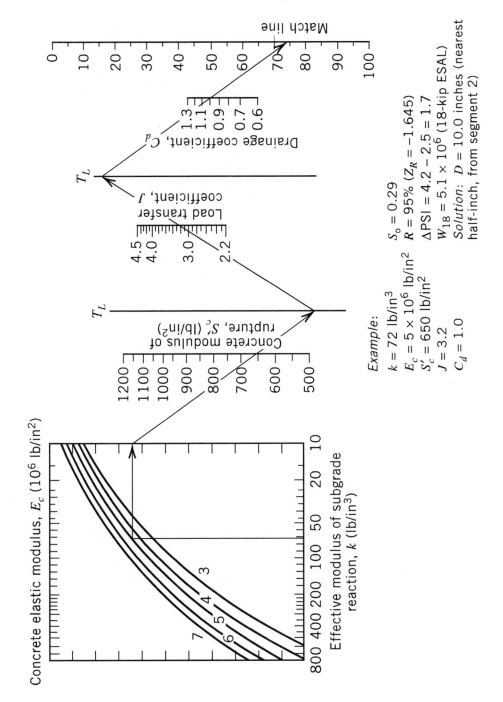

Figure 4.7 Segment 1 of the design chart for rigid pavement based on the use of mean values for each input variable.
Redrawn from *AASHTO Guide for Design of Pavement Structures*, The American Association of State Highway and Transportation Officials, Washington, DC, 1993. Used by permission.

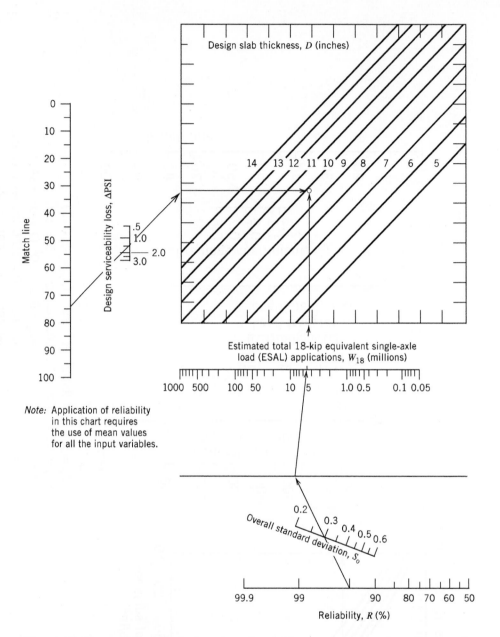

Figure 4.8 Segment 2 of the design chart for rigid pavements based on the use of mean values for each input variable.

Redrawn from *AASHTO Guide for Design of Pavement Structures*, The American Association of State Highway and Transportation Officials, Washington, DC, 1993. Used by permission.

SOLUTION

Because the axle-load equivalency factors presented in Tables 4.6, 4.7, and 4.8 are a function of the slab thickness (D), we have to assume a D value to start the problem (later we will arrive at a slab thickness and check to make sure that it is consistent with our assumed value). A typical assumption is to let $D = 10$ inches. Given this, the 18-kip–equivalent single-axle load (18-kip ESAL) for cars, pickups, and light vans is

2-kip single-axle equivalent = 0.0002 (Table 4.6)

This gives an 18-kip ESAL total of 0.0004 for each vehicle. For single-unit trucks,

10-kip single-axle equivalent   = 0.081 (Table 4.6)

22-kip tandem-axle equivalent = 0.305 (Table 4.7)

This gives an 18-kip ESAL total of 0.386 for single-unit trucks. For tractor semi-trailer trucks,

12-kip single-axle equivalent   = 0.175 (Table 4.6)

18-kip tandem-axle equivalent = 0.132 (Table 4.7)

50-kip triple-axle equivalent   = 3.020 (Table 4.8)

This gives an 18-kip ESAL total of 3.327 for tractor semi-trailer trucks.

Given the computed 18-kip ESAL, the daily traffic on this highway produces an 18-kip ESAL total of 1449.27 (0.0004 × 20,000 + 0.386 × 200 + 3.327 × 410). Traffic (total axle accumulations) over the 20-year design period will be

$$1449.27 \times 365 \times 20 = 10{,}579{,}671 \text{ 18-kip ESAL}$$

With an initial PSI of 4.4 and a TSI of 2.5, ΔPSI = 1.9. Solving Eq. 4.4 for D (using an equation solver on a calculator or computer) with $Z_R = -1.645$ (which corresponds to $R = 95\%$, as shown in Table 4.4) gives $D = 9.21$ inches. (Figs. 4.7 and 4.8 can also be used to arrive at an approximate solution for D.) Note that this value differs from the slab thickness assumed to derive the axle-load equivalency factors. Recomputing the axle-load equivalency factors with $D = 9$ inches (for Tables 4.6, 4.7, and 4.8) for cars, pickups, and light vans gives

2-kip single-axle equivalent = 0.0002 (Table 4.6)

This gives an 18-kip ESAL total of 0.0004 (same as before) for each vehicle. For single-unit trucks,

10-kip single-axle equivalent   = 0.082 (Table 4.6)

22-kip tandem-axle equivalent = 0.308 (Table 4.7)

This gives an 18-kip ESAL total of 0.390 (up from 0.386) for single-unit trucks. For tractor semi-trailer trucks,

12-kip single-axle equivalent   = 0.176 (Table 4.6)

18-kip tandem-axle equivalent = 0.133 (Table 4.7)

50-kip triple-axle equivalent   = 2.940 (Table 4.8)

This gives an 18-kip ESAL total of 3.249 (down from 3.327) for tractor semi-trailer trucks. Given the computed 18-kip ESAL, the daily traffic on this highway produces an 18-kip ESAL of 1418.09 ($0.0004 \times 20,000 + 0.390 \times 200 + 3.249 \times 410$). Traffic (total axle accumulations) over the 20-year design period will be

$$1418.09 \times 365 \times 20 = 10,352,057 \text{ 18-kip ESAL}$$

Again, solving Eq. 4.4 for D gives $D = 9.17$ inches. (Figs. 4.7 and 4.8 can also be used to arrive at an approximate solution for D.) This is very close to the assumed $D = 9.0$ inches and is only a minor change from the 9.21 inches previously obtained. To be conservative, we would round up to the nearest 0.5 inch and make the slab 9.5 inches.

4.7 DESIGN-LANE LOADS

The final point to be covered with regard to pavement design (for both flexible and rigid pavements) relates to the case where there are multiple lanes of a highway (such as an interstate) in one direction. Because traffic tends to be distributed among the lanes, in some instances, the pavement can be designed using a fraction of the total directional W_{18}. However, because traffic tends to concentrate in the right lane (particularly heavy vehicles), this fraction is not as simple as dividing W_{18} by the number of lanes. In equation form,

$$\text{design-lane } W_{18} = PDL \times \text{directional } W_{18} \tag{4.5}$$

where

W_{18} = 18-kip–equivalent single-axle load (ESAL) and

PDL = proportion of directional W_{18} assumed to be in the design lane.

AASHTO-recommended values for PDL are given in Table 4.10.

As an example, suppose the computed directional W_{18} is an 18-kip ESAL of 10,000,000 and there are three lanes in the direction of travel. If the highway is conservatively designed, Table 4.10 shows that 80% of the axle loads can be assumed to be in the design lane ($PDL_{\text{3-lanes}} = 0.8$). So the design W_{18} would be 8,000,000 ($0.8 \times 10,000,000$), and this value would be used in the equations and nomographs. This design procedure applies to both flexible and rigid pavements.

Table 4.10 Proportion of Directional W_{18} Assumed to Be in the Design Lane

Number of directional lanes	Proportion of directional W_{18} in the design lane (*PDL*)
1	1.00
2	0.80–1.00
3	0.60–0.80
4	0.50–0.75

EXAMPLE 4.4   **RIGID-PAVEMENT DESIGN WITH TRAFFIC DISTRIBUTION BY LANE**

In 1996, a rigid pavement on a northbound section of interstate highway was designed with a 12-inch PCC slab, an E_c of 6×10^6 lb/in^2, a concrete modulus of rupture of 800 lb/in^2, a load transfer coefficient of 3.0, an initial PSI of 4.5, and a TSI of 2.5. The overall standard deviation was 0.45, the modulus of subgrade reaction was 190 lb/in^3, and a reliability of 95% was used along with a drainage coefficient of 1.0. The pavement was designed for a 20-year life, and traffic was assumed to be composed entirely of tractor semi-trailer trucks with one 16-kip single axle, one 20-kip single axle, and one 35-kip tandem axle (the effect of all other vehicles was ignored). The interstate has four northbound lanes and was conservatively designed. How many tractor semi-trailer trucks, per day, were assumed to be traveling in the northbound direction?

SOLUTION

Given that the slab thickness D is 12 inches, for the tractor semi-trailer trucks we have

$$
\begin{aligned}
\text{16-kip single-axle equivalent} \quad &= 0.599 \text{ (Table 4.6)} \\
\text{20-kip single-axle equivalent} \quad &= 1.590 \text{ (Table 4.6)} \\
\text{35-kip tandem-axle equivalent} \quad &= 2.245 \text{ (Table 4.7)}
\end{aligned}
$$

Note that the value of 2.245 for the 35,000-lb tandem-axle linear interpolation uses 34-kip and 36-kip values [(1.97 + 2.52)/2]. Summing these axle equivalents gives 4.434 18-kip ESAL per truck.

With an initial PSI of 4.5 and a TSI of 2.5, ΔPSI = 2.0. Solving Eq. 4.4 for W_{18} with $Z_R = -1.645$ (which corresponds to R = 95% as shown in Table 4.4) gives W_{18} = 39,740,309 18-kip ESAL (Figs. 4.7 and 4.8 can also be used to arrive at an approximate solution for W_{18}). Thus, the total daily truck traffic in the design lane is

$$
\begin{aligned}
\text{traffic} &= \frac{39{,}740{,}309 \text{ 18-kip ESAL}}{365 \text{ days / year} \times 20 \text{ years} \times 4.434 \text{ 18-kip ESAL / truck}} \\
&= 1227.76 \text{ trucks / day}
\end{aligned}
$$

To determine the total directional volume (total number of northbound trucks), we note from Table 4.10 that the *PDL* for a conservative design on a four-lane highway is 0.75, and the application of Eq. 4.5 gives

$$\text{directional } W_{18} = \frac{\text{design-lane } W_{18}}{PDL_{\text{4-lanes}}}$$

$$= \frac{1227.76}{0.75}$$

$$= 1637.01 \text{ trucks} / \text{day}$$

EXAMPLE 4.5   FLEXIBLE AND RIGID PAVEMENT DESIGN-LIFE EQUIVALENCE

A new flexible pavement was designed for four lanes of traffic (conservatively designed for load distribution among the lanes). The design is for a total directional daily traffic of 967 10-kip single axles and 1935 30-kip tandem axles. The pavement has an 8-in hot-mix asphaltic (HMA) surface, 10-in dense-graded crushed stone base and a 10-in crushed stone subbase (the drainage coefficients are 0.9 for the base and 0.78 for the subbase). The soil CBR is 2, the reliability used was 95%, the overall standard deviation was 0.4, initial PSI was 4.5, and the TSI was 2.5. Determine the required slab thickness for a rigid pavement designed to last the same number of years as the flexible pavement, but with only three lanes (instead of four) in the design direction (again, conservatively designing for the load distribution among the lanes). The design is to use the same truck traffic, reliability, soil, initial PSI, TSI, and overall standard deviations as the flexible pavement. In addition, the rigid pavement is to have a modulus of rupture of 800 lb/in^2, a concrete modulus of elasticity of 5.5×10^6 lb/in^2, a load transfer coefficient of 3.0, and a drainage coefficient of 0.9.

SOLUTION

We will first need to determine the design life of the existing flexible pavement and then use this value to determine the required slab thickness of the rigid pavement.

The flexible pavement's SN is determined from Eq. 4.3, using Table 4.5 to find $a_1 = 0.44$, $a_2 = 0.18$ and $a_3 = 0.11$, and with $D_1 = 8$, $D_2 = 10$, $D_3 = 10$, $M_2 = 0.9$ and $M_3 = 0.78$ (all given) as

$$SN = a_1D_1 + a_2D_2M_2 + a_3D_3M_3$$

$$SN = 0.44(8) + 0.18(10)(0.9) + 0.11(10)(0.9) = 5.998 \approx 6$$

With a CBR = 2 (given), Eq. 4.2 is applied to give $M_R = 3000$ (1500×2). For other elements required to solve Eq. 4.1 (or alternatively Fig. 4.5), $\Delta PSI = 2$ (initial PSI of 4.5 minus the TSI of 2.5), $S_o = 0.4$ (given), and $Z_R = -1.645$ (which corresponds to $R = 95\%$, as shown in Table 4.4). With these values, applying Eq. 4.1 (or using Fig. 4.5) gives $W_{18} = 5,703,439$.

For the daily axle loads, the equivalency factors (reading axle equivalents from Tables 4.1 and 4.2 while using SN = 6) are

10-kip single-axle equivalent = 0.08 (Table 4.1)

30-kip tandem-axle equivalent = 0.633 (Table 4.2)

Thus the total daily 18-kip ESAL is

$$\text{Daily } W_{18} = 0.08(967) + 0.633(1935) = 1302.215 \text{ 18-kip ESAL/day}$$

From Table 4.10, the *PDL* for a conservative design on a four-lane highway is 0.75, so applying Eq. 4.4 gives

$$\text{design-lane } W_{18} = PDL \times \text{directional } W_{18} = 976.66 \; W_{18} / \text{day}$$

or 356,481 W_{18}/year (976.66 × 365). So the design life of the flexible pavement is:

$$\text{design life (in years)} = \frac{\text{design } W_{18}}{\text{daily } W_{18}}$$

$$= \frac{5,703,439}{356,481}$$

$$= 16 \text{ years}$$

For the rigid pavement, we have that all of the design parameters are the same and are given S'_c = 800 lb/in^2, E_c = 5.5 × 10^6 lb/in^2, J = 3.0, C_d = 0.9. In addition, the modulus of subgrade reaction, k, is 100 lb/in^3 from Table 4.9 (with CBR = 2). Note that the accumulated W_{18} for the rigid pavement will not be the same as the daily axle loads for the flexible pavement because the load equivalency factors will be different. To determine the accumulated W_{18} for the rigid pavement, we assume a slab thickness of 10 inches (D = 10 inches) and read axle equivalents from Tables 4.6 and 4.7 as

$$\text{10-kip single-axle equivalent} = 0.08 \text{ (Table 4.6)}$$
$$\text{30-kip tandem-axle equivalent} = 1.14 \text{ (Table 4.7)}$$

Thus the total daily 18-kip ESAL is

$$\text{Daily } W_{18} = 0.08(967) + 1.14(1935) = 2283.26 \text{ 18-kip ESAL/day}$$

From Table 4.10, the *PDL* for a conservative design on a three-lane highway is 0.8, so applying Eq. 4.4 gives a required slab thickness of <u>10.28 inches</u> (Figs. 4.7 and 4.8 can also be used to arrive at an approximate solution for slab thickness). For design, this can be rounded up to 10.5 inches. It can be shown by the reader that going back to the load equivalency factors and assuming 10.5 inches instead of the previously assumed 10 inches will result in the same, correct slab-thickness solution of 10.5 inches.

EXAMPLE 4.6   **FLEXIBLE AND RIGID PAVEMENT DESIGN-LIFE COMPARISON**

A roadway is determined to have 400 18-kip single axles, 200 24-kip tandem axles and 100 40-kip triple axles per day. The subgrade CBR is 2 and the roadway pavement is designed for an overall standard deviation of 0.4, a reliability of 99% and the initial PSI is 4.5 and the TSI is 2.5. One newly constructed section of this roadway is a rigid pavement designed with a 9-inch slab with a modulus of rupture of 700 lb/in^2, a modulus of elasticity of 4.0×10^6 lb/in^2, and a joint transfer coefficient of 3.0. Another newly constructed section of the same roadway is a flexible pavement with a 5-in HMA surface, 10-in dense-graded crushed stone base and a 9-in crushed-stone subbase. If the roadway has four lanes in each direction and is conservatively designed, which of the pavement sections will last longer and by how many years (all drainage coefficients are 1.0)?

SOLUTION

We first determine the total design W_{18} for the flexible pavement. To do this, the flexible pavement's SN is computed from Eq. 4.3, using Table 4.5 to find $a_1 = 0.44$, $a_2 = 0.18$ and $a_3 = 0.11$, and with $D_1 = 5$, $D_2 = 10$, $D_3 = 9$, $M_2 = 1.0$ and $M_3 = 1.0$ (all given) as

$$SN = a_1D_1 + a_2D_2M_2 + a_3D_3M_3$$

$$SN = 0.44(5) + 0.18(10)(1.0) + 0.11(9)(1.0) = 4.99 \approx 5$$

With a CBR = 2 (given), Eq. 4.2 is applied to give $M_R = 3000$ (1500×2). For other elements required to solve Eq. 4.1 (or alternatively Fig. 4.5), $S_o = 0.4$ (given), $Z_R = -2.326$ (which corresponds to $R = 99\%$, as shown in Table 4.4), and $\Delta PSI = 2$ (initial PSI of 4.5 minus the TSI of 2.5). With these values, applying Eq. 4.1 (or using Fig. 4.5) gives $W_{18} = 775,133$. To determine the total truck traffic, the equivalency factors (reading axle equivalents from Tables 4.1, 4.2 and 4.3 while using SN = 5) are

18-kip single-axle equivalent   = 1.0 (Table 4.1)

24-kip tandem-axle equivalent = 0.260 (Table 4.2)

40-kip triple-axle equivalent   = 0.487 (Table 4.3)

which gives a total W_{18} of 500.7 18-kip ESAL/day ($1.0 \times 400 + 0.260 \times 200 + 0.487 \times 100$). From Table 4.10, the *PDL* for a conservative design on a four-lane highway is 0.75, so applying Eq. 4.4 gives the design-lane $W_{18} = 375.53$ 18-kip ESAL/day (0.75×500.7). So the design life for the flexible pavement is

$$\text{design life (in years)} = \frac{\text{total design-lane } W_{18}}{\text{daily design-lane } W_{18} \times 365 \text{ days / yr}}$$

$$= \frac{775,133}{375.53 \times 365}$$

$$= 5.655 \text{ years}$$

For the rigid pavement, we are given a slab thickness of 9 inches ($D = 9$ inches) $S'_c = 900$ lb/in^2, $E_c = 4.0 \times 10^6$ lb/in^2, $J = 3.0$, $C_d = 1.0$ and the modulus of subgrade reaction,

k, is 100 lb/in³ from Table 4.9 (with CBR = 2), and all other design parameters are the same as those for the flexible pavement. With these values, applying Eq. 4.4 (or using Figs. 4.7 and 4.8) gives W_{18} = 2,364,522. To determine the total truck traffic, the equivalency factors (reading axle equivalents from Tables 4.7, 4.8, and 4.9 while using $D = 9$) are

$$18\text{-kip single-axle equivalent} \quad = 1.0 \text{ (Table 4.6)}$$

$$24\text{-kip tandem-axle equivalent} = 0.444 \text{ (Table 4.7)}$$

$$40\text{-kip triple-axle equivalent} \quad = 1.17 \text{ (Table 4.8)}$$

which gives a total W_{18} of 605.8 18-kip ESAL/day ($1.0 \times 400 + 0.444 \times 200 + 1.17 \times 100$). As in the flexible-pavement case, from Table 4.10, the *PDL* for a conservative design on a four-lane highway is 0.75, so applying Eq. 4.4 gives the design-lane W_{18} = 454.35 18-kip ESAL/day (0.75×605.8). So the design life for the rigid pavement is

$$\text{design life (in years)} = \frac{\text{total design-lane } W_{18}}{\text{daily design-lane } W_{18} \times 365 \text{ days / yr}}$$

$$= \frac{2,364,522}{454.35 \times 365}$$

$$= 14.258 \text{ years}$$

So the rigid pavement will last 8.603 years longer ($14.258 - 5.655$).

4.8   MEASURING PAVEMENT QUALITY AND PERFORMANCE

The design procedure for pavements originally focused on the PSI as a measure of pavement quality. However, the PSI is based on the opinions of a panel of experts (as discussed in Section 4.4.1), which can introduce some variability into their determination. As a result, efforts have been undertaken to develop quantitative measures of pavement condition that provide additional insights into pavement quality and performance and that correlate with the traditional pavement serviceability index. Some factors that are regularly measured by highway pavement agencies now include the International Roughness Index, friction measurements, and rut depth.

4.8.1   International Roughness Index

The International Roughness Index (IRI) has become the most popular measure for evaluating the condition of pavements. The IRI evolved out of a study commissioned by the World Bank [Sayers et al., 1986] to establish uniformity of the physical measurement of pavement roughness. The IRI is determined by measuring vertical movements in a standardized vehicle's suspension per unit length of roadway. Units of IRI are reported in inches per mile (in/mi). The higher the value of the IRI, the rougher the road. To get some sense for how the IRI relates to pavement condition assessments and PSI, Tables 4.11 and 4.12 provide IRI and PSI values corresponding to what is considered poor, mediocre, fair, good, and very good for Interstate and non-Interstate highways [Federal Highway Administration, 2006]. Note that, due to the higher design standards and performance expectations, interstate highways are held to a higher standard for fair, mediocre, and poor pavement assessments.

4.8.2   Friction Measurements

Another important measurement of pavement performance is the surface friction. This is critical because low friction values can increase stopping distances and the probability of accidents. Given the variability of pavement surfaces, weather conditions, and tire characteristics, determining pavement friction over the range of possible values is not an easy task. To estimate friction, a standardized test is conducted under wet conditions using either a treaded or smooth tire. Although other speeds are sometimes used, the standard test is generally conducted at 40 mi/h using a friction-testing trailer in which the wheel is locked on the wetted road surface, and the torque developed from this wheel locking is used to measure a friction number. The friction number resulting from this test gives an approximation of the coefficient of road adhesion under wet conditions (as shown in Table 2.4) and is multiplied by 100 to produce a value between 0 and 100. The friction number with a treaded tire (FN_t) attempts to measure pavement microtexture, which is a function of the aggregate quality and composition. The friction number with a smooth tire (FN_s) provides a measure of pavement macrotexture, which is critical in providing a water drainage escape path between the pavement and tire.

A number of factors influence the friction number, such as changes in traffic volumes or traffic composition, surface age (friction has been found to increase quickly after construction, then as time passes, to level off and eventually decline), seasonal changes (in northern states, the friction number tends to be highest in the spring and lowest in the fall), and speed (the measured value tends to decrease as the test speed increases).

Table 4.11 Relationship between the International Roughness Index (IRI) and Perceptions of Pavement Quality for Interstate Highways

Pavement condition	Present Serviceability Index	Measured International Roughness Index in/mi
Very good	≥ 4.0	< 60
Good	3.5 – 3.9	61–94
Fair	3.1 – 3.4	95–119
Mediocre	2.6 – 3.0	120–170
Poor	≤ 2.5	> 170

Table 4.12 Relationship between the International Roughness Index (IRI) and Perceptions of Pavement Quality for Non-Interstate Highways

Pavement condition	Present Serviceability Index	Measured International Roughness Index in/mi
Very good	≥ 4.0	< 60
Good	3.5 – 3.9	61–94
Fair	3.1 – 3.4	95–170
Mediocre	2.6 – 3.0	171–220
Poor	≤ 2.5	> 220

Also, the friction number measured with the treaded tire tends to be greater than that measured with the smooth tire (usually by a value of about 20), but the difference decreases as the surface texture becomes rougher [Li et al., 2003].

In terms of safety, the amount of friction needed to minimize safety-related problems depends on prevailing traffic and geometric conditions. Guidelines used by some states suggest that values of $FN_t < 30$ or $FN_s < 15$ indicate that poor friction may be contributing to wet-weather accidents. Other state agencies have simply put in place guidelines for minimum friction requirements. For example, in Indiana, the minimum friction value is based on the smooth tire test at 40 mi/h, and a pavement with $FN_s < 20$ is considered in need of surfacing work to improve friction (generally resurfacing).

4.8.3   Rut Depth

Rut depth, which is a measure of pavement surface deformation in the wheel paths, can affect roadway safety because the ruts accumulate water and increase the possibility of vehicle hydroplaning (which results in the tire skimming over a film of water, greatly reducing braking and steering effectiveness). Because of its potential impact on vehicle control, rut depths are regularly measured on many highways to determine if pavement rutting has reached critical values that would require resurfacing or other pavement treatments. Virtually all states measure rut depth using automated equipment that seeks to determine the difference in surface elevation of the pavement in the wheel path relative to the pavement that is not in the wheel path. The critical values of rut depth can vary from one highway agency to the next. Usually, rut depths are considered unacceptably high when their values reach between 0.5 and 1.0 inches, indicating that corrective action is warranted.

4.8.4   Cracking

For flexible pavements, four types of cracking are usually monitored: longitudinal-fatigue cracking, transverse cracking, alligator cracking, and reflection cracking. Longitudinal-fatigue cracking is a surface-down cracking that occurs due to material fatigue in the wheel path. Such cracking can accelerate over time and require significant repairs to protect against water penetration into the flexible pavement structure. Transverse cracking is generally the result of low temperatures that cause fractures across the traffic lanes (resulting in an increase in pavement roughness). Alligator-fatigue cracking is a consequence of material fatigue in the wheel path, generally starting from the bottom of the asphalt layer. Such material fatigue creates a patch of connected cracks that resembles the skin of an alligator (as with other types of cracks, these can accelerate quickly over time and generate the need for maintenance to protect the integrity of the pavement structure). Finally, reflection cracking occurs when HMA overlays are placed over exiting pavement structures that had alligator-fatigue cracking, or other indications of pavement distress, and these old distresses manifest themselves in new distresses in the overlay. This results in surface cracking that increases surface roughness and the need for maintenance to protect water intrusion into the pavement structure.

For rigid pavements, transverse cracking is a common measure of pavement distress. Such cracking can be the result of slab fatigue and can be initiated either at the surface or base of the slab. The spacing and width of transverse cracks, and the potential impact of severe cracking on the structural integrity of the pavement, are critical measures of rigid-pavement distress.

4.8.5   Faulting

For traditional JPCP (Jointed Plain Concrete Pavements) rigid pavements, joint faulting (characterized by different slab elevations) is a critical measure of pavement distress. Faulting is an indicator of erosion or fatigue of the layers beneath the slab and reflects a failure of the load-transfer ability of the pavement between adjacent slabs. Faulting is associated with increased roughness and will be reflected in International Roughness Index measurements.

4.8.6   Punchouts

For Continuously Reinforced Concrete Pavements (CRCP) rigid pavements (those built without expansion/contraction joints), fatigue damage at the top of the slab is often measured by punchouts, which occur when the close spacing of transverse cracks cause in high tensile stresses that result in portions of the slab being broken into pieces. Punchouts are associated with increased roughness and are reflected in International Roughness Index measurements.

4.9   MECHANISTIC-EMPIRICAL PAVEMENT DESIGN

The Mechanistic-Empirical Pavement Design Guide (AASHTO 2008) is one of the more recent tools for the design and rehabilitation of pavement structures. The Mechanistic-Empirical Pavement Design Guide was developed to improve on the traditional pavement design procedures presented earlier in this chapter (*AASHTO Guide for Design of Pavement Structures*, 1993) by providing the ability to predict multiple pavement-performance measures (such as rut depth, various types of cracking, joint faulting, International Roughness Index, etc.) and providing a direct link among pavement elements (materials, structural design, construction, traffic, climate and pavement management practices).

Unlike the traditional pavement-design procedures presented earlier in this chapter, the Mechanistic-Empirical Pavement Design Guide is quite complex and must be done using a software package (the software package is referred to simply as MEPDG, standing for Mechanistic-Empirical Pavement Design Guide). The design of pavements with MEPDG is an iterative process that can be summarized as follows:

1. The design engineer first selects a pavement structure (layer thicknesses, etc.), often using the traditional AASHTO approach (*AASHTO Guide for Design of Pavement Structures*, 1993).

2. Various inputs needed for MEPDG pavement assessment are then gathered and classified in the following broad topic groupings (please note that this is a much more time-intensive effort than the traditional AASHTO pavement design approach demonstrated earlier in this chapter):

   a. General project information. For this, factors such as base/subbase construction month, pavement construction month, and month that the pavement is open to traffic are needed because these factors can affect pavement-performance criteria.

   b. Design criteria and reliability level. For design criteria, the level of tolerable distress such as cracking, faulting, International Roughness Index are needed (these criteria roughly replace the TSI in the traditional AASHTO pavement design). The reliability level needed is similar to that currently used in the traditional AASHTO process.

   c. Site conditions and factors. Here, information is needed on truck traffic (including axle-load distributions, speed limit to account for the effect of truck speed on pavement distress, and monthly and hourly distributions of truck-travel), climate (including hourly temperature, precipitation, wind speed, relative humidity, and cloud cover), and detailed soil information (strength, variability, etc.).

   d. Material properties. Detailed information on new-pavement material properties is needed. This information is along the lines of the structural coefficient values and concrete-strength measurements used in the traditional AASHTO pavement design (although at a significantly higher level of detail).

3. With the above, the MEPDG software can then be run and software outputs will include calculated changes in pavement layer properties, various distresses (such as rut depth, cracking, and faulting), and the International Roughness Index over the design life of the pavement. The designer can then determine if the criteria for a successful pavement design have been met (critical distresses do not cross values that can be considered a failure of the pavement over its design life). If these criteria are not met, the pavement design is altered and the process is continued until an acceptable pavement design is achieved.

Currently, the use of the mechanistic-empirical pavement design process and the MEPDG software is increasing; however, many highway and transportation agencies still use the traditional AASHTO pavement-design approach (*AASHTO Guide for Design of Pavement Structures*, 1993).

4.10 PRACTICE PROBLEMS

**PRACTICE
PROBLEM 4.1**

FLEXIBLE-PAVEMENT DESIGN LIFE AND NUMBER OF LANES

The initial design of northbound direction on a highway consists of a two-lane flexible pavement that has a SN of 5 and is designed for a subgrade CBR of 6, an overall standard deviation of 0.35, and a reliability of 90%. The initial PSI is 4.6 and the final PSI is 2.5. The pavement is designed for 3000 passes per day of a truck that has a 12,000 lb single axle and two 32,000 lb tandem axles. If the pavement is conservatively designed, how many years longer will the pavement last if a third lane is added.

SOLUTION

Note: Open boxes in equations "☐" are to be completed by the reader

This problem shows the effect that the number of lanes can have on the design life of a pavement. Although the number of axle loads may not change when lanes are added or subtracted, the proportion of the axle loads in the design lane will change because axle loads will be distributed over more lanes and fewer will be in the design lanes (as shown in Table 4.10).

 To solve the problem, we must first determine the design-lane W_{18} used in the initial design of the pavement by applying Eq. 4.1 (or by using Fig. 4.5). To apply Eq. 4.1, the soil resilient modulus is determined from Eq. 4.2 as,

$$M_R = 1500 \times \text{CBR} = 1500 \times \boxed{} = \boxed{},$$

the ΔPSI is determined as,

$$\Delta\text{PSI} = \text{initial PSI} - \text{final PSI (TSI)} = \boxed{} - \boxed{} = \boxed{},$$

and Z_R is determined from Table 4.4 to be $\boxed{}$ with $R = 90\%$ (given in the problem). With these values, and the other values given in the problem, solving Eq. 4.1 for W_{18} (which is the design-lane value used in the existing pavement design) gives,

$$\log_{10}W_{18} = Z_R S_o + 9.36\left[\log_{10}(\text{SN}+1)\right] - 0.20$$
$$+ \frac{\log_{10}\left[\Delta\text{PSI}/2.7\right]}{0.40 + \left[1094/(\text{SN}+1)^{5.19}\right]}$$
$$+ 2.32\log_{10}M_R - 8.07$$

$$= \boxed{} \times \boxed{} + 9.36\left[\log_{10}\left(\boxed{}+1\right)\right] - 0.20$$
$$+ \frac{\log_{10}\left[\boxed{}/2.7\right]}{0.40 + \left[1094\Big/\left(\boxed{}+1\right)^{5.19}\right]}$$
$$+ 2.32\log_{10}\boxed{} - 8.07$$

$$\log_{10}W_{18} = \boxed{}, \text{ so } W_{18} = 33{,}142{,}926$$

Initially, the pavement is designed for two lanes in the design direction and, since the problem states that it is conservatively designed, this means that the $PDL_{2\text{-lanes}}$ used in Eq. 4.5 is 1.00 from Table 4.10. With the third lane added in the design direction, the $PDL_{3\text{-lanes}}$ will become ☐ (as shown in Table 4.10 for a conservative design).

To determine the 18-kip–equivalent single-axle loads (EASL) per day, which will be needed to estimate how much longer the pavement will last with the additional lane, the equivalency factors (reading axle equivalents from Tables 4.1 and 4.2 while using SN = 5) are,

$$12\text{-kip single-axle equivalent } = \boxed{} \text{ (Table 4.1)}$$
$$32\text{-kip tandem-axle equivalent } = \boxed{} \text{ (Table 4.2).}$$

Each truck has one 12-kip single axle and two 32-kip tandem axles so the equivalent axle loads per truck is,

$$\boxed{} + 2 \times \boxed{} = 1.903 \text{ 18-kip ESAL/truck.}$$

So the total daily 18-kip ESAL is

$$\text{daily } W_{18} = \boxed{} \times \boxed{} \text{ trucks/day} = \boxed{} \text{ 18-kip ESAL/day.}$$

With this, the number of years the northbound two-lane road would last is,

$$\text{years}_{2\text{-lanes}} = \frac{W_{18} \text{ as designed}}{PDL_{2\text{-lanes}} \times 365 \text{ days / year} \times 18\text{-kip EASL / day}} = \frac{\boxed{}}{\boxed{} \times 365 \times \boxed{}}$$

$$= \boxed{} \text{ years.}$$

Similarly, the number of years the northbound three-lane road would last is,

$$\text{years}_{3\text{-lanes}} = \frac{W_{18} \text{ as designed}}{PDL_{3\text{-lanes}} \times 365 \text{ days / year} \times 18\text{-kip EASL / day}} = \frac{\boxed{}}{\boxed{} \times 365 \times \boxed{}}$$

$$= \boxed{} \text{ years.}$$

So the pavement will last,

$$\text{years}_{3\text{-lanes}} - \text{years}_{2\text{-lanes}} = \boxed{} - \boxed{} = 3.98 \text{ years longer}$$

PRACTICE PROBLEM 4.2

RIGID PAVEMENT DESIGN LIFE AND RELIABILITY

For the conditions in Practice Problem 4.1 above, suppose a rigid pavement was designed with a 12-inch slab (modulus of rupture of 800 lb/in², a modulus of elasticity of 6 million lb/in², a joint transfer coefficient of 3.2, and a drainage coefficient of 1.0) for the two-lane northbound direction described in Practice Problem 4.1 (all other traffic and PSI information is the same as in that problem). How confident would you be that the pavement would have a PSI above 2.5 after 30 years?

SOLUTION

Note: Open boxes in equations "☐" are to be completed by the reader

The intent of this problem is to show how the concept of reliability is used in pavement design, and also to underscore the fact that load equivalency factors are not the same for flexible- and rigid-pavement designs even when the axle loads are the same.

To solve this problem, we will need to apply Eq. 4.4 and solve for the reliability z-statistic Z_R, and then use Table 4.4 to get the percent probability R which will tell us how confident we would be that the pavement would have a PSI above 2.5 after 30 years. To apply Eq. 4.4, many of the terms are given as part of the problem statement, and others will be the same as those in Practice Problem 4.1 (ΔPSI = 2.1, CBR = 6, etc.). The terms that need to be determined are W_{18} (which will differ since the pavement is now rigid instead of flexible), and the modulus of subgrade reaction k which is a term not used in flexible-pavement design.

To determine W_{18}, we start by computing the 18-kip–equivalent single-axle loads (EASL) per day (noting that these are different than the ones previously determined for the flexible pavement in Practice Problem 4.1) from the equivalency factors (reading axle equivalents from Tables 4.6 and 4.7 while using D = 12 inches) are

$$\text{12-kip single-axle equivalent} \quad = \boxed{} \text{ (Table 4.6)}$$
$$\text{32-kip tandem-axle equivalent} \quad = \boxed{} \text{ (Table 4.7)}.$$

As stated in Practice Problem 4.1, each truck has one 12-kip single axle and two 32-kip tandem axles so the equivalent axle loads per truck is,

$$\boxed{} + 2 \times \boxed{} = 3.194 \text{ 18-kip ESAL/truck.}$$

Note that this value is substantially larger than the 1.903 18-kip ESAL/truck found for the flexible-pavement design in Practice Problem 4.1. With this, after 30 years, the design-lane W_{18} is,

design-lane $W_{18} = PDL_{\text{2-lanes}} \times$ design life in years $\times$ days/year $\times$ trucks/day $\times$ ESAL/truck

design-lane $W_{18} = \boxed{} \times \boxed{} \times 365 \times \boxed{} \times \boxed{}$

design-lane W_{18} = 104,922,900.

The modulus of subgrade reaction k is determined to be $\boxed{}$ lb/in³ by interpolation between CBR values of 2 and 10 in Table 4.9. With all values needed for the application of Eq. 4.4 now known, Eq. 4.4 can be rearranged to solve for the reliability z-statistic Z_R,

$$
Z_R = \begin{bmatrix} \log_{10}W_{18} - 7.35\left[\log_{10}(D+1)\right] - 0.06 \\[2mm] -\dfrac{\log_{10}\left[\Delta PSI/3.0\right]}{1+\left[1.624\times10^7/(D+1)^{8.46}\right]} \\[4mm] -(4.22-0.32\mathrm{TSI})\log_{10}\left(\dfrac{S_c'C_d\left[D^{0.75}-1.132\right]}{215.63\times\square\left\{D^{0.75}-\left[18.42/(E_c/k)^{0.25}\right]\right\}}\right) \end{bmatrix} \times S_o^{-1}
$$

$$
= \begin{bmatrix} \log_{10}\square - 7.35\left[\log_{10}(\square+1)\right] - 0.06 \\[2mm] -\dfrac{\log_{10}\left[\square/3.0\right]}{1+\left[1.624\times10^7/(\square+1)^{8.46}\right]} \\[4mm] -(4.22-0.32\times\square)\log_{10}\left(\dfrac{\square\times\square\left[\square^{0.75}-1.132\right]}{215.63\times\square\left\{\square^{0.75}-\left[18.42/(\square/\square)^{0.25}\right]\right\}}\right) \end{bmatrix} \times \square^{-1}
$$

Solving this equation gives $Z_R = -0.613$. With this value, going to Table 4.4 we see the probability R is 73%. Thus we are <u>73%</u> confident that the pavement will have a PSI above 2.5 after 30 years.

PRACTICE PROBLEM 4.3

RIGID PAVEMENT DESIGN LIFE AND INITIAL SERVICEABILITY

A westbound direction of a rigid pavement consists of three lanes of traffic and it is conservatively designed. The pavement is 10 inches thick, has a joint transfer coefficient of 2.8, reliability of 90%, overall standard deviation of 0.4, subgrade CBR of 55, modulus of elasticity of 5,000,000 lb/in², modulus of rupture of 850 lb/in², and a drainage coefficient of 0.9. The westbound direction is designed for daily traffic of 200,000 2-kip single axles, 10,000 12-kip single axles, and 20,000 28-kip tandem axles. If the TSI is 2.5, what is the minimum initial serviceability index for the pavement to have a 20-year design life?

SOLUTION

Note: Open boxes in equations "$\square$" are to be completed by the reader

This problem illustrates how the initial serviceability index (PSI) affects the design life of the pavement.

To begin, the design-lane W_{18} needs to be determined. This is done by first converting existing axle loads into an 18-kip–equivalent single-axle load (ESAL). Reading axle equivalents from Tables 4.6 and 4.7 while using $D = 10$ (inches) gives,

2-kip single-axle equivalent	= $\square$	(Table 4.6)
12-kip single-axle equivalent	= $\square$	(Table 4.6)
24-kip tandem-axle equivalent	= $\square$	(Table 4.7).

So,

$$\text{directional } W_{18} = \begin{bmatrix} (2\text{-kip single-axle equivalent}) \times (2\text{-kip single-axle loads per day}) \\ + (12\text{-kip single-axle equivalent}) \times (12\text{-kip single-axle loads per day}) \\ + (24\text{-kip tandem-axle equivalent}) \times (24\text{-kip tandam-axle loads per day}) \end{bmatrix}$$
$$\times\ 20 \text{ years} \times 365 \text{ days/year}$$

$$= \begin{bmatrix} \boxed{} \times 200{,}000 \\ + \boxed{} \times \boxed{} \\ + \boxed{} \times \boxed{} \end{bmatrix} \times 20 \text{ years} \times 365 \text{ days/year} = 38{,}872{,}500$$

With $PDL = \boxed{}$ for a conservatively designed highway with three directional lanes (see Table 4.10), Eq. 4.5 is applied for the design-lane W_{18} as,

$$\text{design-lane } W_{18} = PDL_{3\text{-lanes}} \times \text{directional } W_{18}$$
$$\text{design-lane } W_{18} = \boxed{} \times \boxed{} = 31{,}098{,}000.$$

With a CBR of 55 (given), Table 4.9 shows that the modulus of subgrade reaction is, by interpolation, $\boxed{}$ lb/in^3. And, with a given reliability of 90% ($R = 90\%$), Table 4.4 shows $Z_R = -1.282$. With these values (and other values provided in the problem description), solving Eq. 4.4 for ΔPSI (using an equation solver on a calculator or computer, or using Figs. 4.7 and 4.8 to arrive at an approximate solution for ΔPSI) we find that ΔPSI $= \boxed{}$. With this, the minimum initial serviceability index for the pavement (initial PSI) can be determined from,

$$\Delta\text{PSI} = \text{initial PSI} - \text{TSI}$$

so, initial PSI = TSI + ΔPSI = $\boxed{}$ + $\boxed{}$ = $\underline{4.334}$.

PRACTICE PROBLEM 4.4

FLEXIBLE PAVEMENT AND SOLVING FOR UNKOWN DESIGN LIFE

A northbound direction on an interstate highway consists of four lanes of flexible pavement. It is conservatively designed with 10-inch sand-mix asphaltic surface, 10-crushed stone base, and a 10-inch crushed stone subbase (all drainage coefficients are 1.0). The subgrade CBR is 3, the overall standard deviation is 0.4, and the reliability is 95%. The initial PSI is 4.6 and the final PSI is 2.5. Daily total northbound road traffic consists of 51,000 cars (each with two 2-kip single axles), 840 buses (each with two 20-kip single axles), and 1000 heavy trucks (each with one 12-kip single axle and two 34-kip tandem axles). How many years was this pavement designed to last?

SOLUTION

Note: Open boxes in equations "$\boxed{}$" are to be completed by the reader

This problem improves the reader's understanding of the relationship between various pavement characteristics and design life by formulating the problem as one where the analysis solves for design life, instead the more conventional approach where the analyst designs the pavement with a given design life. By looking at the pavement problem in this way, the reader can gain a better understanding of the pavement-design process.

The number of years the pavement will last is equal to the W_{18} used to design the pavement (the W_{18} used in Eq. 4.1, which is the total W_{18} in the design lane until pavement reaches the TSI) divided by the W_{18} accumulated annually (which is determined from annual axle loads). To determine the W_{18} from Eq. 4.1, we apply Eq. 4.2 to get,

$$M_R = 1500 \times \text{CBR} = 1500 \times \boxed{} = \boxed{}.$$

From Table 4.4 with R = 95%, $Z_R = \boxed{}$, and

$$\Delta\text{PSI} = \text{initial PSI} - \text{TSI} = \boxed{} - \boxed{} = \boxed{}.$$

Next, the SN of the existing pavement is computed. This is done by first determining the structural coefficients of the flexible-pavement layers, which are:

sand-mix asphalt surface	$a_1 = \boxed{}$	(Table 4.5)
crushed stone base	$a_2 = \boxed{}$	(Table 4.5)
crushed stone subbase	$a_3 = \boxed{}$	(Table 4.5).

With these structural coefficients, and the layer depths and drainage coefficients given in the problem, Eq. 4.3 can be applied to give,

$$\text{SN} = a_1 D_1 + a_2 D_2 M_2 + a_3 D_3 M_3$$

$$\text{SN} = \boxed{} \times \boxed{} + \boxed{} \times \boxed{} \times 1 + \boxed{} \times 10 \times \boxed{} = 6.0$$

With these values, Eq. 4.1 is solved as,

$$\log_{10} W_{18} = Z_R S_o + 9.36 \left[\log_{10} (\text{SN} + 1) \right] - 0.20$$

$$+ \frac{\log_{10} \left[\Delta\text{PSI} / 2.7 \right]}{0.40 + \left[1094 / (\text{SN} + 1)^{5.19} \right]}$$

$$+ 2.32 \log_{10} M_R - 8.07$$

$$= \boxed{} \times \boxed{} + 9.36 \left[\log_{10} \left(\boxed{} + 1 \right) \right] - 0.20$$

$$+ \frac{\log_{10} \left[\boxed{} / 2.7 \right]}{0.40 + \left[1094 / \left(\boxed{} + 1 \right)^{5.19} \right]}$$

$$+ 2.32 \log_{10} \boxed{} - 8.07$$

$$\log_{10} W_{18} = \boxed{}, \text{ so } W_{18} = 16{,}303{,}758$$

Now, to determine the accumulated W_{18} per day in the design lane, the directional daily 18-kip–equivalent single-axle loads (ESALs) need to be computed. With SN = 6, and the four axle type/weight combinations given in the problem, we have,

2-kip single-axle equivalent	= ☐	(Table 4.1)
12-kip single-axle equivalent	= ☐	(Table 4.1).
20-kip single-axle equivalent	= ☐	(Table 4.1)
34-kip tandem-axle equivalent	= ☐	(Table 4.2).

With 51,000 cars per day having two 2-kip single axles each, 840 buses per day having two 20-kip single axles each, and 1,000 heavy trucks per day having one 12-kip single axle and two 34-kip tandem axles we have,

$$\text{daily directional } W_{18} = \begin{bmatrix} (\text{2-kip single-axle equivalent}) \times (\text{2-kip single-axle loads per day}) \\ + (\text{12-kip single-axle equivalent}) \times (\text{12-kip single-axle loads per day}) \\ + (\text{20-kip single-axle equivalent}) \times (\text{20-kip single-axle loads per day}) \\ + (\text{34-kip tandem-axle equivalent}) \times (\text{34-kip tandam-axle loads per day}) \end{bmatrix}$$

$$= \boxed{} \times 102{,}000 + \boxed{} \times 1{,}000 + \boxed{} \times 1{,}680 + \boxed{} \times \boxed{}$$

$$= 4960.4 \text{ 18-kip EASL per day.}$$

To calculate the daily design-lane W_{18}, Eq. 4.5 is written as,

$$\text{daily design-lane } W_{18} = PDL_{\text{4-lanes}} \times \text{daily directional } W_{18}.$$

From Table 4.10, with a conservative design, $PDL_{\text{4-lanes}} = \boxed{}$, so

$$\text{daily design-lane } W_{18} = \boxed{} \times \boxed{} = 3720.3 \text{ 18-kip EASL per day}$$

or 1,357,909.5 18-kip EASL per year (3720.3 × 365).

With these values, the number of years the pavement will last is given by,

$$\text{years} = \frac{W_{18} \text{ as designed}}{W_{18} \text{ per year}} = \frac{\boxed{}}{\boxed{}} = \boxed{} \text{ years.}$$

NOMENCLATURE FOR CHAPTER 4

a_1, a_2, a_3	structural-layer coefficients for wearing surface, base, and subbase	M_R	soil resilient modulus
C_d	drainage coefficient for rigid-pavement design	PDL	proportion of directional W_{18} assumed in the design lane
CBR	California bearing ratio	PSI	present serviceability index
D	slab thickness, AASHTO design equation	R	reliability
		S_c'	concrete modulus of rupture
D_1, D_2, D_3	structural-layer thicknesses for wearing surface, base, and subbase	S_o	overall standard deviation for AASHTO design equations structural number
E_c	concrete modulus of elasticity	SN	structural number
FN_s	friction number with a smooth tire	TSI	terminal serviceability index
FN_t	friction number with a treaded tire	W_{18}	18-kip (80.1-kN)–equivalent single-axle load
IRI	International Roughness Index		
k	modulus of subgrade reaction	Z_R	reliability for AASHTO design equations
M_2, M_3	drainage coefficients for base and subbase	ΔPSI	change in the present serviceability index

REFERENCES

AASHTO (American Association of State Highway and Transportation Officials). *AASHTO Guide for Design of Pavement Structures*. Washington, DC: AASHTO, 1993.

AASHTO (American Association of State Highway and Transportation Officials). *Mechanistic-Empirical Pavement Design Guide—A Manual of Practice*. Washington, DC: AASHTO, 2008.

Carey, W., and P. Irick. *The Pavement Serviceability-Performance Concept*. Highway Research Board Special Report 61E, AASHO Road Test, 1962.

Federal Highway Administration. "2006 Status of the Nation's Highways, Bridges, and Transit: Conditions and Performance; Report to Congress", United States Department of Transportation, Washington, DC, 2007.

Li, S., S. Noureldin and K. Zhu. *Upgrading the INDOT Pavement Friction Testing Program*. Final report FHWA/IN/JTRP-2003/23, West Lafayette, Indiana, 2003.

Sayers, M., T. Gillespie, and W. Paterson. *Guidelines for the Conducting and Calibrating Road Roughness Measurements*. World Bank Technical Paper No. 46. The World Bank, Washington, DC, 1986.

Yoder, E. J., and M. W. Witczak. *Principles of Pavement Design*, 2nd ed. New York: Wiley, 1991.

Chapter 5

Fundamentals of Traffic Flow
and Queuing Theory

5.1 INTRODUCTION

While the primary function of a highway is mobility (measured by various performance measures such as vehicle speeds), safety must also play a prominent role. Many of the safety-related aspects of highway design were discussed in Chapter 3. Starting with the current chapter, emphasis will now turn to understanding elements of highway traffic flow and queuing, and ultimately on to other measures of highway performance and mobility (Chapters 6 and 7).

The analysis of vehicle traffic (including traffic flow and queuing theory) provides the basis for measuring the operating performance of highways. In undertaking such an analysis, the various dimensions of traffic, such as number of vehicles per unit time (flow), vehicle types, vehicle speeds, and the variation in traffic flow over time, must be addressed because they all influence highway design (the selection of the number of lanes, pavement types, and geometric design) and highway operations (selection of traffic control devices, including signs, markings, and traffic signals), both of which impact the performance of the highway. In light of this, it is important for the analysis of traffic to begin with theoretically consistent quantitative techniques that can be used to model traffic flow, speed, and temporal fluctuations. The intent of this chapter is to focus on models of traffic flow and queuing, thus providing the groundwork for quantifying measures of performance (and levels of service, which will be discussed in Chapters 6 and 7).

5.2 TRAFFIC STREAM PARAMETERS

Traffic streams (the flow of vehicles on a highway) can be characterized by a number of different operational performance measures. Before commencing a discussion of the specific measures, it is important to provide definitions for the contexts in which these measures apply. A traffic stream that operates free from the influence of such traffic control devices as signals and stop signs is classified as uninterrupted flow. This type of traffic flow is influenced primarily by roadway characteristics and the interactions of the vehicles in the traffic stream. Freeways, multilane highways, and two-lane highways often operate under uninterrupted flow conditions. Traffic streams that operate under the influence of signals and stop signs are classified as interrupted flow. Although all the concepts in this chapter are generally applicable to both types of flow, there are some additional complexities involved with the analysis of traffic flow at signalized and unsignalized intersections. Chapter 7 will address the additional complexities relating to the analysis of traffic flow at signalized intersections. For details on the analysis of traffic flow at unsignalized intersections, refer to other sources [Transportation Research Board 1975, 2016]. It should be noted that environmental conditions (day vs. night, sunny vs. rainy, etc.) can also affect the flow of traffic, but these issues are beyond the scope of this book.

5.2.1 Traffic Flow, Speed, and Density

Traffic flow, speed, and density are variables that form the underpinnings of traffic analysis. To begin the study of these variables, the basic definitions of traffic flow, speed, and density must be presented. Traffic flow is defined as

$$q = \frac{n}{t} \tag{5.1}$$

where

q = traffic flow in vehicles per unit time,

n = number of vehicles passing some designated roadway point during time t, and

t = duration of time interval.

Flow is often measured over the course of an hour, in which case the resulting value is typically referred to as volume. Thus, when the term "volume" is used, it is generally understood that the corresponding value is in units of vehicles per hour (veh/h). The definition of flow is more generalized to account for the measurement of vehicles over any period of time. In practice, the analysis flow rate is usually based on the peak 15-minute flow within the hour of interest. This aspect will be described in more detail in Chapter 6.

Aside from the total number of vehicles passing a point in some time interval, the amount of time between the passing of successive vehicles (or time between the arrival of successive vehicles) is also of interest. The time between the passage of the front bumpers of successive vehicles, at some designated highway point, is known as the time headway. The time headway is related to t, as defined in Eq. 5.1, by

$$t = \sum_{i=1}^{n} h_i \tag{5.2}$$

where

t = duration of time interval,

h_i = time headway of the ith vehicle (the elapsed time between the arrivals of vehicles i and $i-1$), and

n = number of measured vehicle time headways at some designated roadway point.

Substituting Eq. 5.2 into Eq. 5.1 gives

$$q = \frac{n}{\sum_{i=1}^{n} h_i}$$ (5.3)

or

$$q = \frac{1}{\bar{h}}$$ (5.4)

where $\bar{h}$ = average time headway ($\Sigma \ h_i/n$) in unit time per vehicle. The importance of time headways in traffic analysis will be given additional attention in later sections of this chapter.

The average traffic speed is defined in two ways. The first is the arithmetic mean of the vehicle speeds observed at some designated point along the roadway. This is referred to as the time-mean speed and is expressed as

$$\bar{u}_t = \frac{\sum_{i=1}^{n} u_i}{n}$$ (5.5)

where

$\bar{u}_t$ = time-mean speed in unit distance per unit time,

u_i = spot speed (the speed of the vehicle at the designated point on the highway, as might be obtained using a radar gun) of the ith vehicle, and

n = number of measured vehicle spot speeds.

The second definition of speed is more useful in the context of traffic analysis and is determined on the basis of the time necessary for a vehicle to travel some known length of roadway. This measure of average traffic speed is referred to as the space-mean speed and is expressed as (assuming that the travel time for all vehicles is measured over the same length of roadway)

$$\bar{u}_s = \frac{l}{\bar{t}}$$ (5.6)

where

$\bar{u}_s$ = space-mean speed in unit distance per unit time,

l = length of roadway used for travel time measurement of vehicles, and

$\bar{t}$ = average vehicle travel time, defined as

$$\bar{t} = \frac{1}{n} \sum_{i=1}^{n} t_i$$ (5.7)

where

t_i = time necessary for vehicle i to travel a roadway section of length l, and

n = number of measured vehicle travel times.

Substituting Eq. 5.7 into Eq. 5.6 yields

$$\bar{u}_s = \frac{l}{\dfrac{1}{n}\displaystyle\sum_{i=1}^{n} t_i} \tag{5.8}$$

or

$$\bar{u}_s = \frac{1}{\dfrac{1}{n}\displaystyle\sum_{i=1}^{n}\left[\dfrac{1}{(l/t_i)}\right]} \tag{5.9}$$

which is the harmonic mean of speed (space-mean speed). Space-mean speed is the speed variable used in traffic models.

EXAMPLE 5.1   TIME- AND SPACE-MEAN SPEEDS

The speeds of five vehicles were measured (with radar) at the midpoint of a 0.5-mile section of roadway. The speeds for vehicles 1, 2, 3, 4, and 5 were 44, 42, 51, 49, and 46 mi/h, respectively. Assuming that all vehicles were traveling at constant speed over this roadway section, calculate the time-mean and space-mean speeds.

SOLUTION

For the time-mean speed, Eq. 5.5 is applied, giving

$$\bar{u}_t = \frac{\displaystyle\sum_{i=1}^{n} u_i}{n}$$

$$= \frac{44 + 42 + 51 + 49 + 46}{5}$$

$$= \underline{\underline{46.4 \ \text{mi/h}}}$$

For the space-mean speed, Eq. 5.9 will be applied. This equation is based on travel time; however, because it is known that the vehicles were traveling at constant speed, we can rearrange this equation to utilize the measured speed, knowing that distance, l, divided by travel time, t, is equal to speed ($l / t_i = u$):

$$\bar{u}_s = \frac{1}{\dfrac{1}{n}\displaystyle\sum_{i=1}^{n}\left[\dfrac{1}{(l/t_i)}\right]} = \frac{1}{\dfrac{1}{n}\displaystyle\sum_{i=1}^{n}\left[\dfrac{1}{u_i}\right]}$$

$$= \cfrac{1}{\cfrac{1}{5}\left[\cfrac{1}{44}+\cfrac{1}{42}+\cfrac{1}{51}+\cfrac{1}{49}+\cfrac{1}{46}\right]}$$

$$= \frac{1}{0.02166}$$

$$= \underline{46.17 \ \text{mi/h}}$$

Note that the space-mean speed will always be lower than the time-mean speed, unless all vehicles are traveling at exactly the same speed, in which case the two measures will be equal.

Finally, traffic density is defined as

$$k = \frac{n}{l} \tag{5.10}$$

where

k = traffic density in vehicles per unit distance,

n = number of vehicles occupying some length of roadway at some specified time, and

l = length of roadway.

The density can also be related to the individual spacing between successive vehicles (measured from front bumper to front bumper). The roadway length, l, in Eq. 5.10 can be defined as

$$l = \sum_{i=1}^{n} s_i \tag{5.11}$$

where

s_i = spacing of the ith vehicle (the distance between vehicles i and $i-1$, measured from front bumper to front bumper), and

n = number of measured vehicle spacings.

Substituting Eq. 5.11 into Eq. 5.10 gives

$$k = \frac{n}{\displaystyle\sum_{i=1}^{n} s_i} \tag{5.12}$$

or

$$k = \frac{1}{\bar{s}} \tag{5.13}$$

where $\bar{s}$ = average spacing ($\Sigma s_i / n$) in unit distance per vehicle.

Time headway and spacing are referred to as microscopic measures because they describe characteristics specific to individual pairs of vehicles within the traffic stream. Measures that describe the traffic stream as a whole, such as flow, average speed, and density, are referred to as macroscopic measures. As indicated by the preceding equations, the microscopic measures can be aggregated and related to the macroscopic measures.

Based on the definitions presented, a simple identity provides the basic relationship among traffic flow, speed (space-mean), and density (denoting space-mean speed, $\bar{u}_s$ as simply u for notational convenience):

$$q = uk \tag{5.14}$$

where

q = flow, typically in units of veh/h,

u = speed (space-mean speed), typically in units of mi/h, and

k = density, typically in units of veh/mi.

EXAMPLE 5.2   SPEED, FLOW, AND DENSITY

Vehicle time headways and spacings were measured at a point along a highway, from a single lane, over the course of an hour. The average values were calculated as 2.5 s/veh for headway and 200 ft/veh for spacing. Calculate the average speed of the traffic.

SOLUTION

To calculate the average speed of the traffic, the fundamental relationship in Eq. 5.14 is used. To begin, the flow and density need to be calculated from the headway and spacing data. Flow is determined from Eq. 5.4 as

$$q = \frac{1}{2.5 \text{ s/veh}}$$
$$= 0.40 \text{ veh/s}$$

or, because the data were collected for an hour,

$$q = 0.40 \text{ veh/s} \times 3600 \text{ s/h}$$
$$= 1440 \text{ veh/h}$$

Density is determined from Eq. 5.13 as

$$k = \frac{1}{200 \text{ ft/veh}}$$
$$= 0.005 \text{ veh/ft}$$

or, applying this spacing over the course of one mile,

$$k = 0.005 \text{ veh/ft} \times 5280 \text{ ft/mi}$$
$$= 26.4 \text{ veh/mi}$$

Now applying Eq. 5.14, after rearranging to solve for speed, gives

$$u = \frac{q}{k}$$

$$= \frac{1440 \text{ veh/h}}{26.4 \text{ veh/mi}}$$

$$= 54.5 \text{ mi/h}$$

Note that the average speed of traffic can be determined directly from the average headway and spacing values, as follows:

$$u = \frac{\bar{s}}{\bar{h}}$$

$$= \frac{200 \text{ ft/veh}}{2.5 \text{ s/veh}}$$

$$= 80 \text{ ft/s } (54.5 \text{ mi/h})$$

5.3 BASIC TRAFFIC STREAM MODELS

While the preceding definitions and relationships provide the basis for the measurement and calculation of traffic stream parameters, it is also essential to understand the interaction of the individual macroscopic measures in order to fully analyze the operational performance of traffic streams. The models that describe these interactions are discussed in the following sections, and it will be shown that Eq. 5.14 serves the important function of linking specific models of traffic into a consistent, generalized model.

5.3.1 Speed-Density Model

The most intuitive starting point for developing a consistent, generalized traffic model is to focus on the relationship between speed and density. To begin, consider a section of highway with only a single vehicle on it. Under these conditions, the density (veh/mi) will be very low and the driver will be able to travel freely at a speed close to the design speed of the highway. This speed is referred to as the free-flow speed because vehicle speed is not inhibited by the presence of other vehicles. As more and more vehicles begin to use a section of highway, the traffic density will increase and the average operating speed of vehicles will decline from the free-flow value as drivers slow to allow for the maneuvers of other vehicles. Eventually, the highway section will become so congested (will have such a high density) that the traffic will come to a stop ($u = 0$), and the density will be determined by the length of the vehicles and the spaces that drivers leave between them. This high-density condition is referred to as the jam density.

One possible representation of the process described above is the linear relationship shown in Fig. 5.1. Mathematically, such a relationship can be expressed as

$$u = u_f \left(1 - \frac{k}{k_j} \right) \qquad (5.15)$$

where

u = space-mean speed in mi/h,
u_f = free-flow speed in mi/h,
k = density in veh/mi, and
k_j = jam density in veh/mi.

The advantage of using a linear representation of the speed–density relationship is that it provides a basic insight into the relationships among traffic flow, speed, and density interactions without clouding these insights by the additional complexity that a nonlinear speed–density relationship introduces. However, it is important to note that field studies have shown that the speed–density relationship tends to be nonlinear at low densities and high densities (those that approach the jam density). In fact, the overall speed–density relationship is better represented by three relationships: (1) a nonlinear relationship at low densities that has speed slowly declining from the free-flow value, (2) a linear relationship over the large medium-density region (speed declining linearly with density as shown in Eq. 5.15), and (3) a nonlinear relationship near the jam density as the speed asymptotically approaches zero with increasing density. For the purposes of exposition, we present only traffic stream models that are based on the assumption of a linear speed–density relationship. Examples of nonlinear speed–density relationships are provided elsewhere [Pipes, 1967; Drew, 1965].

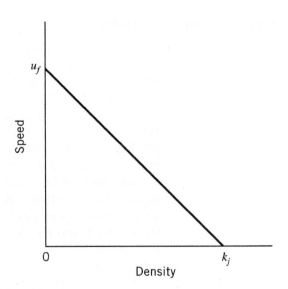

Figure 5.1 Illustration of a typical linear speed–density relationship.

5.3.2   Flow-Density Model

Using the assumption of a linear speed–density relationship as shown in Eq. 5.15, a parabolic flow-density model can be obtained by substituting Eq. 5.15 into Eq. 5.14:

$$q = u_f\left(k - \frac{k^2}{k_j}\right) \tag{5.16}$$

where all terms are as defined previously.

The general form of Eq. 5.16 is shown in Fig. 5.2. Note in this figure that the maximum flow rate, q_{cap}, represents the highest rate of traffic flow that the highway is capable of handling. This is referred to as the traffic flow at capacity, or simply the capacity of the roadway. The traffic density that corresponds to this capacity flow rate is k_{cap}, and the corresponding speed is u_{cap}. Equations for q_{cap}, k_{cap}, and u_{cap} can be derived by differentiating Eq. 5.16, because at maximum flow

$$\frac{dq}{dk} = u_f\left(1 - \frac{2k}{k_j}\right) = 0 \tag{5.17}$$

and because the free-flow speed (u_f) is not equal to zero,

$$k_{cap} = \frac{k_j}{2} \tag{5.18}$$

Substituting Eq. 5.18 into Eq. 5.15 gives

$$u_{cap} = u_f\left(1 - \frac{k_j}{2k_j}\right)$$
$$= \frac{u_f}{2} \tag{5.19}$$

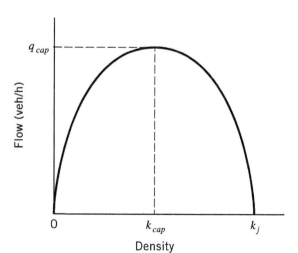

Figure 5.2 Illustration of the parabolic flow–density relationship.

and using Eq. 5.18 and Eq. 5.19 in Eq. 5.14 gives

$$q_{cap} = u_{cap} k_{cap}$$
$$= \frac{u_f k_j}{4} \tag{5.20}$$

5.3.3   Speed-Flow Model

Again returning to the linear speed–density model (Eq. 5.15), a corresponding speed-flow model can be developed by rearranging Eq. 5.15 to

$$k = k_j \left(1 - \frac{u}{u_f} \right) \tag{5.21}$$

and by substituting Eq. 5.21 into Eq. 5.14,

$$q = k_j \left(u - \frac{u^2}{u_f} \right) \tag{5.22}$$

The speed-flow model defined by Eq. 5.22 again gives a parabolic function, as shown in Fig. 5.3. Note that Fig. 5.3 shows that two speeds are possible for flows, q, up to the highway's capacity, q_{cap} (this follows from the two densities possible for given flows as shown in Fig. 5.2). It is desirable, for any given flow, to keep the average space-mean speed on the upper portion of the speed-flow curve (above u_{cap}). When speeds drop below u_{cap}, traffic is in a highly congested and unstable condition.

All of the flow, speed, and density relationships and their interactions are graphically represented in Fig. 5.4.

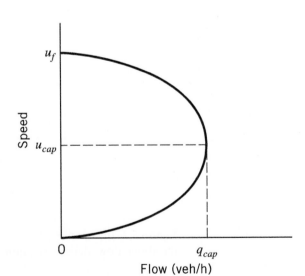

Figure 5.3 Illustration of the parabolic speed–flow relationship.

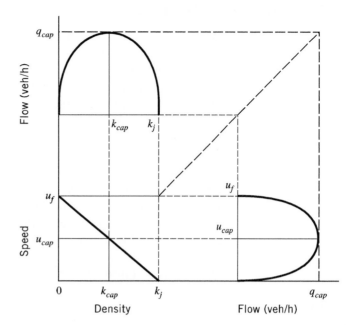

Figure 5.4 Flow–density, speed–density, and speed–flow relationships (assuming a linear speed–density model).

EXAMPLE 5.3   APPLICATION OF SPEED–FLOW–DENSITY RELATIONSHIPS

A section of highway is known to have a free-flow speed of 55 mi/h and a capacity of 3300 veh/h. In a given hour, 2100 vehicles were counted at a specified point along this highway section. If the linear speed–density relationship shown in Eq. 5.15 applies, what would you estimate the space-mean speed of these 2100 vehicles to be?

SOLUTION

The jam density is first determined from Eq. 5.20 as

$$k_j = \frac{4q_{cap}}{u_f}$$

$$= \frac{4 \times 3300}{55}$$

$$= 240.0 \ \text{veh/mi}$$

Rearranging Eq. 5.22 to solve for u,

$$\frac{k_j}{u_f} u^2 - k_j u + q = 0$$

Substituting,

$$\frac{240.0}{55} u^2 - 240.0u + 2100 = 0$$

which gives $u = \underline{\underline{44.08 \ \text{mi/h}}}$ or $\underline{\underline{10.92 \ \text{mi/h}}}$. Both of these speeds are feasible, as shown in Fig. 5.3.

5.4 MODELS OF TRAFFIC FLOW

With the basic relationships among traffic flow, speed, and density formalized, attention can now be directed toward a more microscopic view of traffic flow. That is, instead of simply modeling the number of vehicles passing a specified point on a highway in some time interval, there is considerable analytic value in modeling the time between the arrivals of successive vehicles (the concept of vehicle time headway presented earlier). The most simplistic approach to vehicle arrival modeling is to assume that all vehicles are equally or uniformly spaced. This results in what is termed a deterministic, uniform arrival pattern. Under this assumption, if the traffic flow is 360 veh/h, the number of vehicles arriving in any 5-minute time interval is 30 and the headway between all vehicles is 10 seconds (because h will equal $3600/q$). However, actual observations show that such uniformity of traffic flow is not always realistic because some 5-minute intervals are likely to have more or less traffic flow than other 5-minute intervals. Thus, a representation of vehicle arrivals that goes beyond the deterministic, uniform assumption is often warranted.

5.4.1 Poisson Model

Models that account for the nonuniformity of flow are derived by assuming that the pattern of vehicle arrivals (at a specified point) corresponds to some random process. The problem then becomes one of selecting a probability distribution that is a reasonable representation of observed traffic arrival patterns. An example of such a distribution is the Poisson distribution (the limitations of which will be discussed later), which is expressed as

$$P(n) = \frac{(\lambda t)^n e^{-\lambda t}}{n!} \tag{5.23}$$

where

$P(n)$ = probability of having n vehicles arrive in time t,
λ = average vehicle flow or arrival rate in vehicles per unit time,
t = duration of the time interval over which vehicles are counted, and
e = base of the natural logarithm ($e = 2.718$).

EXAMPLE 5.4   VEHICLE ARRIVALS AS A POISSON PROCESS

An observer counts 360 veh/h at a specific highway location. Assuming that the arrival of vehicles at this highway location is Poisson distributed, estimate the probabilities of having 0, 1, 2, 3, 4, and 5 or more vehicles arriving over a 20-second time interval.

SOLUTION

The average arrival rate, λ, is 360 veh/h, or 0.1 vehicles per second (veh/s). Using this in Eq. 5.23 with $t = 20$ seconds, the probabilities of having exactly 0, 1, 2, 3, and 4 vehicles arrive are

$$P(0) = \frac{(0.1 \times 20)^0 \, e^{-0.1(20)}}{0!} = \underline{\underline{0.135}}$$

$$P(1) = \frac{(0.1 \times 20)^1 \, e^{-0.1(20)}}{1!} = \underline{\underline{0.271}}$$

$$P(2) = \frac{(0.1 \times 20)^2 \, e^{-0.1(20)}}{2!} = \underline{\underline{0.271}}$$

$$P(3) = \frac{(0.1 \times 20)^3 \, e^{-0.1(20)}}{3!} = \underline{\underline{0.180}}$$

$$P(4) = \frac{(0.1 \times 20)^4 \, e^{-0.1(20)}}{4!} = \underline{\underline{0.090}}$$

For five or more vehicles,

$$P(n \geq 5) = 1 - P(n < 5)$$
$$= 1 - 0.135 - 0.271 - 0.271 - 0.180 - 0.090$$
$$= \underline{\underline{0.053}}$$

A histogram of these probabilities is shown in Fig. 5.5.

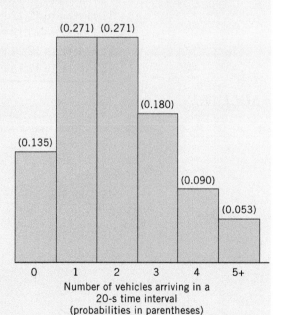

Figure 5.5 Histogram of the Poisson distribution for $\lambda = 0.1$ vehicles per second.

EXAMPLE 5.5   VEHICLE ARRIVALS AS A POISSON PROCESS WITH DETAILED VEHICLE-ARRIVAL DATA

Traffic data are collected in 60-second intervals at a specific highway location as shown in Table 5.1. Assuming the traffic arrivals are Poisson distributed and continue at the same rate as that observed in the 15 time periods shown, what is the probability that six or more vehicles will arrive in each of the next three 60-second time intervals (12:15 P.M. to 12:16 P.M., 12:16 P.M. to 12:17 P.M., and 12:17 P.M. to 12:18 P.M.)?

Table 5.1 Observed Traffic Data for Example 5.5

Time period	Observed number of vehicles
12:00 P.M. to 12:01 P.M.	3
12:01 P.M. to 12:02 P.M.	5
12:02 P.M. to 12:03 P.M.	4
12:03 P.M. to 12:04 P.M.	10
12:04 P.M. to 12:05 P.M.	7
12:05 P.M. to 12:06 P.M.	4
12:06 P.M. to 12:07 P.M.	8
12:07 P.M. to 12:08 P.M.	11
12:08 P.M. to 12:09 P.M.	9
12:09 P.M. to 12:10 P.M.	5
12:10 P.M. to 12:11 P.M.	3
12:11 P.M. to 12:12 P.M.	10
12:12 P.M. to 12:13 P.M.	9
12:13 P.M. to 12:14 P.M.	7
12:14 P.M. to 12:15 P.M.	6

SOLUTION

Table 5.1 shows that a total of 101 vehicles arrive in the 15-minute period from 12:00 P.M. to 12:15 P.M. Thus, the average arrival rate, λ, is 0.112 veh/s (101/900). As in Example 5.4, Eq. 5.23 is applied to find the probabilities of exactly 0, 1, 2, 3, 4, and 5 vehicles arriving.

Applying Eq. 5.23, with $\lambda = 0.112$ veh/s and $t = 60$ seconds, the probabilities of having 0, 1, 2, 3, 4, and 5 vehicles arriving in a 60-second time interval are (using $\lambda t = 6.733$)

$$P(0) = \frac{(6.733)^0 e^{-6.733}}{0!} = \underline{\underline{0.0012}}$$

$$P(1) = \frac{(6.733)^1 e^{-6.733}}{1!} = \underline{\underline{0.008}}$$

$$P(2) = \frac{(6.733)^2 e^{-6.733}}{2!} = \underline{\underline{0.027}}$$

$$P(3) = \frac{(6.733)^3 e^{-6.733}}{3!} = \underline{\underline{0.0606}}$$

$$P(4) = \frac{(6.733)^4 \, e^{-6.733}}{4!} = \underline{\underline{0.102}}$$

$$P(5) = \frac{(6.733)^5 \, e^{-6.733}}{5!} = \underline{\underline{0.137}}$$

The summation of these probabilities is the probability that 0 to 5 vehicles will arrive in any given 60-second time interval, which is

$$P(n \le 5) = \sum_{i=0}^{5} P(n)$$

$$= 0.0012 + 0.008 + 0.027 + 0.0606 + 0.102 + 0.137$$

$$= 0.3358$$

So 1 minus $P(n \le 5)$ is the probability that six or more vehicles will arrive in any 60-second time interval, which is

$$P(n \ge 6) = 1 - P(n \le 5)$$

$$= 1 - 0.3358$$

$$= 0.6642$$

The probability that six or more vehicles will arrive in three successive time intervals (t_1, t_2, and t_3) is simply the product of probabilities, which is

$$P(n \ge 6) \text{ for three successive time periods} = \prod_{t_i=1}^{3} P(n \ge 6)$$

$$= (0.6642)^3$$

$$= \underline{\underline{0.293}}$$

The assumption of Poisson vehicle arrivals also implies a distribution of the time intervals between the arrivals of successive vehicles (time headway). To show this, note that the average arrival rate is

$$\lambda = \frac{q}{3600} \tag{5.24}$$

where

λ = average vehicle arrival rate in veh/s,

q = flow in veh/h, and

3600 = number of seconds per hour.

Substituting Eq. 5.24 into Eq. 5.23 gives

$$P(n) = \frac{(qt/3600)^n \, e^{-qt/3600}}{n!} \tag{5.25}$$

Note that the probability of having no vehicles arrive in a time interval of length t, $P(0)$, is equivalent to the probability of a vehicle headway, h, being greater than or equal to the time interval t. So from Eq. 5.25,

$$P(0) = P(h \geq t)$$
$$= e^{-qt/3600} \tag{5.26}$$

This distribution of vehicle headways is known as the negative exponential distribution and is often simply referred to as the exponential distribution.

EXAMPLE 5.6   HEADWAYS AND THE NEGATIVE EXPONENTIAL DISTRIBUTION

Consider the traffic situation in Example 5.4 (360 veh/h). Again assume that the vehicle arrivals are Poisson distributed. What is the probability that the headway between successive vehicles will be less than 8 seconds, and what is the probability that the headway between successive vehicles will be between 8 and 10 seconds?

SOLUTION

By definition, $P(h < t) = 1 - P(h \geq t)$. This expression gives the probability that the headway will be less than 8 seconds as

$$P(h < t) = 1 - e^{-qt/3600}$$
$$= 1 - e^{-360(8)/3600}$$
$$= \underline{\underline{0.551}}$$

To determine the probability that the headway will be between 8 and 10 seconds, compute the probability that the headway will be greater than or equal to 10 seconds:

$$P(h \geq t) = e^{-qt/3600}$$
$$= e^{-360(10)/3600}$$
$$= \underline{\underline{0.368}}$$

So the probability that the headway will be between 8 and 10 seconds is $\underline{\underline{0.081}}$ $(1 - 0.551 - 0.368)$.

To help in visualizing the shape of the exponential distribution, Fig. 5.6 shows the probability distribution implied by Eq. 5.26, with the flow, q, equal to 360 veh/h as in Example 5.4.

5.4.2   Limitations of the Poisson Model

Empirical observations have shown that the assumption of Poisson-distributed traffic arrivals is most realistic in lightly congested traffic conditions. As traffic flows become heavily congested or when traffic signals cause cyclical traffic stream disturbances, other distributions of traffic flow become more appropriate. The primary limitation of the Poisson model of vehicle arrivals is

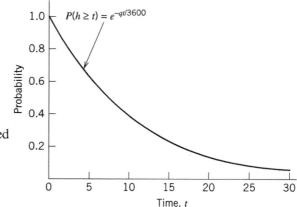

Figure 5.6 Exponentially distributed probabilities of headways greater than or equal to t, with $q = 360$ veh/h.

the constraint imposed by the Poisson distribution that the mean of period observations equals the variance. For example, the mean of period-observed traffic in Example 5.5 is 6.733 and the corresponding variance, σ^2, is 7.210. Because these two values are close, the Poisson model was appropriate for this example. If the variance is significantly greater than the mean, the data are said to be overdispersed, and if the variance is significantly less than the mean, the data are said to be underdispersed. In either case the Poisson distribution is no longer appropriate, and another distribution should be used. Such distributions are discussed in detail in more specialized sources [Transportation Research Board, 1975; Poch and Mannering, 1996; Lord and Mannering, 2010].

5.5   QUEUING THEORY AND TRAFFIC FLOW ANALYSIS

The formation of traffic queues during congested periods is a source of considerable delay and results in a loss of highway performance. Under extreme conditions, queuing delay can account for 90% or more of a motorist's total trip travel time. Given this, it is essential in traffic analysis to develop a clear understanding of the characteristics of queue formation and dissipation along with mathematical formulations that can predict queuing-related elements.

As is well known, the problem of queuing is not unique to traffic analysis. Many non-transportation fields, such as the design and operation of industrial plants, retail stores, service-oriented industries, and computer networks, must also give serious consideration to the problem of queuing. The impact of queues on performance and productivity in manufacturing, retailing, and other fields has led to numerous theories of queuing behavior (the process by which queues form and dissipate). As will be shown, the models of traffic flow presented earlier (uniform, deterministic arrivals and Poisson arrivals) will form the basis for studying traffic queues within the more general context of queuing theory.

5.5.1   Dimensions of Queuing Models

The purpose of traffic queuing models is to provide a means to estimate important measures of highway performance, including vehicle delay and traffic queue lengths. Such estimates are critical to roadway design (the required length of left-turn bays and the number of lanes at intersections) and traffic operations control, including the timing of traffic signals at intersections.

Queuing models are derived from underlying assumptions regarding arrival patterns, departure characteristics, and queue disciplines. Traffic arrival patterns were explored in Section 5.4, where, given an average vehicle arrival rate (λ), two possible distributions of the time between the arrival of successive vehicles were considered:

1. Equal time intervals (derived from the assumption of uniform, deterministic arrivals)
2. Exponentially distributed time intervals (derived from the assumption of Poisson-distributed arrivals)

In addition to vehicle arrival assumptions, the derivation of traffic queuing models requires assumptions relating to vehicle departure characteristics. Of particular interest is the distribution of the amount of time it takes a vehicle to depart—for example, the time to pass through an intersection at the beginning of a green signal, the time required to pay a toll at a toll booth, or the time a driver takes before deciding to proceed after stopping at a stop sign. As was the case for arrival patterns, given an average vehicle departure rate (denoted as μ, in vehicles per unit time), the assumption of a deterministic or exponential distribution of departure times is appropriate.

Another important aspect of queuing models is the number of available departure channels. For most traffic applications only one departure channel will exist, such as a highway lane or group of lanes passing through an intersection. However, multiple departure channels are encountered in some traffic applications, such as at toll booths on turnpikes and at entrances to bridges.

The final necessary assumption relates to the queue discipline. In this regard, two options have been popularized in the development of queuing models: first-in, first-out (FIFO), indicating that the first vehicle to arrive is the first to depart, and last-in, first-out (LIFO), indicating that the last vehicle to arrive is the first to depart. For virtually all traffic-oriented queues, the FIFO queuing discipline is the more appropriate of the two.

Queuing models are often identified by three alphanumeric values. The first value indicates the arrival rate assumption, the second value gives the departure rate assumption, and the third value indicates the number of departure channels. For traffic arrival and departure assumptions, the uniform, deterministic distribution is denoted D and the exponential distribution is denoted M. Thus a $D/D/1$ queuing model assumes deterministic arrivals and departures with one departure channel. Similarly, an $M/D/1$ queuing model assumes exponentially distributed arrival times, deterministic departure times, and one departure channel.

5.5.2   *D/D/1* Queuing

The case of deterministic arrivals and departures with one departure channel ($D/D/1$ queue) is an excellent starting point in understanding queuing models because of its simplicity. The $D/D/1$ queue lends itself to an intuitive graphical or mathematical solution that is best illustrated by an example.

EXAMPLE 5.7   *D/D/*1 **QUEUING WITH CONSTANT ARRIVAL AND DEPARTURE RATES**

Vehicles arrive at an entrance to a recreational park. There is a single gate (at which all vehicles must stop), where a park attendant distributes a free brochure. The park opens at 8:00 A.M., at which time vehicles begin to arrive at a rate of 480 veh/h. After 20 minutes, the arrival flow rate declines to 120 veh/h, and it continues at that level for the remainder of the day. If the time required to distribute the brochure is 15 seconds, and assuming *D/D/*1 queuing, describe the operational characteristics of the queue.

SOLUTION

Begin by putting arrival and departure rates into common units of vehicles per minute:

$$\lambda = \frac{480 \text{ veh/h}}{60 \text{ min/h}} = 8 \text{ veh/min} \quad \text{for } t \leq 20 \text{ min}$$

$$\lambda = \frac{120 \text{ veh/h}}{60 \text{ min/h}} = 2 \text{ veh/min} \quad \text{for } t > 20 \text{ min}$$

$$\mu = \frac{60 \text{ s/min}}{15 \text{ s/veh}} = 4 \text{ veh/min} \quad \text{for all } t$$

Equations for the total number of vehicles that have arrived and departed up to a specified time, *t*, can now be written. Define *t* as the number of minutes after the start of the queuing process (in this case the number of minutes after 8:00 A.M.). The total number of vehicle arrivals at time *t* is equal to

$$8t \quad \text{for } t \leq 20 \text{ min}$$

and

$$160 + 2(t - 20) \quad \text{for } t > 20 \text{ min}$$

Similarly, the number of vehicle departures is

$$4t \quad \text{for all } t$$

The preceding equations can be illustrated graphically as shown in Fig. 5.7. When the arrival curve is above the departure curve, a queue condition exists. The point at which the arrival curve meets the departure curve is the moment when the queue dissipates (no more queue exists). In this example, the point of queue dissipation can be determined graphically by inspection of Fig. 5.7, or analytically by equating appropriate arrival and departure equations, that is,

$$160 + 2(t - 20) = 4t$$

Solving for *t* gives *t* = 60 minutes. Thus the queue that began to form at 8:00 A.M. will dissipate 60 minutes later (9:00 A.M.), at which time 240 vehicles will have arrived and departed (4 veh/min × 60 min).

Another aspect of interest is individual vehicle delay. Under the assumption of a FIFO queuing discipline, the delay of an individual vehicle is given by the horizontal distance between arrival and departure curves starting from the time of the vehicle's arrival in the queue. So, by inspection of Fig. 5.7, the 160th vehicle to arrive will have the longest delay, 20 minutes (the longest horizontal distance between arrival and departure curves), and vehicles arriving after the 239th vehicle will encounter no queue delay because the queue will have dissipated and the departure rate will continue to exceed the arrival rate. It follows that with the LIFO queuing discipline, the first vehicle to arrive would have to wait until the entire queue clears (60 minutes of delay).

The total length of the queue at a specified time, expressed as the number of vehicles, is given by the vertical distance between arrival and departure curves at that time. For example, at 10 minutes after the start of the queuing process (8:10 A.M.) the queue is 40 vehicles long, and the longest queue (longest vertical distance between arrival and departure curves) will occur at $t = 20$ minutes and is 80 vehicles long (see Fig. 5.7).

Total vehicle delay, defined as the summation of the delays for the individual vehicles, is given by the total area between the arrival and departure curves (see Fig. 5.7) and, in this case, is in units of vehicle-minutes. In this example, the area between the arrival and departure curves can be determined by summing triangular areas, giving total delay, D_t, as

$$D_t = \frac{1}{2}(80 \times 20) + \frac{1}{2}(80 \times 40)$$
$$= \underline{\underline{2400 \text{ veh-min}}}$$

Finally, because 240 vehicles encounter queuing delay (as previously determined), the average delay per vehicle is 10 minutes (2400 veh-min/240 veh), and the average queue length is 40 vehicles (2400 veh-min/60 min).

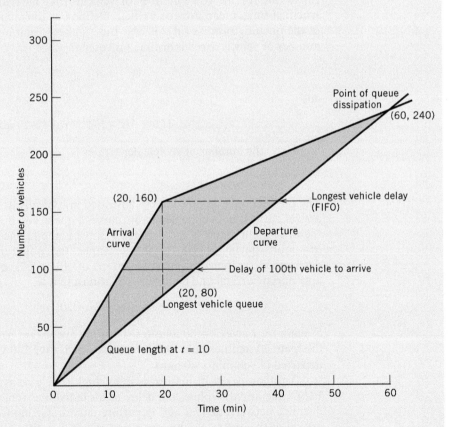

Figure 5.7 *D/D/*1 queuing diagram for Example 5.7.

EXAMPLE 5.8   *D/D/*1 QUEUING WITH TIME-VARYING ARRIVAL RATE AND CONSTANT DEPARTURE RATE

Vehicles start arriving at an entrance to a national park at 5:45 A.M. and a park booth collecting entrance fees opens at 6:00 A.M. and process vehicles at a deterministic rate of 20 veh/min. It is estimated that 20% of the arriving vehicles have priority passes that allow them to access a separate processing booth that also opens at 6:00 A.M. but processes vehicles at a slower deterministic rate of 15 veh/min. The deterministic arrival rate of vehicles is a function of time such that $\lambda(t) = 17.2 - 0.2t$ where t is in minutes after 5:45 A.M. Determine the average delay per vehicle for the non–priority-pass vehicles and the priority-pass vehicles (starting with their arrival at 6:45 A.M.) until their respective queues clear, assuming *D/D/*1 queuing.

SOLUTION

This problem has time-dependent deterministic arrivals and departure rates that do not vary over time. Begin by computing the arrivals of non–priority-pass vehicles. Because 20% of all arrivals are priority-pass vehicles, the non–priority-pass vehicle arrivals will be (starting at 5:45 A.M.):

$$= \int_0^t 17.2 - 0.2t \; dt$$

$$= 17.2t - 0.1t^2 - 0.2\left(17.2t - 0.1t^2\right)$$

$$= 17.2t - 0.1t^2 - 3.44t + 0.02t^2$$

$$= 13.76t - 0.08t^2$$

The time required to clear the queue of the non–priority-pass vehicles (with a departure rate of 20 veh/min) is determined as:

$$13.76t - 0.08t^2 = 20(t - 15)$$

$$-0.08t^2 - 6.24t + 300 = 0$$

$$t = 33.60 \, \text{min}$$

At $t = 33.60$ minutes, the total number of non–priority-pass vehicle arrivals is

$$= 13.76t - 0.08t^2$$

$$= 13.76(33.60) - 0.08(33.60)^2$$

$$= 372.02 \, \text{veh}$$

So the total delay for the non–priority-pass vehicles, D_{np} will be the area under the arrival function minus the area under the departure function, which will be a simple triangle with a height of 372.02 vehicles and a base of 18.06 min (33.60 − 15):

$$D_{np} = \int_0^{33.60} 13.76t - 0.08t^2 \; dt - \frac{1}{2}(33.60 - 15)(372.02)$$

$$= 6.88(33.60)^2 - 0.0267(33.60)^3 - 3459.786$$

$$= 3294.65 \text{ veh-min}$$

So the average delay per vehicle for the non–priority-pass vehicles is <u>8.86 min</u> (3294.65/372.02).

    For the priority-pass vehicles, the arrival function is 20% of the total vehicle arrival function or,

$$= 0.2 \int_0^t 17.2 - 0.2t \; dt$$

$$= 0.2\left(17.2t - 0.1t^2\right)$$

$$= 3.44t - 0.02t^2$$

The time required to clear the queue of the priority-pass vehicles (with a departure rate of 15 veh/min) is:

$$3.44t - 0.02t^2 = 15(t - 15)$$

$$-0.02t^2 - 11.56t + 225 = 0$$

$$t = 18.84 \text{ min}$$

At t = 18.84 minutes, the total number of priority-pass vehicle arrivals is

$$= 3.44t - 0.02t^2$$

$$= 3.44(18.84) - 0.02(18.84)^2$$

$$= 57.71 \text{ veh}$$

So the total delay for the priority-pass vehicles, D_p will again be the area under the arrival function minus the area under the departure function, which will be a simple triangle with a height of 57.71 vehicles and a base of 3.84 min (18.84 − 15):

$$D_p = \int_0^{18.84} 3.44t - 0.02t^2 \; dt - \frac{1}{2}(18.84 - 15)(57.71)$$

$$= 1.72(18.84)^2 - 0.0067(18.84)^3 - 110.803$$

$$= 454.90 \text{ veh-min}$$

So the average delay per vehicle for the priority-pass vehicles is <u>7.88 min</u> (454.9/57.71). Thus the priority pass saves an average of 0.98 min, or about 59 s.

EXAMPLE 5.9   *D/D/*1 **QUEUING WITH TIME-VARYING ARRIVAL AND DEPARTURE RATES**

After observing arrivals and departures at a highway toll booth over a 60-minute time period, an observer notes that the arrival and departure rates (or service rates) are deterministic, but instead of being uniform, they change over time according to a known function. The arrival rate is given by the function $\lambda(t) = 2.2 + 0.17t - 0.0032t^2$, and the departure rate is given by $\mu(t) = 1.2 + 0.07t$, where t is in minutes after the beginning of the observation period and $\lambda(t)$ and $\mu(t)$ are in vehicles per minute. Determine the total vehicle delay at the toll booth and the longest queue, assuming *D/D/*1 queuing.

SOLUTION

Note that this problem is an example of a time-dependent deterministic queue because the deterministic arrival and departure rates change over time. Begin by computing the time to queue dissipation by equating vehicle arrivals and departures:

$$\int_0^t 2.2 + 0.17t - 0.0032t^2 \, dt = \int_0^t 1.2 + 0.07t \, dt$$

$$2.2t + 0.085t^2 - 0.00107t^3 = 1.2t + 0.035t^2$$

$$-0.00107t^3 + 0.05t^2 + t = 0$$

This gives $t = 61.8$ minutes. Therefore, the total vehicle delay (the area between the arrival and departure functions) is

$$D_t = \int_0^{61.8} 2.2t + 0.085t^2 - 0.00107t^3 \, dt - \int_0^{61.8} 1.2t + 0.035t^2 \, dt$$

$$= 1.1t^2 + 0.0283t^3 - 0.0002675t^4 - 0.6t^2 - 0.0117t^3 \ \Big|_0^{61.8}$$

$$= -0.0002675(61.8)^4 + 0.0166(61.8)^3 + 0.5(61.8)^2$$

$$= \underline{1925.8 \text{ veh-min}}$$

The queue length (in vehicles) at any time t is given by the function

$$Q(t) = \int_0^t 2.2 + 0.17t - 0.0032t^2 \, dt - \int_0^t 1.2 + 0.07t \, dt$$

$$= -0.00107t^3 + 0.05t^2 + t$$

Solving for the time at which the maximum queue length occurs yields

$$\frac{dQ(t)}{dt} = -0.00321t^2 + 0.1t + 1 = 0$$

$$t = \underline{39.12 \text{ min}}$$

Substituting with $t = 39.12$ minutes gives the maximum queue length:

$$Q(39.12) = -0.00107t^3 + 0.05t^2 + t \Big|_0^{39.12}$$

$$= -0.00107(39.12)^3 + 0.05(39.12)^2 + 39.12$$

$$= \underline{\underline{51.58 \text{ veh}}}$$

EXAMPLE 5.10   DETERMINING A REQUIRED DEPARTURE RATE

A parking garage has a single processing booth where cars pay for parking. The garage opens at 6:00 A.M. and vehicles start arriving at 6:00 A.M. at a deterministic rate of $\lambda(t) = 6.1 - 0.22t$ where $\lambda(t)$ is in vehicles per minute and t is in minutes after 6:00 A.M. What is the minimum constant departure rate (from 6:00 A.M. on) needed to ensure that the queue length does not exceed 10 vehicles?

SOLUTION

Let μ be the unknown departure rate (in veh/min) so that the queue length (in vehicles) at any time t is:

$$Q(t) = \int_0^t 6.1 - 0.22t \, dt - \int_0^t \mu \, dt$$

$$= 6.1t - 0.11t^2 - \mu t$$

Using this and solving for the time at which the maximum queue length occurs gives

$$\frac{dQ(t)}{dt} = 6.1 - 0.22t - \mu = 0$$

$$t = \frac{6.1 - \mu}{0.22}$$

Substituting this value of t into the queue-length equation with a maximum $Q(t) = 10$ vehicles as specified in the problem gives

$$Q(t) = 10 = 6.1\left(\frac{6.1-\mu}{0.22}\right) - 0.11\left(\frac{6.1-\mu}{0.22}\right)^2 - \mu\left(\frac{6.1-\mu}{0.22}\right)$$

which gives $2.271\mu^2 - 27.727\mu + 74.566 = 0$. Solving for μ gives possible solutions as 8.21 veh/min and 4 veh/min, so the minimum departure rate would be $\underline{\underline{4 \text{ veh/min}}}$.

5.5.3   *M*/*D*/1 Queuing

The assumption of exponentially distributed times between the arrivals of successive vehicles (Poisson arrivals) will, in some cases, give a more realistic representation of traffic flow than the assumption of uniformly distributed arrival times. Therefore, the *M*/*D*/1 queue (exponentially distributed arrivals, deterministic departures, and one departure channel) has some important applications within the traffic analysis field. Although a graphical solution to an *M*/*D*/1 queue is difficult, a mathematical solution is straightforward. Defining a new term (traffic intensity) for the ratio of average arrival to departure rates as

$$\rho = \frac{\lambda}{\mu} \tag{5.27}$$

where

ρ = traffic intensity, unitless,
λ = average arrival rate in vehicles per unit time, and
μ = average departure rate in vehicles per unit time,

and assuming that ρ is less than 1, it can be shown that for an *M*/*D*/1 queue the following queuing performance equations apply:

$$\overline{Q} = \frac{\rho^2}{2(1-\rho)} \tag{5.28}$$

$$\overline{w} = \frac{\rho}{2\mu(1-\rho)} \tag{5.29}$$

$$\overline{t} = \frac{2-\rho}{2\mu(1-\rho)} \tag{5.30}$$

where

$\overline{Q}$ = average length of queue in vehicles,
$\overline{w}$ = average waiting time in the queue, in unit time per vehicle,
$\overline{t}$ = average time spent in the system (the summation of average waiting time in the queue and average departure time), in unit time per vehicle, and
other terms are as defined previously.

It is important to note that under the assumption that the traffic intensity is less than 1 ($\lambda < \mu$), the *D*/*D*/1 queue will predict no queue formation. However, a queuing model that is derived based on random arrivals or departures, such as the *M*/*D*/1 queuing model, will predict queue formations under such conditions. Also, note that the *M*/*D*/1 queuing model presented here is based on steady-state conditions (constant average arrival and departure rates), with randomness arising from the assumed probability distribution of arrivals. This contrasts with the time-varying deterministic queuing case, presented in Example 5.9, in which arrival and departure rates changed over time but randomness was not present.

EXAMPLE 5.11   M/D/1 QUEUING: PARK-ENTRANCE APPLICATION

Consider the entrance to the recreational park described in Example 5.7. However, let the average arrival rate be 180 veh/h and Poisson distributed (exponential times between arrivals) over the entire period from park opening time (8:00 A.M.) until closing at dusk. Compute the average length of queue (in vehicles), average waiting time in the queue, and average time spent in the system, assuming *M/D/*1 queuing.

SOLUTION

Putting arrival and departure rates into common units of vehicles per minute gives

$$\lambda = \frac{180 \text{ veh/h}}{60 \text{ min/h}} = 3 \text{ veh/min} \quad \text{for all } t$$

$$\mu = \frac{60 \text{ s/min}}{15 \text{ s/veh}} = 4 \text{ veh/min} \quad \text{for all } t$$

and

$$\rho = \frac{\lambda}{\mu} = \frac{3 \text{ veh/min}}{4 \text{ veh/min}} = 0.75$$

For the average length of queue (in vehicles), Eq. 5.28 is applied:

$$\overline{Q} = \frac{0.75^2}{2(1-0.75)}$$

$$= \underline{1.125 \text{ veh}}$$

For average waiting time in the queue, Eq. 5.29 gives

$$\overline{w} = \frac{0.75}{2(4)(1-0.75)}$$

$$= \underline{0.375 \text{ min/veh}}$$

For average time spent in the system [queue time plus departure (service) time], Eq. 5.30 is used:

$$\overline{t} = \frac{2-0.75}{2(4)(1-0.75)}$$

$$= \underline{0.625 \text{ min/veh}}$$

or, alternatively, because the departure (service) time is $1/\mu$ (the 0.25 minutes it takes the park attendant to distribute the brochure),

$$\bar{t} = \bar{w} + \frac{1}{\mu}$$

$$= 0.375 + \frac{1}{4}$$

$$= \underline{\underline{0.625 \text{ min/veh}}}$$

5.5.4   *M/M/*1 Queuing

A queuing model that assumes one departure channel and exponentially distributed departure times in addition to exponentially distributed arrival times (an *M/M/*1 queue) is applicable in some traffic applications. For example, exponentially distributed departure patterns might be a reasonable assumption at a toll booth, where some arriving drivers have the correct toll and can be processed quickly, and others do not have the correct toll, producing a distribution of departures about some mean departure rate. Under standard *M/M/*1 assumptions, it can be shown that the following queuing performance equations apply (again assuming that ρ is less than 1):

$$\bar{Q} = \frac{\rho^2}{1-\rho} \tag{5.31}$$

$$\bar{w} = \frac{\lambda}{\mu(\mu-\lambda)} \tag{5.32}$$

$$\bar{t} = \frac{1}{\mu-\lambda} \tag{5.33}$$

where

$\bar{Q}$ = average length of queue in vehicles,

$\bar{w}$ = average waiting time in the queue, in unit time per vehicle,

$\bar{t}$ = average time spent in the system ($\bar{w} + 1/\mu$), in unit time per vehicle, and

other terms are as defined previously.

EXAMPLE 5.12   M/M/1 QUEUING: PARKING-LOT APPLICATION

Assume that the park attendant in Examples 5.7 and 5.11 takes an average of 15 seconds to distribute brochures, but the distribution time varies depending on whether park patrons have questions relating to park operating policies. Given an average arrival rate of 180 veh/h as in Example 5.11, compute the average length of queue (in vehicles), average waiting time in the queue, and average time spent in the system, assuming *M/M/*1 queuing.

SOLUTION

Using the average arrival rate, departure rate, and traffic intensity as determined in Example 5.11, the average length of queue is (from Eq. 5.31)

$$\overline{Q} = \frac{0.75^2}{1 - 0.75}$$
$$= 2.25 \text{ veh}$$

the average waiting time in the queue is (from Eq. 5.32)

$$\overline{w} = \frac{3}{4(4-3)}$$
$$= 0.75 \text{ min/veh}$$

and the average time spent in the system is (from Eq. 5.33)

$$\overline{t} = \frac{1}{4-3}$$
$$= 1 \text{ min/veh}$$

5.5.5   *M/M/N* Queuing

A more general formulation of the *M/M/*1 queue is the *M/M/N* queue, where *N* is the total number of departure channels. *M/M/N* queuing is a reasonable assumption at toll booths on turnpikes or at toll bridges, where there is often more than one departure channel available (more than one toll booth open). A parking lot is another example, with *N* being the number of parking stalls in the lot and the departure rate, μ, being the exponentially distributed times of parking duration. *M/M/N* queuing is also frequently encountered in non-transportation applications such as checkout lines at retail stores, security checks at airports, and so on.

The following equations describe the operational characteristics of *M/M/N* queuing. Note that unlike the equations for *M/D/*1 and *M/M/*1, which require that the traffic intensity, ρ, be less than 1, the following equations allow ρ to be greater than 1 but apply only when ρ/N (which is called the utilization factor) is less than 1.

$$P_0 = \frac{1}{\displaystyle\sum_{n_c=0}^{N-1} \frac{\rho^{n_c}}{n_c!} + \frac{\rho^N}{N!(1-\rho/N)}} \tag{5.34}$$

$$P_n = \frac{\rho^n P_0}{n!} \qquad \text{for } n \le N \tag{5.35}$$

$$P_n = \frac{\rho^n P_0}{N^{n-N} N!} \quad \text{for } n \geq N \tag{5.36}$$

$$P_{n>N} = \frac{P_0 \rho^{N+1}}{N!N(1-\rho/N)} \tag{5.37}$$

where

P_0 = probability of having no vehicles in the system,

P_n = probability of having n vehicles in the system,

$P_{n>N}$ = probability of waiting in a queue (the probability that the number of vehicles in the system is greater than the number of departure channels),

n = number of vehicles in the system,

N = number of departure channels,

n_c = departure channel number, and

ρ = traffic intensity (λ/μ).

$$\overline{Q} = \frac{P_0 \, \rho^{N+1}}{N!N}\left[\frac{1}{(1-\rho/N)^2}\right] \tag{5.38}$$

$$\overline{w} = \frac{\rho + \overline{Q}}{\lambda} - \frac{1}{\mu} \tag{5.39}$$

$$\overline{t} = \frac{\rho + \overline{Q}}{\lambda} \tag{5.40}$$

where

$\overline{Q}$ = average length of queue (in vehicles),

$\overline{w}$ = average waiting time in the queue, in unit time per vehicle,

$\overline{t}$ = average time spent in the system, in unit time per vehicle, and

other terms are as defined previously.

EXAMPLE 5.13   *M/M/N* QUEUING: TOLL-BRIDGE APPLICATION

At an entrance to a toll bridge, four toll booths are open. Vehicles arrive at the bridge at an average rate of 1200 veh/h, and at the booths, drivers take an average of 10 seconds to pay their tolls. Both the arrival and departure rates can be assumed to be exponentially distributed. How would the average queue length, time in the system, and probability of waiting in a queue change if a fifth toll booth were opened?

SOLUTION

Using the equations for *M/M/N* queuing, we first compute the four-booth case. Note that $\mu = 6$ veh/min and $\lambda = 20$ veh/min, and therefore $\rho = 3.333$. Also, because $\rho/N = 0.833$ (which is less than 1), Eqs. 5.34 to 5.40 can be used. The probability of having no vehicles in the system with four booths open (using Eq. 5.34) is

$$P_0 = \cfrac{1}{1 + \cfrac{3.333}{1!} + \cfrac{3.333^2}{2!} + \cfrac{3.333^3}{3!} + \cfrac{3.333^4}{4!(0.1667)}}$$

$$= 0.0213$$

The average queue length is (from Eq. 5.38)

$$\overline{Q} = \frac{0.0213(3.333)^5}{4!4}\left[\frac{1}{(0.1667)^2}\right]$$

$$= 3.287 \text{ veh}$$

The average time spent in the system is (from Eq. 5.40)

$$\overline{t} = \frac{3.333 + 3.287}{20}$$

$$= 0.331 \text{ min/veh}$$

And the probability of having to wait in a queue is (from Eq. 5.37)

$$P_{n>N} = \frac{0.0213(3.333)^5}{4!4(0.1667)}$$

$$= 0.548$$

With a fifth booth open, the probability of having no vehicles in the system is (from Eq. 5.34)

$$P_0 = \cfrac{1}{1 + \cfrac{3.333}{1!} + \cfrac{3.333^2}{2!} + \cfrac{3.333^3}{3!} + \cfrac{3.333^4}{4!} + \cfrac{3.333^5}{5!(0.3333)}}$$

$$= 0.0318$$

The average queue length is (from Eq. 5.38)

$$\overline{Q} = \frac{0.0318(3.333)^6}{5!5}\left[\frac{1}{(0.3333)^2}\right]$$

$$= 0.654 \text{ veh}$$

The average time spent in the system is (from Eq. 5.40)

$$\overline{t} = \frac{3.333 + 0.654}{20}$$

$$= 0.199 \text{ min/veh}$$

And the probability of having to wait in a queue is (from Eq. 5.37)

$$P_{n>N} = \frac{0.0318(3.333)^6}{5!5(0.3333)}$$

$$= 0.218$$

So opening a fifth booth would reduce the average queue length by 2.633 veh (3.287 − 0.654), the average time in the system by 0.132 min/veh (0.331 − 0.199), and the probability of waiting in a queue by 0.330 (0.548 − 0.218).

EXAMPLE 5.14   *M/M/N* **QUEUING: PARKING-LOT APPLICATION**

A convenience store has four available parking spaces. The owner predicts that the duration of customer shopping (the time that a customer's vehicle will occupy a parking space) is exponentially distributed with a mean of 6 minutes. The owner knows that in the busiest hour customer arrivals are exponentially distributed with a mean arrival rate of 20 customers per hour. What is the probability that a customer will not find an open parking space when arriving at the store?

SOLUTION

Putting mean arrival and departure rates in common units gives $\mu = 10$ veh/h and $\lambda = 20$ veh/h. So $\rho = 2.0$, and because $\rho/N = 0.5$ (which is less than 1), Eqs. 5.34 to 5.40 can be used. The probability of having no vehicles in the system with four parking spaces available (using Eq. 5.34) is

$$P_0 = \frac{1}{1 + \dfrac{2}{1!} + \dfrac{2^2}{2!} + \dfrac{2^3}{3!} + \dfrac{2^4}{4!(0.5)}}$$

$$= 0.1304$$

Thus the probability of not finding an open parking space upon arrival is (from Eq. 5.37)

$$P_{n>N} = \frac{0.1304(2)^5}{4!4(0.5)}$$

$$= \underline{0.087}$$

5.6   TRAFFIC ANALYSIS AT HIGHWAY BOTTLENECKS

Some of the most severe congestion problems occur at highway bottlenecks, which are defined as a portion of highway with a lower capacity (q_{cap}) than the incoming section of highway. This reduction in capacity can originate from a number of sources, including a decrease in the number of highway lanes and reduced shoulder widths (which tend to cause drivers to slow and thus effectively reduce highway capacity, as will be discussed in Chapter 6). There are two general types of traffic bottlenecks—those that are recurring and those that are incident induced. Recurring bottlenecks exist where the highway itself limits capacity—for example, by a physical reduction in the number of lanes. Traffic congestion at such bottlenecks results from recurring traffic flows that exceed the vehicle capacity of the highway in the bottleneck area. In contrast, incident-induced bottlenecks occur as a result of vehicle breakdowns or accidents that effectively reduce highway capacity by restricting the through movement of traffic. Because incident-induced bottlenecks are unanticipated and temporary in nature, they have features that distinguish them from recurring bottlenecks, such as the possibility that the capacity resulting from an incident-induced bottleneck may change over time. For example, an accident may initially stop traffic flow completely, but as the wreckage is cleared, partial capacity (one lane open) may be provided for a period of time before full capacity is eventually restored. A feature shared by recurring and incident-induced bottlenecks is the adjustment in traffic flow that may occur as travelers choose other routes and/or different trip departure times, to avoid the bottleneck area, in response to visual information or traffic advisories.

The analysis of traffic flow at bottlenecks can be undertaken using the queuing models discussed in Section 5.5. The most intuitive approach to analyzing traffic congestion at bottlenecks is to assume $D/D/1$ queuing.

EXAMPLE 5.15 $D/D/1$ QUEUING: HIGHWAY BOTTLENECK APPLICATION

An incident occurs on a freeway that has a capacity in the northbound direction, before the incident, of 4000 veh/h and a constant flow of 2900 veh/h during the morning commute (no adjustments to traffic flow result from the incident). At 8:00 A.M., a traffic accident closes the freeway to all traffic. At 8:12 A.M., the freeway is partially opened with a capacity of 2000 veh/h. Finally, the wreckage is removed, and the freeway is restored to full capacity (4000 veh/h) at 8:31 A.M. Assume $D/D/1$ queuing to determine time of queue dissipation, longest queue length, total delay, average delay per vehicle, and longest wait of any vehicle (assuming FIFO).

SOLUTION

Let μ be the full-capacity departure rate and μ_r be the restrictive partial-capacity departure rate. Putting arrival and departure rates in common units of vehicles per minute, we have

$$\mu = \frac{4000 \text{ veh/h}}{60 \text{ min/h}} = 66.67 \text{ veh/min}$$

$$\mu_r = \frac{2000 \text{ veh/h}}{60 \text{ min/h}} = 33.33 \text{ veh/min}$$

$$\lambda = \frac{2900 \text{ veh/h}}{60 \text{ min/h}} = 48.33 \text{ veh/min}$$

The arrival rate is constant over the entire time period, and the total number of vehicles is equal to λt, where t is the number of minutes after 8:00 A.M. The total number of departing vehicles is

$$
\begin{array}{ll}
0 & \text{for } t \leq 12 \text{ min} \\
\mu_r(t-12) & \text{for } 12 \text{ min} < t \leq 31 \text{ min} \\
633.33 + \mu(t-31) & \text{for } t > 31 \text{ min}
\end{array}
$$

Note that the value of 633.33 in the departure rate function for $t > 31$ is based on the preceding departure rate function [$33\frac{1}{3}$ (31 − 12)]. These arrival and departure rates can be represented graphically as shown in Fig. 5.8. As discussed in Section 5.5, for $D/D/1$ queuing, the queue will dissipate at the intersection point of the arrival and departure curves, which can be determined as

$$\lambda t = 633.33 + \mu(t-31) \quad \text{or} \quad t = \underline{78.16 \text{ min}} \text{ (just after 9:18 A.M.)}$$

At this time a total of 3777.5 vehicles (48.33 × 78.16) will have arrived and departed (for the sake of clarity, fractions of vehicles are used). The longest queue (longest vertical distance between arrival and departure curves) occurs at 8:31 a.m. and is

$$
\begin{aligned}
Q_{max} &= \lambda t - \mu_r (t-12) \\
&= 48.33(31) - 33.33(19) \\
&= \underline{865 \text{ veh}}
\end{aligned}
$$

Total vehicle delay is (using equations for triangular and trapezoidal areas to calculate the total area between the arrival and departure curves)

$$
\begin{aligned}
D_t &= \frac{1}{2}(12)(580) + \frac{1}{2}(580+1498.33)(19) - \frac{1}{2}(19)(633.33) \\
&\quad + \frac{1}{2}(1498.33-633.33)(78.16-31) \\
&= \underline{37,604.2 \text{ veh-min}}
\end{aligned}
$$

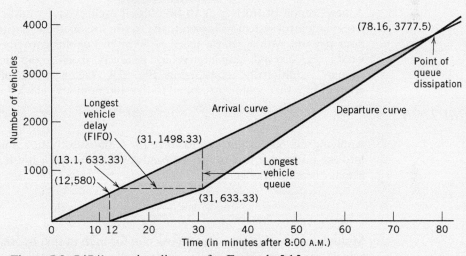

Figure 5.8 *D/D/*1 queuing diagram for Example 5.15.

The average delay per vehicle is <u>9.95 min</u> (37,604.2/3777.5). The longest wait of any vehicle (the longest horizontal distance between the arrival and departure curves), assuming a FIFO queuing discipline, will be the delay time of the 633.33rd vehicle to arrive. This vehicle will arrive 13.1 minutes (633.33/48.33) after 8:00 A.M. and will depart at 8:31 A.M., being delayed a total of <u>17.9 min</u>.

5.7   IMPACT OF AUTONOMOUS VEHICLES

While there has been tremendous progress in recent years with respect to the technology that allows a vehicle to drive autonomously, understanding the impact that such vehicles will have on traffic operations and speed-flow relationships is not well understood.

The traffic stream models of Section 5.3 are theoretical realizations based on empirical observations of traffic flow. Consequently, these models of traffic flow are founded on human driving behavior. This driving behavior, as well as the automobile, has changed significantly since the relationships of Section 5.3 were first proposed. Consequently, traffic stream models based on current observations of traffic flow, while similar to the models of Section 5.3, do consist of some significant differences, as will be shown in Chapter 6.

With the notion of a traffic stream comprised entirely of computer-controlled vehicles, the possibility exists to essentially define the traffic stream relationships as desired to achieve a much more efficient use of existing roadway space.

EXAMPLE 5.16   FREEWAY TRAFFIC FLOW WITH AUTONOMOUS VEHICLES

A new section of freeway is to be utilized exclusively by autonomous vehicles. The freeway control system will operate the traffic stream at a capacity of 3600 vehicles per hour per lane with a vehicle speed of 70 mi/h. For this capacity flow rate and speed, what is the corresponding inter-vehicle headway, spacing, and the density of the traffic stream? Additionally, considering Fig. 5.4, sketch a set of speed–flow–density relationships that could correspond to this autonomous-vehicle traffic stream.

SOLUTION

Applying Eq. 5.4, rearranged to solve for $\bar{h}$ and substituting a unit conversion factor into the numerator to convert the headway from units of h/veh to units of s/veh, the average headway per lane is

$$\bar{h} = \frac{3600 \text{ s/h}}{3600 \text{ veh/h}} = \underline{\underline{1.0 \text{ s/veh}}}$$

Multiplying speed (with a unit conversion for mi/h to ft/s) by headway gives the inter-vehicle spacing as

$$\bar{s} = (70 \times 5280/3600) \times 1.0 = \underline{102.67 \text{ ft/veh}}$$

Applying Eq. 5.13, rearranged to solve for k and substituting a unit conversion factor into the numerator to convert the density from units of veh/ft to units of veh/mi, the density per lane is

$$k = \frac{5280 \text{ ft/mi}}{102.67 \text{ ft/veh}} = \underline{\underline{51.4 \text{ veh/mi}}}$$

Note that applying Eq. 5.14 provides a check on our previous calculations

$$q = uk$$
$$= 70 \times 51.4 \approx 3600 \text{ veh/h}$$

In considering the speed–flow–density relationships for this section of freeway, it needs to be recognized that the autonomous vehicle stream will be controlled to ensure that there will be no breakdown (i.e., the point at which traffic flow transitions from uncongested to congested conditions). This breakdown point is represented by the apex of the parabolic speed–flow and flow–density relationships shown in Fig. 5.4, and is regarded as the capacity flow rate value.

Furthermore, since vehicle speeds and spacing will be precisely regulated, the parabolic shape of the speed–flow and flow–density relationships of Figure 5.4 will be avoided. Currently, the parabolic shapes of the speed–flow and flow–density relationships reflect the phenomenon of drivers reducing their speed as more vehicles use the roadway, and reducing their inter-vehicle spacing, as reflected in the speed–density relationship of Fig. 5.4. With autonomous vehicles, a potential realization of the speed–flow–density relationships is shown in Fig. 5.9. Note that in these relationships, there is no jam density (k_j) and speed at capacity and free-flow speed are the same.

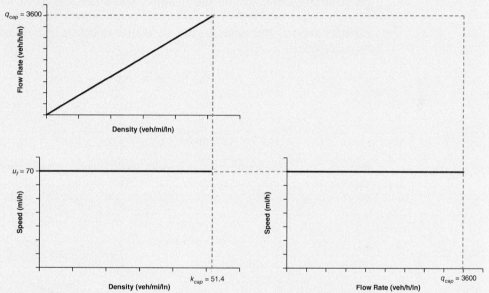

Figure 5.9. Flow–density, speed–density, and speed–flow relationships for autonomous vehicle stream.

Due to uncertainties relating to the technologies involved, it is currently impossible to project the traffic-stream parameter values that will ultimately be realized from autonomous vehicle traffic streams. However, there is clearly great potential for such technologies to lead to significant improvements in traffic operations efficiency. Still, it must also be pointed out that other considerations may ultimately place limits on the traffic stream operational improvements realized from autonomous vehicles. For example, fuel consumption and emissions output considerations could result in limits on speed, human tolerances for deceleration and the physics of vehicle braking could result in limits to speed and/or vehicle spacing, and so on.

5.8 PRACTICE PROBLEMS

PRACTICE PROBLEM 5.1

CAPACITY AND SPEED–FLOW RELATIONSHIPS

A section of highway has a speed–flow relationship of the form $q = au^2 + bu$, where q is the traffic flow in veh/h, u is the space-mean speed of traffic in mi/h, and a and b are unknown parameters. It is known that at capacity (which is 3100 veh/h) the space-mean speed of traffic is 28 mi/h. Determine the speed when the flow is 1500 veh/h and the free-flow speed.

SOLUTION

Note: Open boxes in equations "☐" are to be completed by the reader

This problem seeks to improve readers' understanding of the speed–flow relationship (which Chapter 6 will show is critical in determining highway mobility performance) by demonstrating elements of the functional form of the relationship. In looking at the problem statement, to find the free-flow speed and the speed when the flow is 1500 veh/h we need to first find the unknown parameters a and b. With flow and speed at capacity known, we can set the first derivative of $q = au^2 + bu$ equal to zero to arrive at capacity values,

$$q_{cap} = au_{cap}^2 + bu_{cap}$$

$$\frac{dq_{cap}}{du_{cap}} = \boxed{} \times a \times \boxed{} + b = 0$$

$$\text{so } b = -\boxed{} \times a \times \boxed{}.$$

With $q_{cap} = 3100$ veh/h and $u_{cap} = 28$ mi/h (both given), substituting these values and the expression for b obtained from the first derivative (above) back into $q = au^2 + bu$ gives,

$$\boxed{} = a \times \boxed{}^2 + \left(-\boxed{} \times \boxed{} \times \boxed{}\right) \times \boxed{}$$

$$a = -3.954.$$

With this, b can be calculated,

$$b = -\boxed{} \times -3.954 \times \boxed{}$$

$$= 221.424.$$

With these values of a and b, the speed with a flow of 1500 veh/h is,

$$q = au^2 + bu$$

$$\boxed{} = \boxed{} \times u^2 + \boxed{} \times u.$$

Note that the equation above, for a given flow, will give two feasible values for speed, one in uncongested flow conditions (the higher speed) and one in congested flow conditions (the lower speed). This is illustrated by the general speed–flow relationship shown in Fig. 5.3. Solving the above equation for speed gives $u = \underline{48.12 \text{ mi/h}}$ in uncongested conditions and $u = \underline{7.88 \text{ mi/h}}$ in congested conditions.

Finally, the free-flow speed is obtained by substituting $q = 0$ into the speed–flow relationship above, giving $u = \underline{56 \text{ mi/h}}$.

| PRACTICE PROBLEM 5.2 | **POISSON ARRIVAL PATTERNS AND VEHICLE TIME HEADWAYS** |

A vehicle pulls out onto a single-lane highway that has a flow rate of 300 veh/h (Poisson distributed). The driver of the vehicle does not look for oncoming traffic. Road conditions and vehicle speeds on the highway are such that it takes 1.7 seconds for an oncoming vehicle to stop once the brakes are applied. Assuming that a standard driver reaction time of 2.5 seconds (see Chapter 3) and ignoring the length of the vehicles (which will have just a small influence on the results), what is the probability that the vehicle pulling out will get in an accident with oncoming traffic?

SOLUTION

Note: Open boxes in equations "☐" are to be completed by the reader

This problem shows how a Poisson process can be used to arrive at time spacings between vehicles. With 1.7 seconds required for a stop and 2.5 seconds required for driver reaction time (for a total of 4.2 seconds), an accident will occur when the vehicle headway is less than 4.2 seconds. Using Eq. 5.26 with $t = 4.2$ s and $q = 300$ veh/h we have,

$$P(h \ge t) = e^{-qt/3600}$$

$$P\left(h \ge \boxed{}\right) = e^{-\left(\boxed{}\right)\left(\boxed{}\right)/3600} = \boxed{}.$$

With this, the probability that the time-headway will be less than 4.2 seconds, and thus the probability of an accident, is,

$$P(h < 4.2) = 1 - P\left(h \ge \boxed{}\right)$$

$$= 1 - \boxed{} = \underline{\underline{0.295}}.$$

| PRACTICE PROBLEM 5.3 | ***D/D/*1 QUEUING WITH TIME-INVARIANT ARRIVALS AND DEPARTURES** |

Vehicles begin arriving at a closed entrance to a parking lot at a rate of six vehicles per minute for the first 15 minutes and then continue to arrive at two vehicles per minute for times after this initial 15-minute period. The parking-lot entrance opens when the vehicle queue reaches 30 vehicles and the parking-lot attendant services one vehicle every 15 seconds. Assuming *D/D/*1 queuing, what is the total delay?

SOLUTION

Note: Open boxes in equations "☐" are to be completed by the reader

To begin, a few key points for the points for the *D/D/*1 queuing diagram need to be determined. First, the number of minutes into the process when parking-lot entrance opens can be calculated as,

$$t_{open} = \frac{\text{number of queued vehicles when open}}{\text{vehicle arrival rate}} = \frac{\boxed{}}{\boxed{}} = 5 \text{ min.}$$

Next, at 15 min into the process the arrival rate changes from 6 veh/min to 2 veh/min. So after the first 15 min total vehicle arrivals are,

$$\text{arrivals}_{15} = 15 \times \boxed{} = 90 \text{ veh.}$$

It is known that the parking-lot attendant services vehicles at a rate of one every 15 sec or, equivalently, 4 veh/min starting 15 min into the process (when parking service begins). So the number of departures after 15 min will be,

$$\text{departures}_{15} = 10 \times \boxed{} = \boxed{} \text{ veh,}$$

which means that the queue at 15 min into the process is,

$$\text{queue}_{15} = \text{arrivals}_{15} - \text{departures}_{15} = \boxed{} - \boxed{} = \boxed{} \text{ veh.}$$

So the equation for the time after 15 min to queue clearance ($t_{>15}$) can be determined from,

$$\text{queue}_{15} + (\text{arrival rate after 15 min})(t_{>15}) = (\text{departure rate after 15 min})(t_{>15})$$

or,

$$t_{>15} = \frac{\text{queue}_{15}}{(\text{departure rate after 15 min}) - (\text{arrival rate after 15 min})}$$

$$= \frac{\boxed{}}{\boxed{} - \boxed{}} = 25 \text{ min.}$$

So the queue will clear 15 min + 25 min or 40 min after the start of the process. For the total delay, we need the area between arrival and departure curves. At the 15th minute, 90 veh have arrived so the area under the arrival curve is a triangle with a height of 90 veh and a base of 15 min. Also at the 15th minute, 40 veh will have departed so the area under the departure curve is a triangle with a height of 40 veh and but the base is now 10 min (15 − 5) since departures do not start until 5 min after the first vehicle arrival. The subtraction of these two triangles will give the area between arrival and departure curves up to 15 min.

For the area between arrival and departure curves after 15 min, note that the difference between total vehicle arrivals and departures at 15 min is 50 veh (90 veh − 40 veh) which forms the base of a triangular area, and the height of this triangle is the time between 15 min and the 40 min clearance time or $t_{>15}$ = 25 min.

With these areas, the total delay D_t is,

$$D_t = \left(\frac{\boxed{} \times \boxed{}}{2} \right) - \left(\frac{\boxed{} \times \boxed{}}{2} \right) + \left(\frac{\boxed{} \times \boxed{}}{2} \right)$$

$$= 1{,}100 \text{ veh-min.}$$

PRACTICE PROBLEM 5.4	***D/D/*1 QUEUING WITH TIME-VARYING ARRIVAL AND DEPARTURE RATES**

The gate entrance to a park opens at 9:00 A.M. At 9:00 A.M. there are 32 vehicles in the queue waiting to enter. Vehicles continue to arrive (from 9:00 A.M. onward) at a rate of $\lambda(t) = 4.2 - 0.05t$ (with $\lambda(t)$ in veh/min and t in minutes after 9:00 A.M.). The gate attendant processes vehicles at a rate of $\mu(t) = 3 + 0.3t$ [with $\mu(t)$ in veh/min and t in minutes after 9:00 A.M.]. Assuming *D/D/*1 queuing, what is the maximum queue length and total vehicle delay from 9:00 A.M. onward?

SOLUTION   **Note: Open boxes in equations "$\boxed{}$" are to be completed by the reader**

To begin, we need to develop equations for arrivals and departures from the equations we are given for arrival and departure rates. If we integrate the arrival and departure rate equations with respect to time, we will have equations for arrivals and departures (in units of vehicles) at any time t (which will be the number of minutes after 9:00 A.M.). To do this for the arrivals we have,

$$= \int_0^t 4.2 - 0.05t \ dt = \boxed{} \times \boxed{} - \boxed{} \times \boxed{}^{\boxed{}} ,$$

and for departures we have,

$$= \int_0^t 3 + 0.3t \ dt = \boxed{} \times \boxed{} + \boxed{} \times \boxed{}^{\boxed{}} .$$

So the queue length at any time $Q(t)$, at any time t after 9:00 A.M., is going to be the arrivals minus the departures from the above functions plus the 32 vehicles waiting at 9:00 A.M.,

$$Q(t) = \left(\boxed{} \times \boxed{} - \boxed{} \times \boxed{}^{\boxed{}} \right) - \left(\boxed{} \times \boxed{} + \boxed{} \times \boxed{}^{\boxed{}} \right) + 32$$

$$= -0.175t^2 + 1.2t + 32$$

Solving for the time at which the maximum queue length occurs by taking the first derivative of the above equation with respect to time yields

$$\frac{dQ(t)}{dt} = -\boxed{} \times t + \boxed{} = 0$$

$$t = 3.43 \text{ min.}$$

Substituting with $t = 3.43$ minutes gives the maximum queue length as

$$Q(t) = -0.175t^2 + 1.2t + 32$$

$$= -0.175 \times \boxed{}^2 + 1.2 \times \boxed{} + 32 = \underline{34.06 \text{ veh.}}$$

Next, to determine total delay, we must first determine when the queue will clear. This will happen when the total number of arrivals equals the total number of departures. Equating the arrival and departure equations from above (with the 32 vehicles initially in the queue added to the arrivals), and solving for time t gives,

$$32 + \left(\boxed{} \times \boxed{} - \boxed{} \times \boxed{}^{\boxed{}} \right) = \left(\boxed{} \times \boxed{} + \boxed{} \times \boxed{}^{\boxed{}} \right),$$

which gives $0 = -\boxed{} \times t^2 + \boxed{} \times t + 32$, or $t = 17.38$ min (time until queue clearance).

With this, the total delay D_t is going to be the area under the arrival curve minus the area under the departure curve (plus the delay of the initial 32 vehicles), so

$$D_t = 32 \times t + \int_0^t \left[\Box \times \Box - \Box \times \Box^{\Box} \right] dt - \int_0^t \left[\Box \times \Box + \Box \times \Box^{\Box} \right] dt$$

$$= 32t + \Box \times t^2 - \Box \times t^3 - \Box \times t^2 - \Box \times t^3 \Big|_0^{17.38}$$

$$= \underline{431.17 \text{ veh-min.}}$$

NOMENCLATURE FOR CHAPTER 5

D	deterministic arrivals or departures	Q_{max}	maximum length of queue
D_t	total vehicle delay	s	vehicle spacing
h	vehicle time headway	t	time
k	traffic density	$\bar{t}$	average time spent in the system
k_j	traffic jam density	u	space-mean speed (also denoted $\bar{u}_s$)
k_{cap}	traffic density at capacity	u_i	spot speed of vehicle i
l	roadway length	u_f	free-flow speed
M	exponentially distributed arrivals or departures	u_{cap}	speed at capacity
		$\bar{u}_s$	space-mean speed (also denoted simply as u)
n	number of vehicles	$\bar{u}_t$	time-mean speed
n_c	departure channel number	$\bar{w}$	average time waiting in the queue
N	total number of departure channels	λ	arrival rate
q	traffic flow	μ	departure rate
q_{cap}	traffic flow at capacity (maximum traffic flow)	ρ	traffic intensity
Q	length of queue		
$\bar{Q}$	average length of queue		

REFERENCES

Drew, D. R. "Deterministic Aspects of Freeway Operations and Control." *Highway Research Record*, 99, 1965.

Lord, D., and F. Mannering. "The Statistical Analysis of Crash-Frequency Data: A Review and Assessment of Methodological Alternatives," *Transportation Research Part A*, vol. 44, no. 5, 2010.

Pipes, L. A. "Car Following Models and the Fundamental Diagram of Road Traffic." *Transportation Research*, vol. 1, no. 1, 1967.

Poch, M., and F. Mannering. "Negative Binomial Analysis of Intersection-Accident Frequencies," *Journal of Transportation Engineering*, vol. 122, no. 2, March/April 1996.

Transportation Research Board. *Traffic Flow Theory: A Monograph.* Special Report 165. Washington, DC: National Research Council, 1975.

Transportation Research Board. *Highway Capacity Manual*. Washington, DC: National Research Council, 2010.

Chapter 6

Highway Capacity and Level-of-Service Analysis

6.1 INTRODUCTION

The underlying objective of traffic analysis is to quantify a roadway's performance with regard to specified traffic volumes. This performance can be measured in terms of travel delay (as the roadway becomes increasingly congested) as well as other factors. The comparative performance of various roadway segments (which is determined from an analysis of traffic) is important because it can be used as a basis to allocate limited roadway construction and improvement funds. The purpose of this chapter is to apply the elements of uninterrupted traffic flow theory covered in Chapter 5 to the practical field analysis of traffic flow and capacity on freeways, multilane highways, and two-lane highways.

The main challenge of such a process is to adapt the theoretical formulations to the wide range of conditions that occur in the field. These diverse field conditions must be taken into account in a traffic analysis methodology, yet the methodology must remain theoretically consistent. For example, in Chapter 5, capacity (q_{cap}) is simply defined as the highest traffic flow rate that the roadway is capable of supporting. For applied traffic analysis, a consistent and reasonably precise method of determining capacity must be developed within this definition. Because it can readily be shown that the capacity of a roadway segment is a function of factors such as roadway type (freeway, multilane highway, or two-lane highway), free-flow speed (*FFS*), number of lanes, and widths of lanes and shoulders, the method of capacity determination clearly must account for a wide variety of physical and operational roadway characteristics.

Additionally, recall that Chapter 5 defines traffic flow on the basis of units of vehicles per hour. Two practical issues arise concerning this unit of measure. First, in many cases, vehicular traffic consists of a variety of vehicle types with substantially different performance characteristics. These performance differentials are likely to be magnified by changing roadway geometrics, such as upgrades or downgrades, which have a differential effect on the acceleration and deceleration capabilities of the various types of vehicles; for example, grades have a greater impact on the performance of large trucks (heavy vehicles) than on automobiles. As a result, traffic must be defined not only in terms of vehicles per unit time but also in terms of vehicle composition, because it is clear that a 1500-veh/h traffic flow consisting of 100% automobiles will differ significantly with regard to operating speed and traffic density from a 1500-veh/h traffic flow that consists of 50% automobiles and 50% heavy vehicles.

The other flow-related concern is the temporal distribution of traffic. In practice, the analysis of roadway traffic usually focuses on the most critical condition, which is the most congested hour within a 24-hour daily period (the temporal distribution of traffic is discussed in more detail in Section 6.7). However, within this most congested peak hour, traffic flow is likely to be nonuniform. It is therefore necessary to arrive at some method of defining and measuring the nonuniformity of flow within the peak hour.

To summarize, the objective of applied traffic analysis is to provide a practical method of quantifying the degree of traffic congestion and to relate this to the overall traffic-related performance of the roadway. The following sections of this chapter discuss and demonstrate accepted standards for applied traffic analysis for the three major types of uninterrupted-flow roadways: freeways, multilane highways, and two-lane highways (one lane in each direction).

6.2   LEVEL-OF-SERVICE CONCEPT

The *Highway Capacity Manual* (HCM), produced by the Transportation Research Board [2016], is a synthesis of the state of the art in methodologies for quantifying traffic operational performance and capacity utilization (congestion level) for a variety of transportation facilities. One of the foundations of the HCM is the concept of level of service (LOS). The LOS represents a qualitative ranking of the traffic operational conditions experienced by users of a facility under specified roadway, traffic, and traffic control (if present) conditions. Current practice designates six levels of service ranging from A to F, with LOS A representing the best operating conditions and LOS F the worst.

A number of operational performance measures, such as speed, flow, and density, can be measured or calculated for any transportation facility. To apply the LOS concept to traffic analysis, it is necessary to select a performance measure that is representative of how motorists actually perceive the quality of service they are receiving on a facility. Motorists tend to evaluate their received quality of service in terms of factors such as speed and travel time, freedom to maneuver, traffic interruptions, and comfort and convenience. Thus, it is important to select a measure that encompasses some or all of these factors. The performance measure that is selected for LOS analysis for a particular transportation facility is referred to as the service measure.

The HCM [Transportation Research Board 2016] defines the LOS categories for freeways and multilane highways as follows:

Level of service A. LOS A represents free-flow conditions (traffic operating at *FFS*, as defined in Chapter 5). Individual users are virtually unaffected by the presence of others in the traffic stream. Freedom to select speeds and to maneuver within the traffic stream is extremely high. The general level of comfort and convenience provided to drivers is excellent.

Level of service B. LOS B also allows speeds at or near *FFS*, but the presence of other users in the traffic stream begins to be noticeable. Freedom to select speeds is relatively unaffected, but there is a slight decline in the freedom to maneuver within the traffic stream relative to LOS A.

Level of service C. LOS C has speeds at or near *FFS*, but the freedom to maneuver is noticeably restricted (lane changes require careful attention on the part of drivers). The general level of comfort and convenience declines significantly at this level. Disruptions in the traffic stream, such as an incident (for example, vehicular accident or disablement), can result in significant queue formation and vehicular delay. In contrast, the effects of incidents at LOS A or LOS B are minimal, with only minor delay in the immediate vicinity of the event.

Level of service D. LOS D represents the conditions where speeds begin to decline slightly with increasing flow. The freedom to maneuver becomes more restricted, and drivers experience reductions in physical and psychological comfort. Incidents can generate lengthy queues because the higher density associated with this LOS provides little space to absorb disruptions in the traffic flow.

Level of service E. LOS E represents operating conditions at or near the roadway's capacity. Even minor disruptions to the traffic stream, such as vehicles entering from a ramp or vehicles changing lanes, can cause delays as other vehicles give way to allow such maneuvers. In general, maneuverability is extremely limited, and drivers experience considerable physical and psychological discomfort.

Level of service F. LOS F describes a breakdown in vehicular flow. Queues form quickly behind points in the roadway where the arrival flow rate temporarily exceeds the departure rate, as determined by the roadway's capacity (see Chapter 5). Such points occur at incidents and on- and off-ramps, where incoming traffic results in capacity being exceeded. Vehicles typically operate at low speeds under these conditions and are often required to come to a complete stop, usually in a cyclic fashion. The cyclic formation and dissipation of queues is a key characterization of LOS F.

A visual perspective of the LOS definitions for freeways is provided in Fig. 6.1. In dealing with LOS, it is important to remember that when the traffic volume is at or near the roadway capacity (which will be shown as a function of the prevailing traffic and physical characteristics of the roadway), the roadway is operating at LOS E. This, however, is not a desirable condition because under LOS E conditions there is considerable driver discomfort, which could increase the likelihood of vehicular crashes and overall delay. In roadway design, the possibility of degradation of LOS to LOS E should be avoided, although this is not always possible due to financial and environmental constraints that may limit the design speed, number of lanes, and other factors affecting roadway capacity.

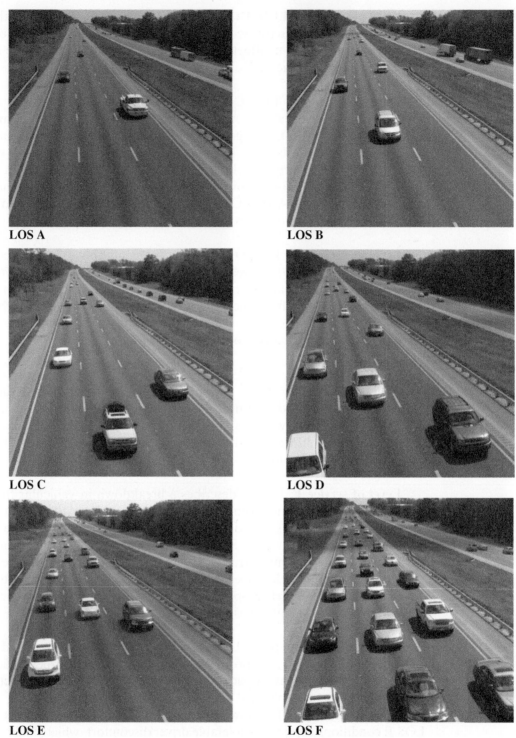

LOS A

LOS B

LOS C

LOS D

LOS E

LOS F

Figure 6.1 Illustration of freeway level of service (A to F).

6.3 LEVEL-OF-SERVICE DETERMINATION

There are several steps in a basic LOS determination for an uninterrupted-flow facility. The remainder of this section describes the general details of each step, as applicable to uninterrupted-flow facility analyses. Facility-specific details of these steps are described in the sections that follow.

6.3.1 Base Conditions and Capacity

The determination of a roadway's LOS begins with the specification of base roadway conditions. Recall that in the introduction to this chapter, the effects of vehicle performance and roadway design characteristics on traffic flow were discussed qualitatively. In practice, the effects of such factors on traffic flow are measured quantitatively, relative to the base traffic and roadway design conditions. For uninterrupted-flow roadways, base conditions can be categorized as those relating to roadway conditions, such as lane widths, lateral clearances, access frequency, and terrain; and traffic stream conditions such as the effects of heavy vehicles (large trucks, buses, and RVs). Base conditions are defined as those conditions that represent unrestrictive geometric and traffic conditions. Additionally, base conditions are assumed to consist of favorable environmental conditions (such as dry roadways).

The capacity of a particular roadway segment will be greatest when all roadway and traffic conditions meet or exceed their base values. Empirical studies have identified the values of these base conditions for which the capacity of a roadway segment is maximized. Values in excess of the base conditions will not increase the capacity of the roadway, but values more restrictive than the base conditions will result in a lower capacity. For example, studies have identified a base lane width of 12 ft. That is, lane widths in excess of 12 ft will not result in increased capacity; however, lane widths less than 12 ft will result in a reduction in capacity. Capacity values for base conditions have been determined for all uninterrupted-flow facility types from field studies. It should be noted that for purposes of LOS analysis, capacity is defined not as the absolute maximum flow rate ever observed for a particular facility type, but rather as the maximum flow rate that can be reasonably expected on a recurring basis.

Because all base conditions for a particular roadway type are seldom realized in practice, methods for converting the measured flow rate into an equivalent analysis flow rate in terms of passenger cars for the given traffic conditions and estimating the actual *FFS* for the given roadway conditions are needed. The following sections describe the procedures for arriving at flow and speed values for given roadway and traffic conditions.

6.3.2 Determine Free-Flow Speed

FFS is a term that was introduced in Chapter 5 as the speed of traffic as the traffic density approaches zero. In practice, *FFS* is governed by roadway design characteristics (horizontal and vertical curves, lane and shoulder widths, and median design), the frequency of access points, the complexity of the driving environment (possible distractions from roadway signs and the like), and posted speed limits.

The *FFS* must be determined given the characteristics of the roadway segment. *FFS* is the mean speed of traffic as measured when flow rates are low to moderate (specific values are given under the individual sections for each roadway type). Ideally, *FFS* should be measured directly in the field at the site of interest. However, if this is not possible or feasible, an alternative method can be employed to arrive at an estimate of *FFS* under the prevailing conditions. This method makes adjustments to a base *FFS* (*BFFS*) depending on the physical characteristics of the roadway segment, such as lane width, shoulder width, and access frequency. This method has the same basic structure for the various roadway types, but contains adjustment factors and values appropriate for each roadway type.

6.3.3   Determine Analysis Flow Rate

One of the fundamental inputs to a traffic analysis is the actual traffic volume on the roadway, in vehicles per hour, which is given the symbol *V*. Generally, the highest volume in a 24-hour period (the peak-hour volume) is used for *V* in traffic analysis computations. However, this hourly volume needs to be adjusted to reflect the temporal variation of traffic demand within the analysis hour and the impacts due to heavy vehicles. To account for these effects, the hourly volume is divided by adjustment factors to obtain an equivalent flow rate in terms of passenger cars per hour (pc/h). Additionally, the flow rate is expressed on a per-lane basis (pc/h/ln) by dividing by the number of lanes in the analysis segment.

6.3.4   Calculate Service Measure(s) and Determine LOS

Once the previous steps have been completed, all that remains is to calculate the value of the service measure and then determine the LOS from the service measure value. For freeways and multilane highways, this is a relatively straightforward task. However, for two-lane highways, there are actually two service measures, and the calculation of these and the subsequent LOS determination are more involved.

6.4   BASIC FREEWAY SEGMENTS

A basic freeway segment is defined as a section of a divided roadway having two or more lanes in each direction, full access control, and traffic that is unaffected by merging or diverging movements near ramps. It is important to note that capacity analysis for divided roadways focuses on the traffic flow in one direction only. This is reasonable because the objective is to measure the highest level of congestion. Due to directional imbalance of traffic flows—for example, morning rush hours having higher volumes going toward the central city and evening rush hours having higher volumes going away from the central city— consideration of traffic volumes in both directions is likely to seriously understate the true level of traffic congestion.

6.4.1   Speed versus Flow Rate Relationship

Central to the analysis of basic freeway segments is the relationship between average traffic stream speed and flow rate. Chapter 5 introduced a theoretical speed–flow relationship. This relationship was parabolic in form, derived from an assumed linear speed–density relationship. Empirical studies of current traffic stream patterns show a speed–flow relationship, that while similar to the theoretical parabolic relationship, is significantly different in one respect. The parabolic model for speed-flow yields a continuous change in speed with changes in flow rate, whereas the empirical relationship shows that traffic stream speed stays constant up to moderate flow rates, and then starts to decline in a curvilinear manner. The HCM [Transportation Research Board 2016] defines the relationship between average speed and flow rate as follows:

$$S = FFS \qquad\qquad v_p \leq BP \qquad\qquad (6.1)$$

and

$$S = FFS - \frac{\left(FFS - \frac{c}{45}\right)\left(v_p - BP\right)^{2.0}}{\left(c - BP\right)^{2.0}} \qquad BP < v_p \leq c \qquad (6.2)$$

where

c = Capacity in pc/h/ln,
v_p = 15-min passenger car equivalent flow rate in pc/h/ln,
BP = Linear to curvilinear flow rate breakpoint value in pc/h/ln, and
45 = Density at capacity in pc/mi/ln, as shown in Table 6.2.

Other terms are as defined previously.

Capacity is given by the following equation:

$$c = \text{Min}\left[2200 + 10\left(FFS - 50\right), 2400\right] \qquad 55 \leq FFS \leq 75 \qquad (6.3)$$

Results from this equation for 5-mi/h increments across the valid range of *FFS* is shown in Table 6.1.

Table 6.1 Relationship between Free-Flow Speed and Capacity on Basic Freeway Segments

Source: Transportation Research Board, *Highway Capacity Manual*, 6th Edition, National Academy of Sciences, Washington, D.C., 2016.

Free-flow speed (mi/h)	Capacity (pc/h/ln)
75	2400
70	2400
65	2350
60	2300
55	2250

The breakpoint, *BP*, value is given by the following equation:

$$BP = \left[1000 + 40\left(75 - FFS\right)\right] \qquad 55 \leq FFS \leq 75 \qquad (6.4)$$

The speed-flow rate curves resulting from the application of Eqs. 6.1 and 6.2, for 5-mi/h increments across the valid range of *FFS* are shown in Fig. 6.2.

Table 6.2 provides the LOS criteria corresponding to traffic density, speed, volume-to-capacity ratio, and maximum flow rate. The maximum service flow rate is simply the maximum flow rate, under base conditions, that can be sustained for a given LOS. This value is based on the speed–flow relationship and the maximum density values given in Table 6.2.

It should be noted that the versions of Eqs. 6.1 and 6.2 in the HCM use the variables FFS_{adj} and c_{adj} instead of FFS and c. This is to accommodate additional factors that might reduce the FFS and/or capacity values from those discussed in this chapter. Examples of these types of factors include temporary work zones, adverse weather, incidents, and drivers unfamiliar with the roadway. In these instances, the FFS and c values as calculated from the factors discussed in this chapter would be further adjusted. Discussion of these topics is beyond the scope of this book; therefore, the FFS and c variables are used in lieu of the FFS_{adj} and c_{adj} variables in Eqs. 6.1 and 6.2.

6.4.2 Base Conditions and Capacity

The base conditions for a basic freeway segment are defined as [Transportation Research Board 2016]

- 12-ft minimum lane widths
- 6-ft minimum right-shoulder clearance between the edge of the travel lane and objects (utility poles, retaining walls, etc.) that influence driver behavior
- Only passenger cars in the traffic stream
- 6-mi or greater interchange ramp spacing
- Level terrain (no grades greater than 2%)

These conditions represent a high operating level, with a FFS of 75.4 mi/h or higher.

The capacity, c, for basic freeway segments, in passenger cars per hour per lane (pc/h/ln), is given in Tables 6.1 and 6.2. From Table 6.2, note that, by definition, the upper boundary of LOS E corresponds to the value of capacity and a v/c of 1.0. Other values of v/c for a specific LOS are obtained by simply dividing the maximum flow rate for that LOS by capacity (the maximum flow rate at LOS E).

6.4.3 Service Measure

The service measure for basic freeway segments is density. Density, as discussed in Chapter 5, is typically measured in terms of passenger cars per mile per lane (pc/mi/ln) and therefore provides a good measure of the relative mobility of individual vehicles in the traffic stream. A low traffic stream density gives individual vehicles the ability to change lanes and speeds with relative ease, while a high density makes it very difficult for individual vehicles to maneuver within the traffic stream. Thus, traffic density is the primary determinant of freeway LOS.

Recall Eq. 5.14 from Chapter 5:

$$q = uk \qquad (6.5)$$

where

q = flow in veh/h,
u = speed in mi/h, and
k = density in veh/mi.

Density is therefore calculated as flow divided by speed. The following sections will describe how to arrive at flow and speed values for the given roadway and traffic conditions. Once the flow and speed values have been determined according to the given conditions, a density can be calculated and then referenced in Table 6.2 or Fig. 6.2 to arrive at a LOS for the freeway segment.

Table 6.2 LOS Criteria for Basic Freeway Segments

Criterion	LOS				
	A	B	C	D	E
FFS = 75 mi/h					
Maximum density (pc/mi/ln)	11	18	26	35	45
Average speed (mi/h)	75.0	73.9	68.3	60.9	53.3
Maximum *v/c*	0.34	0.55	0.74	0.89	1.00
Maximum flow rate (pc/h/ln)	825	1330	1775	2130	2400
FFS = 70 mi/h					
Maximum density (pc/mi/ln)	11	18	26	35	45
Average speed (mi/h)	70.0	70.0	66.7	60.4	53.3
Maximum *v/c*	0.32	0.53	0.72	0.88	1.00
Maximum flow rate (pc/h/ln)	770	1260	1735	2115	2400
FFS = 65 mi/h					
Maximum density (pc/mi/ln)	11	18	26	35	45
Average speed (mi/h)	65.0	65.0	64.0	58.9	52.2
Maximum *v/c*	0.30	0.50	0.71	0.88	1.00
Maximum flow rate (pc/h/ln)	715	1170	1665	2060	2350
FFS = 60 mi/h					
Maximum density (pc/mi/ln)	11	18	26	35	45
Average speed (mi/h)	60.0	60.0	60.0	57.1	51.1
Maximum *v/c*	0.29	0.47	0.68	0.87	1.00
Maximum flow rate (pc/h/ln)	660	1080	1560	2000	2300
FFS = 55 mi/h					
Maximum density (pc/mi/ln)	11	18	26	35	45
Average speed (mi/h)	55.0	55.0	55.0	54.7	50.0
Maximum *v/c*	0.27	0.44	0.64	0.85	1.00
Maximum flow rate (pc/h/ln)	605	990	1430	1915	2250

Note: Maximum flow rate values are rounded to the nearest 5 passenger cars.

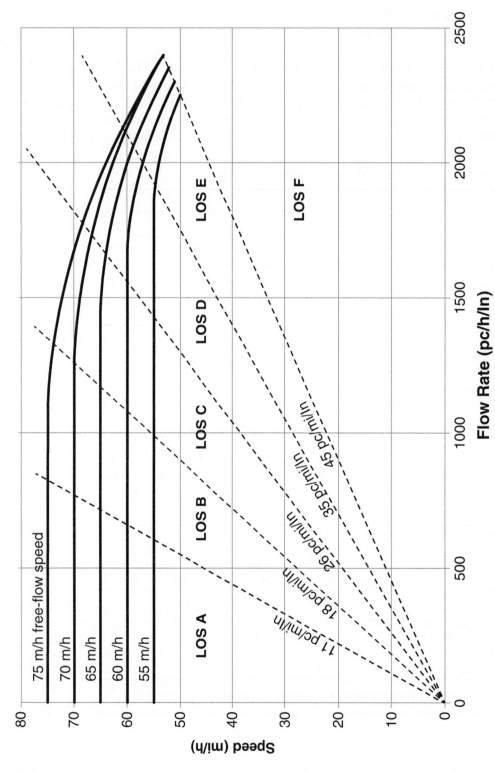

Figure 6.2 Basic freeway segment speed-flow curves and level-of-service criteria. (pickup from current edition)
Reproduced with permission of the Transportation Research Board, *Highway Capacity Manual*, 6th Edition, Copyright, National Academy of Sciences, Washington, D.C. 2016. Exhibit 12-7, p. 12-11.

6.4.4   Determine Free-Flow Speed

For basic freeway segments, *FFS* is the mean speed of passenger cars operating in flow rates up to 1300 passenger cars per hour per lane (pc/h/ln). If *FFS* is to be estimated rather than measured, the following equation can be used. It accounts for the roadway characteristics of lane width, right-shoulder lateral clearance, and ramp density.

$$FFS = 75.4 - f_{LW} - f_{RLC} - 3.22TRD^{0.84} \qquad (6.6)$$

where

FFS = estimated free-flow speed in mi/h,

f_{LW} = adjustment for lane width in mi/h,

f_{RLC} = adjustment for lateral clearance in mi/h,

$3.22TRD^{0.84}$ = adjustment for total ramp density in mi/h (with *TRD* in ramps/mi).

The constant value of 75.4 in Eq. 6.6 is considered to be the base free-flow speed (*BFFS*) and applies to freeways in urban and rural areas. The following sections describe the procedures for estimating the adjustment factor values.

Lane Width Adjustment

When lane widths are narrower than the base 12 ft, the adjustment factor f_{LW} is used to reflect the impact on *FFS*. Such an adjustment is needed because narrow lanes cause traffic to slow as a result of reduced psychological comfort and limits on driver maneuvering and accident avoidance options. Thus, *FFS* under these conditions is less than the value that would be observed if base lane widths were provided. The adjustment factors used in current practice are presented in Table 6.3.

Table 6.3 Adjustment for Lane Width

Lane width (ft)	Reduction in free-flow speed, f_{LW} (mi/h)
12	0.0
11	1.9
10	6.6

Reproduced with permission of the Transportation Research Board, *Highway Capacity Manual*, 6th Edition, Copyright, National Academy of Sciences, Washington, D.C. 2016. Exhibit 12-20, pp. 12–29.

Lateral Clearance Adjustment

When obstructions are closer than 6 ft (at the right shoulder) from the traveled pavement, the adjustment factor f_{RLC} is used to reflect the impact on *FFS*. Again, these conditions lead to reduced psychological comfort for the driver and consequently reduced speeds. An obstruction is a right-side object that can either be continuous (such as a retaining wall or barrier) or periodic (such as light posts or utility poles). Table 6.4 provides corrections for obstructions on the right side of the roadway.

Table 6.4 Adjustment for Right-Shoulder Lateral Clearance

Right-shoulder lateral clearance (ft)	Reduction in free-flow speed, f_{RLC} (mi/h), lanes in one direction			
	2	3	4	≥5
≥6	0.0	0.0	0.0	0.0
5	0.6	0.4	0.2	0.1
4	1.2	0.8	0.4	0.2
3	1.8	1.2	0.6	0.3
2	2.4	1.6	0.8	0.4
1	3.0	2.0	1.0	0.5
0	3.6	2.4	1.2	0.6

Reproduced with permission of the Transportation Research Board, *Highway Capacity Manual*, 6th Edition, Copyright, National Academy of Sciences, Washington, D.C. 2016. Exhibit 12-21, pp. 12–29.

Total Ramp Density Adjustment

Ramp density provides a measure of the impact of merging and diverging traffic on *FFS*. Total ramp density is the number of on- and off-ramps (in one direction) within a distance of three miles upstream and three miles downstream of the midpoint of the analysis segment, divided by six miles.

6.4.5   Determine Analysis Flow Rate

The analysis flow rate is calculated using the following equation:

$$v_p = \frac{V}{PHF \times N \times f_{HV}} \qquad (6.7)$$

where

$$
\begin{aligned}
v_p &= \text{15-min passenger car equivalent flow rate in pc/h/ln,} \\
V &= \text{hourly volume in veh/h,} \\
PHF &= \text{peak-hour factor,} \\
N &= \text{number of lanes, and} \\
f_{HV} &= \text{heavy-vehicle adjustment factor.}
\end{aligned}
$$

The adjustment factors *PHF* and f_{HV} are described next.

Peak-Hour Factor

As previously mentioned, vehicle arrivals during the period of analysis [typically the highest hourly volume within a 24-h period (peak hour)] will likely be nonuniform. To account for this varying arrival rate, the peak 15-min vehicle arrival rate within the analysis hour is usually used for practical traffic analysis purposes. The peak-hour factor has been developed for this purpose, and is defined as the ratio of the hourly volume to the maximum 15-min flow rate expanded to an hourly volume, as follows:

$$PHF = \frac{V}{V_{15} \times 4} \tag{6.8}$$

where

PHF = peak-hour factor,
V = hourly volume for hour of analysis,
V_{15} = maximum 15-min volume within hour of analysis, and
4 = number of 15-min periods per hour.

Equation 6.8 indicates that the further the *PHF* is from unity, the more *peaked* or nonuniform the traffic flow is during the hour. For example, consider two roads both of which have a peak-hour volume, *V*, of 1800 veh/h. The first road has 600 vehicles arriving in the highest 15-min interval, and the second road has 500 vehicles arriving in the highest 15-min interval. The first road has a more nonuniform flow, as indicated by its *PHF* of 0.75 [1800/(600 × 4)], which is further from unity than the second road's *PHF* of 0.90 [1800/(500 × 4)].

Heavy-Vehicle Adjustment

Large trucks, buses, and recreational vehicles have performance characteristics (slow acceleration and inferior braking) and dimensions (length, height, and width) that have an adverse effect on roadway capacity. Recall that base conditions stipulate that no heavy vehicles are present in the traffic stream, and when prevailing conditions indicate the presence of such vehicles, the adjustment factor f_{HV} is used to translate the traffic stream from base to prevailing conditions. The f_{HV} correction term is found using a two-step process. The first step is to determine the passenger car equivalent (PCE) for each heavy vehicle in the traffic stream. These values represent the number of passenger cars that would consume the same amount of roadway capacity as a single heavy vehicle. Heavy vehicles are classified as either single-unit trucks (SUTs) or tractor-trailers (TTs). Buses and recreational vehicles are treated as SUTs. The passenger car equivalent values, denoted as E_T, are a function of roadway grades because steep grades will tend to magnify the poor performance of heavy vehicles as well as the sight distance problems caused by their larger dimensions (the visibility afforded to drivers in vehicles following heavy vehicles). For segments of freeway that contain a mix of grades, an extended segment analysis can be used as long as no single grade is steep enough or long enough to significantly impact the overall operations of the segment. As a guideline, an extended segment analysis can be used for freeway segments where no single grade that is less than 3% is more than 0.5 mi long, or no single grade that is 3% or greater is longer than 0.25 mi. If an extended segment analysis is used, the terrain must be generally classified according to the following definitions [Transportation Research Board 2016]:

Level terrain. Any combination of horizontal and vertical alignment permitting heavy vehicles to maintain approximately the same speed as passenger cars. This generally includes short grades of no more than 2%.

Rolling terrain. Any combination of horizontal and vertical alignment that causes heavy vehicles to reduce their speed substantially below those of passenger cars but does not cause heavy vehicles to operate at their limiting speed [$F_{net}(V) \neq 0$] for the given terrain for any significant length of time or at frequent intervals due to high grade resistance, as illustrated in Fig. 2.2.

The passenger car equivalency factor values for an extended segment analysis can be obtained from Table 6.5.

Table 6.5 Passenger Car Equivalents (PCEs) for Extended Freeway Segments

	Type of terrain	
Factor	Level	Rolling
E_T	2.0	3.0

Reproduced with permission of the Transportation Research Board, *Highway Capacity Manual*, 6th Edition, Copyright, National Academy of Sciences, Washington, D.C. 2016. Exhibit 12-25, pp. 12–35.

Any grade that does not meet the conditions for an extended segment analysis, must be analyzed as a separate segment because of its significant impact on traffic operations. In these cases, grade-specific PCE values must be used. Tables 6.6–6.8 provide these values. These tables correspond to heavy vehicle percentage splits (%SUTs/%TTs) of 30/70, 50/50, 70/30, respectively. These tables assume average weight-to-horsepower ratios of 100 and 150 lb/hp for SUTs and TTs, respectively. Linear interpolation should be used with these tables as necessary.

Note that the equivalency values presented in these tables increase with increasing grade and length of grade, but decrease with increasing heavy vehicle percentage. This decrease with increasing percentage is due to the fact that heavy vehicles tend to group together as their percentages increase on steep, extended grades, thus decreasing their adverse impact on the traffic stream.

Negative grades (downgrades) also have an impact on equivalency factors because the comparatively poor braking characteristics of heavy vehicles have a more deleterious effect on the traffic stream than the level-terrain case. For negative grades, use the level grade PCE values from Tables 6.6–6.8.

Sometimes it is necessary to determine the cumulative effect on traffic operations of several significant grades in succession. For this situation, a distance-weighted average may be used if all grades are less than 4% or the total combined length of the grades is less than 4000 ft. For example, a 2% upgrade for 1000 ft followed immediately by a 3% upgrade for 2000 ft would use the equivalency factor for a 2.67% upgrade [(2 × 1000 + 3 × 2000)/3000] for 3000 ft or 0.568 mi. For information on additional analysis situations involving composite grades, refer to the *Highway Capacity Manual* [Transportation Research Board, 2016]. These situations include combining two or more successive grades when the grades exceed 4% or the combined length is greater than 4000 ft, determining the length of a grade that starts or ends on a vertical curve, and determining the point of greatest traffic impact in a series of grades (for example, if a long 5% grade were immediately followed by a 2% grade, the end of the 5% grade would be used, as this would be the point of minimum vehicle speed).

Table 6.6 Passenger Car Equivalents (E_T) for Heavy Vehicles (30% SUTs/70% TTs) on Specific Grades

Grade (%)	Length (mi)	Percentage of heavy vehicles								
		2	4	5	6	8	10	15	20	≥25
≤ 0	≤ 0.125	2.62	2.37	2.30	2.24	2.17	2.12	2.04	1.99	1.97
	0.375	2.62	2.37	2.30	2.24	2.17	2.12	2.04	1.99	1.97
	0.625	2.62	2.37	2.30	2.24	2.17	2.12	2.04	1.99	1.97
	0.875	2.62	2.37	2.30	2.24	2.17	2.12	2.04	1.99	1.97
	1.25	2.62	2.37	2.30	2.24	2.17	2.12	2.04	1.99	1.97
	≥ 1.5	2.62	2.37	2.30	2.24	2.17	2.12	2.04	1.99	1.97
2	≤ 0.125	2.62	2.37	2.30	2.24	2.17	2.12	2.04	1.99	1.97
	0.375	3.76	2.96	2.78	2.65	2.48	2.38	2.22	2.14	2.09
	0.625	4.47	3.33	3.08	2.91	2.68	2.54	2.34	2.23	2.17
	0.875	4.80	3.50	3.22	3.03	2.77	2.61	2.39	2.28	2.21
	1.25	5.00	3.60	3.30	3.09	2.83	2.66	2.42	2.30	2.23
	≥ 1.5	5.04	3.62	3.32	3.11	2.84	2.67	2.43	2.31	2.23
2.5	≤ 0.125	2.62	2.37	2.30	2.24	2.17	2.12	2.04	1.99	1.97
	0.375	4.11	3.14	2.93	2.78	2.58	2.46	2.28	2.19	2.13
	0.625	5.04	3.62	3.32	3.11	2.84	2.67	2.43	2.31	2.23
	0.875	5.48	3.85	3.51	3.27	2.96	2.77	2.50	2.36	2.28
	1.25	5.73	3.98	3.61	3.36	3.03	2.83	2.54	2.40	2.31
	≥ 1.5	5.80	4.02	3.64	3.38	3.05	2.84	2.55	2.41	2.32
3.5	≤ 0.125	2.62	2.37	2.30	2.24	2.17	2.12	2.04	1.99	1.97
	0.375	4.88	3.54	3.25	3.05	2.80	2.63	2.41	2.29	2.22
	0.625	6.34	4.30	3.87	3.58	3.20	2.97	2.64	2.48	2.38
	0.875	7.03	4.66	4.16	3.83	3.39	3.12	2.76	2.57	2.46
	1.25	7.44	4.87	4.33	3.97	3.50	3.22	2.82	2.62	2.50
	≥ 1.5	7.53	4.92	4.38	4.01	3.53	3.24	2.84	2.63	2.51
4.5	≤ 0.125	2.62	2.37	2.30	2.24	2.17	2.12	2.04	1.99	1.97
	0.375	5.80	4.02	3.64	3.38	3.05	2.84	2.55	2.41	2.32
	0.625	7.90	5.11	4.53	4.14	3.63	3.32	2.90	2.68	2.55
	0.875	8.91	5.64	4.96	4.50	3.92	3.56	3.07	2.82	2.67
	≥ 1.0	9.19	5.78	5.08	4.60	3.99	3.62	3.11	2.85	2.70
5.5	≤ 0.125	2.62	2.37	2.30	2.24	2.17	2.12	2.04	1.99	1.97
	0.375	6.87	4.58	4.10	3.77	3.35	3.09	2.73	2.55	2.44
	0.625	9.78	6.09	5.33	4.82	4.16	3.76	3.21	2.93	2.77
	0.875	11.20	6.83	5.94	5.33	4.56	4.09	3.45	3.12	2.93
	≥ 1.0	11.60	7.04	6.11	5.47	4.67	4.18	3.51	3.17	2.97
≥ 6	≤ 0.125	2.62	2.37	2.30	2.24	2.17	2.12	2.04	1.99	1.97
	0.375	7.48	4.90	4.36	3.99	3.52	3.23	2.83	2.63	2.51
	0.625	10.87	6.66	5.79	5.21	4.46	4.01	3.39	3.08	2.89
	0.875	12.54	7.54	6.51	5.81	4.94	4.40	3.67	3.30	3.08
	≥ 1.0	13.02	7.78	6.71	5.99	5.07	4.51	3.75	3.37	3.14

Table 6.7 Passenger Car Equivalents (E_T) for Heavy Vehicles (50% SUTs/50% TTs) on Specific Grades

Grade (%)	Length (mi)	Percentage of heavy vehicles								
		2	4	5	6	8	10	15	20	≥ 25
≤ 0	≤ 0.125	2.67	2.38	2.31	2.25	2.16	2.11	2.02	1.97	1.93
	0.375	2.67	2.38	2.31	2.25	2.16	2.11	2.02	1.97	1.93
	0.625	2.67	2.38	2.31	2.25	2.16	2.11	2.02	1.97	1.93
	0.875	2.67	2.38	2.31	2.25	2.16	2.11	2.02	1.97	1.93
	1.25	2.67	2.38	2.31	2.25	2.16	2.11	2.02	1.97	1.93
	≥ 1.5	2.67	2.38	2.31	2.25	2.16	2.11	2.02	1.97	1.93
2	≤ 0.125	2.67	2.38	2.31	2.25	2.16	2.11	2.02	1.97	1.93
	0.375	3.76	2.95	2.77	2.64	2.47	2.36	2.20	2.11	2.06
	0.625	4.32	3.24	3.01	2.84	2.63	2.49	2.29	2.19	2.12
	0.875	4.57	3.37	3.11	2.93	2.70	2.55	2.33	2.22	2.15
	1.25	4.71	3.45	3.17	2.99	2.74	2.58	2.36	2.24	2.17
	≥ 1.5	4.74	3.47	3.19	3.00	2.75	2.59	2.36	2.24	2.17
2.5	≤ 0.125	2.67	2.38	2.31	2.25	2.16	2.11	2.02	1.97	1.93
	0.375	4.10	3.13	2.92	2.77	2.57	2.44	2.26	2.16	2.10
	0.625	4.84	3.52	3.23	3.03	2.77	2.61	2.38	2.26	2.18
	0.875	5.17	3.69	3.37	3.15	2.87	2.69	2.43	2.30	2.22
	1.25	5.36	3.79	3.45	3.22	2.92	2.73	2.47	2.33	2.24
	≥ 1.5	5.40	3.81	3.47	3.24	2.93	2.74	2.47	2.33	2.25
3.5	≤ 0.125	2.67	2.38	2.31	2.25	2.16	2.11	2.02	1.97	1.93
	0.375	4.89	3.54	3.25	3.05	2.79	2.62	2.39	2.26	2.19
	0.625	6.05	4.15	3.75	3.47	3.11	2.89	2.58	2.42	2.32
	0.875	6.58	4.43	3.97	3.66	3.26	3.01	2.67	2.49	2.39
	1.25	6.88	4.58	4.10	3.77	3.35	3.09	2.72	2.53	2.42
	≥ 1.5	6.95	4.62	4.13	3.80	3.37	3.10	2.73	2.54	2.43
4.5	≤ 0.125	2.67	2.38	2.31	2.25	2.16	2.11	2.02	1.97	1.93
	0.375	5.83	4.03	3.65	3.39	3.05	2.84	2.55	2.39	2.30
	0.625	7.53	4.92	4.38	4.01	3.53	3.24	2.83	2.62	2.50
	0.875	8.32	5.34	4.72	4.29	3.75	3.42	2.97	2.73	2.59
	≥ 1.0	8.53	5.45	4.81	4.37	3.81	3.47	3.00	2.76	2.62
5.5	≤ 0.125	2.67	2.38	2.31	2.25	2.16	2.11	2.02	1.97	1.93
	0.375	6.97	4.63	4.14	3.81	3.38	3.11	2.74	2.55	2.43
	0.625	9.37	5.89	5.16	4.68	4.05	3.67	3.14	2.88	2.72
	0.875	10.49	6.48	5.65	5.09	4.37	3.93	3.34	3.03	2.85
	≥ 1.0	10.80	6.64	5.78	5.20	4.46	4.01	3.39	3.08	2.89
≥ 6	≤ 0.125	2.67	2.38	2.31	2.25	2.16	2.11	2.02	1.97	1.93
	0.375	7.64	4.98	4.43	4.05	3.56	3.26	2.85	2.64	2.51
	0.625	10.45	6.45	5.63	5.07	4.36	3.92	3.33	3.03	2.85
	0.875	11.78	7.16	6.20	5.56	4.74	4.24	3.56	3.22	3.01
	≥ 1.0	12.15	7.35	6.36	5.69	4.85	4.33	3.62	3.27	3.05

Table 6.8 Passenger Car Equivalents (E_T) for Heavy Vehicles (70% SUTs/30% TTs) on Specific Grades

Grade (%)	Length (mi)	Percentage of heavy vehicles								
		2	4	5	6	8	10	15	20	≥ 25
≤ 0	≤ 0.125	2.39	2.18	2.12	2.07	2.01	1.96	1.89	1.85	1.83
	0.375	2.39	2.18	2.12	2.07	2.01	1.96	1.89	1.85	1.83
	0.625	2.39	2.18	2.12	2.07	2.01	1.96	1.89	1.85	1.83
	0.875	2.39	2.18	2.12	2.07	2.01	1.96	1.89	1.85	1.83
	1.25	2.39	2.18	2.12	2.07	2.01	1.96	1.89	1.85	1.83
	≥ 1.5	2.39	2.18	2.12	2.07	2.01	1.96	1.89	1.85	1.83
2	≤ 0.125	2.67	2.32	2.23	2.17	2.08	2.03	1.94	1.89	1.86
	0.375	3.63	2.82	2.64	2.52	2.35	2.25	2.10	2.02	1.97
	0.625	4.12	3.08	2.85	2.69	2.49	2.36	2.18	2.08	2.02
	0.875	4.37	3.21	2.96	2.78	2.56	2.42	2.22	2.11	2.05
	1.25	4.53	3.29	3.02	2.84	2.60	2.45	2.24	2.13	2.07
	≥ 1.5	4.58	3.31	3.04	2.86	2.61	2.46	2.25	2.14	2.07
2.5	≤ 0.125	2.75	2.36	2.27	2.20	2.11	2.04	1.95	1.90	1.87
	0.375	4.01	3.02	2.80	2.65	2.46	2.33	2.16	2.06	2.01
	0.625	4.66	3.35	3.08	2.88	2.64	2.48	2.26	2.15	2.08
	0.875	4.99	3.52	3.21	3.00	2.73	2.56	2.32	2.19	2.12
	1.25	5.20	3.64	3.30	3.08	2.79	2.60	2.35	2.22	2.14
	≥ 1.5	5.26	3.67	3.33	3.10	2.80	2.62	2.36	2.23	2.15
3.5	≤ 0.125	2.93	2.45	2.34	2.26	2.16	2.09	1.98	1.92	1.89
	0.375	4.86	3.46	3.16	2.96	2.69	2.53	2.30	2.18	2.10
	0.625	5.88	3.99	3.59	3.32	2.98	2.76	2.46	2.31	2.22
	0.875	6.40	4.26	3.81	3.51	3.12	2.88	2.55	2.38	2.28
	1.25	6.74	4.43	3.96	3.63	3.21	2.96	2.60	2.42	2.32
	≥ 1.5	6.83	4.48	3.99	3.66	3.24	2.98	2.62	2.44	2.33
4.5	≤ 0.125	3.13	2.56	2.43	2.34	2.21	2.13	2.01	1.95	1.91
	0.375	5.88	3.99	3.59	3.32	2.98	2.76	2.46	2.31	2.22
	0.625	7.35	4.75	4.22	3.85	3.39	3.10	2.71	2.51	2.39
	0.875	8.11	5.15	4.54	4.13	3.60	3.27	2.83	2.61	2.47
	≥ 1.0	8.33	5.27	4.63	4.21	3.66	3.33	2.87	2.64	2.50
5.5	≤ 0.125	3.37	2.69	2.53	2.42	2.28	2.19	2.05	1.98	1.94
	0.375	7.09	4.62	4.11	3.76	3.31	3.04	2.66	2.47	2.36
	0.625	9.13	5.68	4.97	4.49	3.88	3.51	3.00	2.74	2.59
	0.875	10.21	6.24	5.43	4.88	4.18	3.76	3.18	2.89	2.71
	≥ 1.0	10.52	6.41	5.57	5.00	4.27	3.83	3.24	2.93	2.75
≥ 6	≤ 0.125	3.51	2.76	2.59	2.47	2.32	2.22	2.08	2.00	1.95
	0.375	7.78	4.98	4.40	4.01	3.51	3.20	2.78	2.56	2.44
	0.625	10.17	6.23	5.42	4.87	4.17	3.75	3.18	2.88	2.71
	0.875	11.43	6.88	5.95	5.32	4.53	4.04	3.39	3.06	2.86
	≥ 1.0	11.81	7.08	6.11	5.46	4.64	4.13	3.45	3.11	2.90

It should be noted that utilizing PCE values to accommodate heavy vehicles in the analysis is not appropriate for mountainous terrain situations. Mountainous terrain is considered to be any combination of horizontal and vertical alignment that causes heavy vehicles to operate at their limiting speed for significant distances or at frequent intervals. For this type of situation, a more accurate analysis procedure is provided in the HCM [Transportation Research Board 2016].

Once the appropriate equivalency factors have been obtained, the following equation is applied to arrive at the heavy-vehicle adjustment factor f_{HV}:

$$f_{HV} = \frac{1}{1 + P_T(E_T - 1)} \tag{6.9}$$

where

f_{HV} = heavy-vehicle adjustment factor,

P_T = proportion of heavy vehicles in the traffic stream, and

E_T = passenger car equivalent for heavy vehicles, from Tables 6.5–6.8.

As an example of how the heavy-vehicle adjustment factor is computed, consider a freeway with a 1.25-mi 3.5% upgrade with a traffic stream having 10% heavy vehicles (70% SUT/30% TT split). Table 6.8 must be used because the grade is too steep and long for Table 6.5 to apply. The corresponding equivalency value from Table 6.8 for this roadway is $E_T = 2.96$. Substituting this E_T value and P_T = 0.1 (for 10% heavy vehicles) into Eq. 6.5 gives f_{HV} = 0.84, or a 16% reduction in effective roadway capacity relative to the base condition of no heavy vehicles in the traffic stream.

6.4.6 Calculate Density and Determine LOS

With all the terms in the previous equations defined, these equations can now be applied to determine freeway LOS and capacity. The final step before LOS can be determined is to calculate the density of the traffic stream. The alternative notation to Eq. 6.5 is shown in Eq. 6.10, which will be used in subsequent example problems (for consistency with the *Highway Capacity Manual*).

$$D = \frac{v_p}{S} \tag{6.10}$$

where

D = density in pc/mi/ln,

v_p = flow rate in pc/h/ln, and

S = average speed in mi/h.

The average speed is found by reading it from the *y*-axis of Fig. 6.2 for the corresponding flow rate (v_p) and *FFS*. Once the density value is calculated, the LOS can be read from Table 6.1 or Fig. 6.2.

Application of the process for determining basic freeway segment capacity and LOS is demonstrated in example problems 6.1 and 6.2.

EXAMPLE 6.1 **BASIC FREEWAY SEGMENT LOS WITH GENERAL TERRAIN CLASSIFICATION**

A six-lane urban freeway (three lanes in each direction) is on level terrain with 11-ft lanes, obstructions 2 ft from the right edge of the traveled pavement, and nine ramps within three miles upstream and three miles downstream of the midpoint of the analysis segment. A directional weekday peak-hour volume of 3000 vehicles is observed, with 810 vehicles arriving in the most congested 15-min period. If the traffic stream has 10% single-unit trucks and 10% tractor-trailer trucks, determine the LOS.

SOLUTION

Determine the *FFS* according to Eq. 6.6.

$$FFS = 75.4 - f_{LW} - f_{RLC} - 3.22TRD^{\,0.84}$$

with

$f_{LW} = 1.9$ mi/h (Table 6.3),
$f_{RLC} = 1.6$ mi/h (Table 6.4), and
$TRD = \dfrac{9}{6} = 1.5$ ramps/mi

$$FFS = 75.4 - 1.9 - 1.6 - 3.22(1.5)^{0.84} = 67.4 \text{ mi/h}$$

Determine the flow rate according to Eq. 6.7:

$$v_p = \frac{V}{PHF \times N \times f_{HV}}$$

with

$PHF = \dfrac{3000}{810 \times 4} = 0.926$
$N = 3$ (given), and
$E_T = 2.0$ (level terrain, Table 6.5).

From Eq. 6.9, we obtain:

$$f_{HV} = \frac{1}{1 + 0.20(2.0 - 1)} = 0.833$$

So, from Eq. 6.7,

$$v_p = \frac{3000}{0.926 \times 3 \times 0.833} = 1296.5 \rightarrow 1297 \text{ pc/h/ln}$$

To determine the average speed (*S*), first calculate the breakpoint value for the speed–flow relationship, from Eq. 6.4:

$$BP = \left[1000 + 40(75 - 67.4)\right] = 1304 \text{ pc/h/ln}$$

Since the v_p of 1297 is less than the *BP* value of 1304, *S* will be equal to the FFS, 67.4 mi/h. In this case, the average speed is still the same as the *FFS* because the flow rate is low enough such that it is still on the linear/flat part of the speed-flow curve.

Now, density can be calculated with Eq. 6.10:

$$D = \frac{1297}{67.4} = 19.2 \text{ pc/mi/ln}$$

From Table 6.2, it can be seen that this corresponds to <u>LOS C</u> (18.0 [max density for LOS B] < 19.2 < 26.0 [max density for LOS C]). Thus, this freeway segment operates at LOS C.

This problem can also be solved graphically by applying Fig. 6.2. Using this figure, draw a vertical line up from 1297 pc/h/ln (on the figure's *x*-axis) and find that this line intersects the 67.4 mi/h *FFS* curve (interpolated between the 65 and 70 mi/h curves) in the LOS C density region (the dashed diagonal lines).

EXAMPLE 6.2   BASIC FREEWAY SEGMENT LOS WITH A SPECIFIC GRADE

Consider the freeway and traffic conditions in Example 6.1. At some point further along the roadway there is a 6% upgrade that is 1.25 mi long. All other characteristics are the same as in Example 6.1. What is the LOS of this portion of the roadway, and how many vehicles can be added before the roadway reaches capacity (assuming that the proportion of vehicle types and the peak-hour factor remain constant)?

SOLUTION

To determine the LOS of this segment of the freeway, we note that all adjustment factors are the same as those in Example 6.1 except f_{HV}, which must now be determined using an equivalency factor, E_T, drawn from the specific-grade tables (in this case Table 6.7, for 50%/50% SUT/TT split). From Table 6.7, $E_T = 3.27$, which gives (from Eq. 6.9)

$$f_{HV} = \frac{1}{1+0.20(3.27-1)} = 0.688$$

So,

$$v_p = \frac{3000}{0.926 \times 3 \times 0.688} = 1569.8 \rightarrow 1570 \text{ pc/h/ln}$$

In this case, the v_p of 1570 is more than the *BP* value of 1304 (compare to Example 6.1), so *S* will not be equal to the *FFS*. Therefore, we must proceed with applying Eq. 6.2. First, calculate the capacity from Eq. 6.3, which gives

$$c = \text{Max}\left[2200+10(67.4-50), 2400\right] = 2374 \text{ pc/h/ln}$$

Substituting the values for *FFS*, *c*, v_p, and *BP* into Eq. 6.2 yields

$$S = 67.4 - \frac{\left(67.4 - \dfrac{2374}{45}\right)(1570-1304)^{2.0}}{(2374-1304)^{2.0}} = 66.5 \text{ mi/h}$$

Now, density can be calculated with Eq. 6.10:

$$D = \frac{1570}{66.5} = 23.6 \text{ pc/mi/ln}$$

which still gives <u>LOS C</u> from Table 6.2 (18.0 [max density for LOS B] < 23.6 < 26.0 [max density for LOS C]). However, traffic conditions are still worse on the upgrade section than the level section because the density is higher.

To determine how many vehicles can be added before capacity is reached, the hourly volume at capacity must be computed. Recall that capacity corresponds to a volume-to-capacity ratio of 1.0 (the threshold between LOS E and LOS F). For a *FFS* of 67.4 mi/h, the capacity is 2374 pc/h/ln. Equation 6.7 is rearranged and used to solve for the hourly volume based upon this capacity:

$$v_p = \frac{V}{PHF \times N \times f_{HV}} \Rightarrow 2374 = \frac{V}{0.926 \times 3 \times 0.688}$$

which gives $V = 4537$ veh/h. This means that about <u>1537 vehicles</u> (4537 − 3000) can be added during the peak hour before capacity is reached. It should be noted that the assumption that the peak-hour factor will remain constant as the roadway approaches capacity is not very realistic. In practice, it is observed that as a roadway approaches capacity, *PHF* gets closer to 1. This implies that the flow rate over the peak hour becomes more uniform.

6.5   MULTILANE HIGHWAY SEGMENTS

Multilane highways are similar to freeways in most respects, except for a few key differences:

- Vehicles may enter or leave the roadway at at-grade intersections and driveways (multilane highways do not have full access control).
- Multilane highways may or may not be divided (by a barrier or median separating opposing directions of flow), whereas freeways are always divided.
- Traffic signals may be present.
- Design standards (such as design speeds) are sometimes lower than those for freeways.
- The visual setting and development along multilane highways are usually more distracting to drivers than in the freeway case.

Multilane highways usually have four or six lanes (both directions), have posted speed limits between 40 and 65 mi/h, and can have physical medians, medians that are two-way left-turn lanes (TWLTLs), or opposing directional volumes that may not be divided by a median at all. Two examples of multilane highways are shown in Fig. 6.3.

The determination of LOS on multilane highways closely mirrors the procedure for freeways. The main differences lie in the speed–flow relationship and some of the *FFS* adjustment factors and their values. The procedure we present is valid only for sections of highway that are not significantly influenced by large queue formations and dissipations resulting from traffic signals (this is generally taken as having traffic signals spaced 2.0 mi apart or more), do not have significant on-street parking, do not have bus stops with high usage, and do not have significant pedestrian activity.

Figure 6.3 Examples of multilane highways.

6.5.1   Speed versus Flow Rate Relationship

Just as for freeways, the speed–flow relationship is central to the analysis of multilane highways. However, unlike freeways, the breakpoint (the transition from a linear to curvilinear flow rate) is not a function of *FFS* but rather is constant at 1400 pc/h/ln. With this, the HCM [Transportation Research Board 2016] defines the relationship between average speed and flow rate as follows:

$$S = FFS \qquad\qquad v_p \le 1400 \qquad \textbf{(6.11)}$$

and

$$S = FFS - \frac{\left(FFS - \dfrac{c}{45}\right)\left(v_p - 1400\right)^{1.31}}{\left(c - 1400\right)^{1.31}} \qquad\qquad 1400 < v_p \le c \qquad \textbf{(6.12)}$$

where

c = Capacity in pc/h/ln,
v_p = 15-min passenger car equivalent flow rate in pc/h/ln,
45 = Density at capacity in pc/mi/ln, as shown in Table 6.10, and
1400 = Linear to curvilinear flow rate breakpoint value in pc/h/ln.

Other terms are as defined previously.

Capacity is given by the following equation:

$$c = \text{Min}\left[1900 + 20\left(FFS - 45\right), 2300\right] \qquad\qquad 45 \le FFS \le 70 \qquad \textbf{(6.13)}$$

Results from this equation for 5-mi/h increments across the valid range of *FFS* are shown in Table 6.9. The speed-flow rate curves resulting from the application of Eqs. 6.11 and 6.12, for 5-mi/h increments across the valid range of *FFS* are shown in Fig. 6.4. Table 6.10 provides the LOS criteria corresponding to traffic density, speed, volume-to-capacity ratio, and maximum flow rate.

Table 6.9 Relationship between Free-Flow Speed and Capacity on Multilane Highway Segments

Source: Transportation Research Board, *Highway Capacity Manual*, 6th Edition, National Academy of Sciences, Washington, D.C. 2016.

Free-flow speed (mi/h)	Capacity (pc/h/ln)
70	2300
65	2300
60	2200
55	2100
50	2000
45	1900

Table 6.10 LOS Criteria for Multilane Highway Segments

Criterion	LOS				
	A	B	C	D	E
FFS = 70 mi/h					
Maximum density (pc/mi/ln)	11	18	26	35	45
Average speed (mi/h)	70.0	70.0	65.4	58.1	51.1
Maximum *v/c*	0.33	0.55	0.74	0.88	1.00
Maximum flow rate (pc/h/ln)	770	1260	1700	2035	2300
FFS = 65 mi/h					
Maximum density (pc/mi/ln)	11	18	26	35	45
Average speed (mi/h)	65.0	65.0	62.7	57.0	51.1
Maximum *v/c*	0.31	0.51	0.71	0.87	1.00
Maximum flow rate (pc/h/ln)	715	1170	1630	1995	2300
FFS = 60 mi/h					
Maximum density (pc/mi/ln)	11	18	26	35	45
Average speed (mi/h)	60.0	60.0	59.0	54.1	48.9
Maximum *v/c*	0.30	0.49	0.70	0.86	1.00
Maximum flow rate (pc/h/ln)	660	1080	1535	1895	2200
FFS = 55 mi/h					
Maximum density (pc/mi/ln)	11	18	26	35	45
Average speed (mi/h)	55.0	55.0	54.8	51.1	46.7
Maximum *v/c*	0.29	0.47	0.68	0.85	1.00
Maximum flow rate (pc/h/ln)	605	990	1425	1790	2100
FFS = 50 mi/h					
Maximum density (pc/mi/ln)	11	18	26	35	45
Average speed (mi/h)	50.0	50.0	50.0	48.0	44.4
Maximum *v/c*	0.28	0.45	0.65	0.84	1.00
Maximum flow rate (pc/h/ln)	550	900	1300	1680	2000
FFS = 45 mi/h					
Maximum density (pc/mi/ln)	11	18	26	35	45
Average speed (mi/h)	45.0	45.0	45.0	44.4	42.2
Maximum *v/c*	0.26	0.43	0.62	0.82	1.00
Maximum flow rate (pc/h/ln)	495	810	1170	1555	1900

Note: Maximum flow rate values are rounded to the nearest 5 passenger cars.

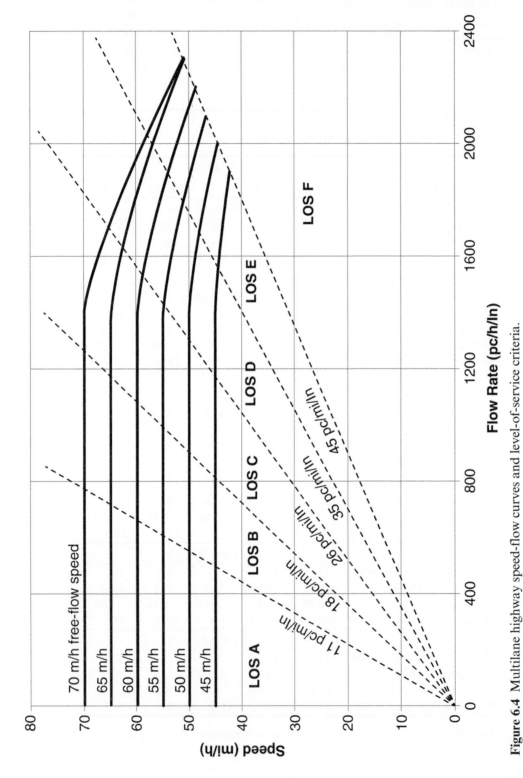

Figure 6.4 Multilane highway speed-flow curves and level-of-service criteria.
Reproduced with permission of the Transportation Research Board, *Highway Capacity Manual*, 6th Edition, Copyright, National Academy of Sciences, Washington, D.C. 2016. Exhibit 12-8, p. 12–11.

6.5.2   Base Conditions and Capacity

The base conditions for multilane highways are defined as [Transportation Research Board 2016]

- 12-ft minimum lane widths
- 12-ft minimum total lateral clearance from roadside objects (right shoulder and median) in the travel direction
- Only passenger cars in the traffic stream
- No direct access points along the roadway
- Divided highway
- Level terrain (no grades greater than 2%)
- Free-flow speed of 60 mi/h or more

As was the case with the freeway LOS analysis, adjustments will have to be made when nonbase conditions are encountered.

The capacity, c, for multilane highway segments, in pc/h/ln, is given in Table 6.9. From Table 6.10, note again that these capacity values correspond to the maximum service flow rate at LOS E and a v/c of 1.0.

6.5.3   Service Measure

Due to the large degree of similarity between multilane highway and freeway facilities, density is also the service measure (performance measure used for determining LOS) for multilane highways. As can be seen in Table 6.10, the LOS density thresholds for multilane highways are the same as for freeways.

6.5.4   Determine Free-Flow Speed

FFS for multilane highways is the mean speed of passenger cars operating in flow rates up to 1400 passenger cars per hour per lane (pc/h/ln). If FFS is to be estimated rather than measured, the following equation can be used, which takes into account the roadway characteristics of lane width, lateral clearance, presence (or lack) of a median, and access frequency:

$$FFS = BFFS - f_{LW} - f_{TLC} - f_M - f_A \qquad (6.14)$$

where

$$
\begin{aligned}
FFS &= \text{estimated free-flow speed in mi/h,} \\
BFFS &= \text{estimated free-flow speed, in mi/h, for base conditions,} \\
f_{LW} &= \text{adjustment for lane width in mi/h,} \\
f_{LC} &= \text{adjustment for lateral clearance in mi/h,} \\
f_M &= \text{adjustment for median type in mi/h, and} \\
f_A &= \text{adjustment for the number of access points along the roadway in mi/h.}
\end{aligned}
$$

As can be seen, this equation closely resembles Eq. 6.6 in the freeway section. Both include adjustments for lane width and lateral clearance, and the access frequency adjustment is similar to the ramp density adjustment. The main difference is that Eq. 6.14 also includes an adjustment for median type. The presence of a physical barrier or wide separation between opposing flows (such as a TWLTL) will lead to higher *FFS* than if there is no separation or physical barrier between opposing flows. This adjustment is not included for freeways since, by definition, all freeways are divided.

As for *BFFS*, many factors can influence the *FFS*, with the posted speed limit often being a significant one. For multilane highways, research has found that *FFS*, under base conditions, are about 7 mi/h higher than the speed limit for 40- and 45-mi/h posted-speed-limit roadways, and about 5 mi/h higher for 50-mi/h and higher posted-speed-limit roadways. The following sections describe the procedures for estimating the adjustment factor values.

Lane Width Adjustment

The same lane width adjustment factor values are used for multilane highways as are used for freeways. Thus, Table 6.3 should be used for multilane highways as well.

Lateral Clearance Adjustment

The adjustment factor for potentially restrictive lateral clearances (f_{TLC}) is determined first by computing the total lateral clearance, which is defined as

$$TLC = LC_R + LC_L \qquad (6.15)$$

where

TLC = total lateral clearance in ft,

LC_R = lateral clearance on the right side of the travel lanes to obstructions (retaining walls, utility poles, signs, trees, etc.), and

LC_L = lateral clearance on the left side of the travel lanes to obstructions.

For undivided highways, there is no adjustment for left-side lateral clearance because this is already taken into account in the f_M term (thus LC_L = 6 ft in Eq. 6.8). If an individual lateral clearance (either left or right side) exceeds 6 ft, 6 ft is used in Eq. 6.15. Finally, highways with TWLTLs are considered to have LC_L equal to 6 ft. Once Eq. 6.15 is applied, the value for f_{TLC} can be determined directly from Table 6.11.

Table 6.11 Adjustment for Lateral Clearance

*Total lateral clearance is the sum of the lateral clearances of the median (if greater than 6 ft, use 6 ft) and shoulder (if greater than 6 ft, use 6 ft). Therefore, for purposes of analysis, total lateral clearance cannot exceed 12 ft.

Reproduced with permission of the Transportation Research Board, *Highway Capacity Manual*, 6th Edition, Copyright, National Academy of Sciences, Washington, D.C. 2016. Exhibit 12-22, p. 12-30.

Total lateral clearance* (ft)	Reduction in free-flow speed, f_{TLC} (mi/h)	
	Four-lane highways	Six-lane highways
12	0.0	0.0
10	0.4	0.4
8	0.9	0.9
6	1.3	1.3
4	1.8	1.7
2	3.6	2.8
0	5.4	3.9

Median Adjustment

Values for the adjustment factor for median type, f_M, are provided in Table 6.12. This table shows that undivided highways have a *FFS* that is 1.6 mi/h lower than divided highways (which include those with two-way left-turn lanes).

Table 6.12 Adjustment for Median Type

Reproduced with permission of the Transportation Research Board, *Highway Capacity Manual*, 6th Edition, Copyright, National Academy of Sciences, Washington, D.C. 2016. Exhibit 12-23, p. 12-30.

Median type	Reduction in free-flow speed, f_M (mi/h)
Undivided highways	1.6
Divided highways (including TWLTLs)	0.0

Access Frequency Adjustment

The final adjustment factor in Eq. 6.14 is for the number of access points per mile, f_A. Access points are defined to include intersections and driveways (on the right side of the highway in the direction being considered) that significantly influence traffic flow, and thus do not generally include driveways to individual residences or service driveways at commercial sites. Adjustment values for access point frequency are provided in Table 6.13.

Table 6.13 Adjustment for Access-Point Frequency

Reproduced with permission of the Transportation Research Board, *Highway Capacity Manual*, 6th Edition, Copyright, National Academy of Sciences, Washington, D.C. 2016. Exhibit 12-24, p. 12-31.

Access points/ mile	Reduction in free-flow speed f_A (mi/h)
0	0.0
10	2.5
20	5.0
30	7.5
≥ 40	10.0

6.5.5   Determine Analysis Flow Rate

The analysis flow rate for multilane highways is determined in the same manner as for freeways, using Eq. 6.7 and the remainder of the procedure outlined in Section 6.4.5.

6.5.6   Calculate Density and Determine LOS

The procedure for calculating density and determining LOS for multilane highways is essentially the same as for freeways (see Section 6.4.6). Equation 6.10 is applied to arrive at a density. However, different speed-flow curves are used for multilane highways. Once the density value is calculated, the LOS can be determined from Table 6.10 or Fig. 6.4.

EXAMPLE 6.3   MULTILANE HIGHWAY FREE-FLOW SPEED

A four-lane undivided highway (two lanes in each direction) has 11-ft lanes, with 4-ft shoulders on the right side. There are seven access points per mile, and the posted speed limit is 50 mi/h. What is the estimated *FFS*?

SOLUTION

This problem can be solved by direct application of Eq. 6.14 to arrive at an estimated *FFS*:

$$FFS = BFFS - f_{LW} - f_{TLC} - f_M - f_A$$

with

$$
\begin{aligned}
BFFS &= 55 \text{ mi/h (assume } FFS = \text{posted speed} + 5 \text{ mi/h)}, \\
f_{LW} &= 1.9 \text{ mi/h (Table 6.3)}, \\
f_{TLC} &= 0.4 \text{ mi/h (Table 6.11, with } TLC = 4 + 6 = 10 \text{ from Eq. 6.15, with } LC_L = 6 \\
&\quad \text{ft because the highway is undivided)}, \\
f_M &= 1.6 \text{ mi/h (Table 6.12), and} \\
f_A &= 1.75 \text{ mi/h (Table 6.13, by interpolation)}.
\end{aligned}
$$

Substitution gives

$$FFS = 55 - 1.9 - 0.4 - 1.6 - 1.75 = \underline{\underline{49.35 \text{ mi/h}}}$$

which means that the more restrictive roadway characteristics relative to the base conditions result in a reduction in *FFS* of 5.65 mi/h.

EXAMPLE 6.4   MULTILANE HIGHWAY LOS

A six-lane divided highway (three lanes in each direction) is on rolling terrain with two access points per mile and has 10-ft lanes, with a 5-ft shoulder on the right side and a 3-ft shoulder on the left side. The peak-hour factor is 0.85, and the directional peak-hour volume is 3400 vehicles per hour. There are 7% single-unit trucks and 3% tractor-trailer trucks in the traffic stream. No speed studies are available, but the posted speed limit is 55 mi/h. Determine the LOS.

SOLUTION

We begin by determining *FFS* by applying Eq. 6.14:

$$FFS = BFFS - f_{LW} - f_{TLC} - f_M - f_A$$

with

$$
\begin{aligned}
BFFS &= 60 \text{ mi/h (assume } FFS = \text{posted speed} + 5 \text{ mi/h)}, \\
f_{LW} &= 6.6 \text{ mi/h (Table 6.3)}, \\
f_{TLC} &= 0.9 \text{ mi/h (Table 6.11, with TLC} = 5 + 3 = 8 \text{ from Eq. 6.15)}, \\
f_M &= 0.0 \text{ mi/h (Table 6.12), and} \\
f_A &= 0.5 \text{ mi/h (Table 6.13, by interpolation)}.
\end{aligned}
$$

Substitution gives

$$FFS = 60.0 - 6.6 - 0.9 - 0.0 - 0.5 = 52.0 \ \text{mi/h}$$

Next we determine the analysis flow rate using Eq. 6.7:

$$v_p = \frac{V}{PHF \times N \times f_{HV}}$$

with

$$V = 3400 \ \text{veh/h (given)},$$
$$PHF = 0.85 \ \text{(given)},$$
$$N = 3 \ \text{(given), and}$$
$$E_T = 3.0 \ \text{(Table 6.5)}.$$

From Eq. 6.9, we find

$$f_{HV} = \frac{1}{1 + 0.10(3.0 - 1)} = 0.833$$

Substitution gives

$$v_p = \frac{3400}{0.85 \times 3 \times 0.833} = 1600.6 \rightarrow 1601 \ \text{pc/h/ln}$$

To determine the average speed (S), first check the v_p value against the BP value. Since 1601 > 1400, S will not be equal to the FFS. Therefore, we must proceed with applying Eq. 6.12. First, calculate the capacity from Eq. 6.13, which gives

$$c = \text{Max}\left[1900 + 20(52 - 45), 2300\right] = 2040 \ \text{pc/h/ln}$$

Substituting the values for FFS, c, and v_p into Eq. 6.12 yields

$$S = 52 - \frac{\left(52 - \dfrac{2040}{45}\right)(1601 - 1400)^{1.31}}{(1601 - 1400)^{1.31}} = 50.5 \ \text{mi/h}$$

Now, density can be calculated with Eq. 6.10:

$$D = \frac{1601}{50.5} = 31.7 \ \text{pc/mi/ln}$$

which gives <u>LOS D</u> from Table 6.10 [26.0 (max density for LOS C) < 31.7 < 35.0 (max density for LOS D)]. This can also be confirmed from Fig. 6.4, for FFS = 52 mi/h (interpolated), where the 1601-pc/h/ln flow rate intersects this curve in the LOS D density region

EXAMPLE 6.5   MULTILANE HIGHWAY CAPACITY

A local manufacturer wishes to open a factory near the segment of highway described in Example 6.4. How many heavy vehicles can be added to the peak-hour directional volume before capacity is reached? (Add only heavy vehicles and assume that the *PHF* remains constant.)

SOLUTION

Note that *FFS* will remain unchanged at 52 mi/h. The capacity for *FFS* = 52 mi/h, as calculated in Example 6.4 is 2040 pc/h/ln. The current number of heavy vehicles in the peak-hour traffic stream is 340 (0.10 × 3400). Let us denote the number of new heavy vehicles added as V_{nt}, so the combination of Eqs. 6.7 and 6.9 gives

$$v_p = \frac{V + V_{nt}}{(PHF)(N)\left[\dfrac{1}{1+\left(\dfrac{340+V_{nt}}{V+V_{nt}}\right)(E_T-1)}\right]}$$

with

$$
\begin{aligned}
v_p &= 2040 \text{ pc/h/ln (capacity)}, \\
V &= 3400 \text{ veh/h (Example 6.5)}, \\
PHF &= 0.85 \text{ (Example 6.4)}, \\
N &= 3 \text{ (Example 6.4), and} \\
E_T &= 3.0 \text{ (Example 6.4)}.
\end{aligned}
$$

$$2040 = \frac{3400 + V_{nt}}{(0.85)(3)\left[\dfrac{1}{1+\left(\dfrac{340+V_{nt}}{3400+V_{nt}}\right)(3.0-1)}\right]}$$

which gives $V_{nt} = 374$, which is the number of heavy vehicles that can be added to the peak-hour volume before capacity is reached.

6.6   TWO-LANE HIGHWAYS

Two-lane highways are defined as roadways with one lane available in each direction. For level of service (LOS) determination, a key distinction between two-lane highways and the freeways and multilane highways previously discussed is that traffic in both directions must now be considered (previously we considered traffic in one direction only). This is because traffic in an opposing direction has a strong influence on LOS. For example, a high opposing traffic volume limits the opportunity to pass slow-moving vehicles (because such a pass requires the passing vehicle to occupy the opposing lane) and thus forces a lower traffic speed—and, as a consequence, a lower LOS. It also follows that any geometric features that restrict passing sight distance (such as sight distance on horizontal and vertical curves) will have an adverse impact on the LOS. Finally, the type of terrain (grades, horizontal curves) plays a more critical role in LOS calculations, relative to freeways and multilane highways, because of the sometimes limited ability to pass slower-moving vehicles on grades in areas where passing is prohibited due to sight distance restrictions or where opposing traffic does not permit safe passing. Some examples of two-lane highways are shown in Fig. 6.5.

(a)

(b)

(c)

(d)

Figure 6.5 Examples of two-lane highways.

This section provides an overview of the analysis procedure for two-lane highways in the *Highway Capacity Manual* [Transportation Research Board 2016].

6.6.1   Analysis Concepts

The analysis methodology considers several roadway and traffic concepts, described as follows.

Segmentation

A two-lane highway is divided into segments for analysis purposes. The ability to pass, lane geometry, grades, lane and shoulder widths, posted speed limits, traffic demands, adjacent land uses, driveways, and other characteristics of the facility should be homogeneous within each analysis segment. Note that segmentation will be different for each direction of the highway because passing zones and other characteristics will start and end in different locations depending on the direction of travel. The results of each segment analysis are then combined into results for the full highway.

The analysis methodology can be used to analyze individual homogeneous segments of a two-lane highway, as well as extended lengths of a two-lane highway that are composed of multiple contiguous segments. The methodology applies to the following segment types:

- *Passing Constrained*: A segment in which passing in the oncoming lane is either prohibited or is effectively negligible due to geometric or sight distance limitations. Traffic operations in a passing constrained segment are a function of analysis direction flow rate, percent heavy vehicles, horizontal and vertical alignment, and segment length (see Figs. 6.5a and 6.5b).

- *Passing Zone*: A segment where passing in the oncoming lane is not restricted. However, to be effective in accommodating passing maneuvers, these zones must provide a minimum length of useable passing distance. The effectiveness of a passing zone in improving traffic operations is a function of analysis and opposing direction flow rates, percent heavy vehicles, horizontal and vertical alignment, and segment length (see Fig. 6.5c, direction with centerline skip-striping).

- *Passing/Climbing Lane*: A segment in which a relatively short length of additional lane is provided in the same travel direction. A passing lane can be effective in dispersing platoons by providing an opportunity for faster vehicles to pass slower vehicles. Passing lanes on significant upgrades (where the speed of large trucks is reduced well below their desired speed) are generally referred to as *climbing lanes*. The effectiveness of a passing or climbing lane in improving traffic operations is a function of analysis direction flow rate, percent heavy vehicles, horizontal and vertical alignment, and segment length (see Fig. 6.5d).

Access Points

Access points are major driveways and side roads where significant traffic enters and/or leaves the two-lane highway within an analysis segment. Access points lower the free-flow speed (*FFS*) for the segment. By lowering the *FFS*, access points will also indirectly affect the average speed and the follower density. Residential driveways and other low-volume driveways and side roads (generally with average daily traffic below 20 vehicles per day) are generally not counted as access points.

Treatment of Terrain and Heavy Vehicles

Unlike other facility analysis methodologies in the HCM, the two-lane highway analysis procedure works directly with traffic stream units of vehicles per hour, rather than passenger cars per hour. That is, passenger car equivalent (PCE) values are not utilized. This approach provides the ability to more accurately represent the performance measure relationships as a function of the more varying horizontal and vertical alignment often encountered on two-lane highways. Additionally, it simplifies the analysis methodology by removing the process of converting units from veh/h to pc/h. The two-lane highway analysis methodology therefore includes the percentage of heavy vehicles as an input to the performance measure estimation models.

In general, a heavy vehicle is defined as any vehicle (or vehicle–trailer unit) with more than four wheels on the ground during normal operation. Heavy vehicles generally consist of large trucks, buses, and recreational vehicles (RVs).

Vertical alignment is accommodated through a classification scheme providing five different levels. Each classification is a function of the segment length and up- or downgrade percentage. The classification levels are based on reductions in heavy-vehicle *FFS*. On an upgrade, the speed reduction is a function of the effect of grade resistance limiting a vehicle's acceleration capability. On a downgrade, the speed reduction is a result of trucks downshifting to avoid the potential of a "runaway" acceleration condition. Vertical Alignment Class 1 corresponds to a minimal reduction in *FFS*, while Vertical Alignment Class 5 corresponds to a crawl-speed condition.

Horizontal alignment is also accommodated through a classification scheme that considers reductions in heavy-vehicle *FFS*. The horizontal classifications are a function of curve radius and superelevation.

Applicable Performance Measures

Three performance measures underpin this analysis methodology: average speed, percent followers, and follower density. Typical agency performance objectives for two-lane highways focus on reduced travel times and reduced pressure to pass (reduced percent and density of followers). The three performance measures were selected to represent the quality of service provided by two-lane highways because they most closely conform to typical agency performance objectives. They also have the advantage of being sensitive to the design and demand features typically considered when planning and designing two-lane highways.

Average Speed Average speed is a key performance measure for two-lane highways. The general relationship between average speed and flow rate is illustrated in Fig. 6.6.

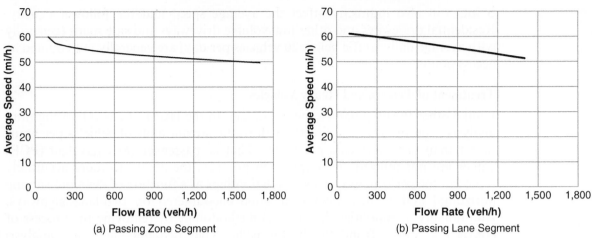

Figure 6.6 Average speed versus directional flow rate.

Percent Followers The measure *percent followers* represents the freedom to maneuver and the comfort and convenience of travel. It can also serve as a proxy for the need to provide passing opportunities. Percent followers is the percentage of vehicles passing a given point on the roadway that are considered to be in a follower state. Being in a follower state is used as a surrogate to indicate a driver's perception that they are being delayed by a slower driver. Higher values are generally indicative of restrictions in passing (very infrequent passing zones or passing lanes, high volumes in the opposing direction).

A critical headway threshold value of 2.5 s is used to determine whether a vehicle is in a follower state. Thus, any vehicle following another vehicle at a headway of 2.5 s or less is considered to be in a follower state and percent followers is simply the percentage of all vehicles with a headway of 2.5 s or less at a given roadway location. The general relationship between percent followers and flow rate is illustrated in Fig 6.7.

Note that even at a capacity flow rate, 100% followers will not be reached. This is because there will always be some gaps between platoons within the traffic stream as a result of different platoon lead vehicle desired speeds.

Follower Density Follower density is defined as the number of vehicles in a follower state per mile per lane. Mathematically, it is the percent followers multiplied by density. This measure is used as the service measure for two-lane highways because of its sensitivity to both traffic demand and geometric alignment variability. On two-lane highways, it is possible for two roadways to have similar values of density but very different levels of percent followers. Likewise, it is possible for two roadways to have similar percent followers but very different values of density. Therefore, service quality is better represented by considering the combination of follower percentage and density. The general relationship between follower density and flow rate is illustrated in Fig. 6.8.

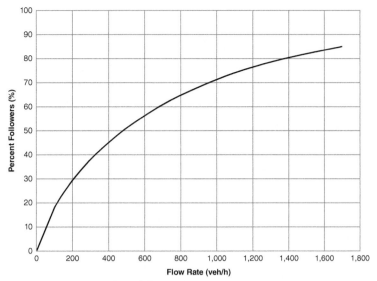

Figure 6.7 Percent followers versus directional flow rate.

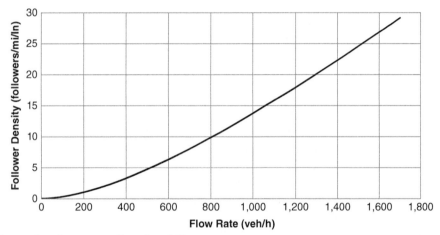

Figure 6.8 Follower density versus directional flow rate.

Analysis Approach The operational performance of a segment, either individually or within a facility, is reported for the end of the segment, rather than corresponding to an aggregated value across the full length of the segment. Thus, the reported values are point estimates of performance and are not necessarily representative of average conditions along an extended length of segment. This approach makes it easier for practitioners to use field-measured point values to directly calculate LOS with the method and to use point values to validate method outputs. Nevertheless, when considering multiple contiguous segments as a facility, the point measure is used as the estimate of the performance of a given segment. From a traveler's perspective, the conditions at the end of a segment (particularly passing zones) probably factor more heavily into their assessment of the quality of service. For passing lane segments, additional calculations are performed, which is described briefly at the end of this section.

Individual segments can be analyzed with this methodology. Additionally, multiple contiguous two-lane highway segments (in the same direction) may be combined to analyze a longer section (with varying characteristics) as a facility.

While this methodology provides estimation equations for all of the key performance measures, it is also possible to assess LOS through direct field measurement of speed, flow rate, and percent followers at a specific point.

Base Conditions and Capacity The base conditions for two-lane highways are defined as:

- 12-ft minimum lane widths
- 6-ft minimum paved shoulder widths
- No restrictions on passing in the oncoming lane
- Only passenger cars in the traffic stream
- No direct access points along the roadway
- No impediments to through traffic due to traffic control or turning vehicles
- Level terrain and straight alignment

Flow rates equivalent to capacity are rarely observed on two-lane highways, except possibly in short segments. Because service quality deteriorates at relatively low demand flow rates, most two-lane highways are upgraded before demand approaches capacity. Thus, the consideration of a capacity value for a two-lane highway operations analysis is rarely a concern. Nevertheless, being able to estimate capacity can be important for some planning situations, such as evacuation planning, special event planning, recreational routes, and evaluating the downstream impacts of incident bottlenecks once cleared.

The capacity of nonpassing lane segments, under base conditions, is 1,700 veh/h/ln. A distinction between the capacities of a Passing Zone and a Passing Constrained segment is not made, as the empirical evidence is currently too limited in this regard.

Passing lane segments can significantly improve traffic flow, especially on steep and long grades. However, the merging behavior of vehicles at the end of the passing lane can become problematic at high flow rates. Higher flow rates reduce the average gap between vehicles, which forces drivers to merge into smaller gaps. This behavior creates shockwaves, as following vehicles must decelerate for the merging vehicles. At some point, breakdown is reached, and the performance of the passing lane degrades below that of a nonpassing lane segment. Maximum flow rates for passing lane segments are a function of the heavy vehicle percentage and the vertical classification. For example, the capacity flow rate at the merge point on a vertical class 5 passing lane segment with 25% or higher trucks is approximately 1100 veh/h.

Level of Service If the demand-to-capacity ratio is less than or equal to 1.0, follower density is used as the service measure for all two-lane highways. However, two sets of LOS thresholds are used to account for differences in driver perception between driving on higher-speed versus lower-speed highways.

On higher-speed two-lane highways ($\geq$ 50 mi/h), absolute speed and delay due to passing restrictions are generally both important to motorists. Higher-speed two-lane highways are most commonly encountered as inter-city connecting routes. Lower-speed two-lane highways (< 50 mi/h) are typically encountered as intra-city routes and in scenic and rural-developed areas. These highways generally have posted speed limits of 35–45 mi/h and have limited passing opportunities. Thus, for two-lane highways in these areas, high speeds are usually not expected and higher percentages of followers are generally tolerated. Consequently, the follower density thresholds for a given LOS are higher for lower-speed highways than higher-speed highways. The LOS follower density thresholds for high-speed two-lane highways range from 2.0 followers/mi (LOS A/B) to 12.0 followers/mi (LOS E/F). For lower-speed two-lane highways, the thresholds range from 2.5 followers/mi (LOS A/B) to 15.0 followers/mi (LOS E/F).

At LOS A, motorists experience operating speeds near the posted speed limit and little difficulty in passing. Platooning is minimal, and follower density is very low. At LOS E, speeds may still be reasonable, but platooning is significant and follower density is high. Passing, if allowed, is essentially impossible. Conditions for LOS B, C, and D represent gradations between the conditions for LOS A and E. LOS F exists whenever demand flow exceeds the capacity of the segment.

Performance Measure Estimation A summary of the models used to estimate the performance measures used in this analysis methodology is provided in this section.   The reader is referred to the HCM for the details on supporting equations, determination of equation coefficient values, and the specific sequence of steps in the analysis procedure.

Average Speed

Average speed is computed from Eq. 6.16:

$$S = \begin{cases} FFS & v_d \leq 100 \\ FFS - m\left(\dfrac{v_d}{1{,}000} - 0.1\right)^p & v_d > 100 \end{cases} \tag{6.16}$$

where

S = average speed in the analysis direction (mi/h),

FFS = free-flow speed (mi/h),

v_d = flow rate in the analysis direction (veh/h),

m = slope coefficient, and

p = power coefficient.

The slope (m) and (p) coefficients are calculated from supporting equations, which include one or more of the following independent variables: analysis direction flow rate, opposing direction flow rate (for passing zone segments), FFS, percent heavy vehicles, segment length, and coefficient values as a function of vertical class (1-5). If horizontal curves are present within the segment alignment, the average speed will be reduced through a supplementary calculation process.

The formulation of Eq. 6.16 allows for the two different shapes of curves as shown in Fig. 6.6—one for passing constrained and passing zone segments and one for passing lane segments. The shape of the speed versus flow rate for passing lane segments (Fig. 6.6b) has a shape more consistent with that for freeways and multilane highways, as expected, since passing lane segments have two lanes and allow for passing without entering the opposing lane.

Percent Followers

The percent followers is computed from Eq. 6.17:

$$PF = 100 \times \left[1 - e^{\left(m \times \left\{ \frac{v_d}{1,000} \right\}^p \right)} \right] \tag{6.17}$$

where

PF = percent followers in the analysis direction (%),

v_d = flow rate in the analysis direction (veh/h),

m = slope coefficient, and

p = power coefficient.

The slope (m) and (p) coefficients are calculated in a similar manner as for average speed.

Follower Density

Follower density is computed from Eq. 6.18:

$$FD = \frac{PF}{100} \times \frac{v_d}{S} \tag{6.18}$$

where

FD = follower density in the analysis direction (followers/mi/ln), and

other terms are as defined previously.

EXAMPLE 6.6   TWO-LANE HIGHWAY PASSING CONSTRAINED SEGMENT

A segment of two-lane highway has the following roadway characteristics:

- Passing constrained in the analysis direction
- Segment Length (L) = 3,960 ft
- Level Terrain (Vertical Class 1)
- FFS = 62.7 mi/h (based on posted speed limit of 55 mi/h)

For these conditions, determine the follower density for an analysis direction flow rate (v_d) of 800 veh/h (this accounts for the measured hourly volume and the PHF) that consists of only passenger cars ($HV\%$ = 0).

SOLUTION

Applying the full analysis methodology of the *Highway Capacity Manual* for these roadway characteristics and no heavy vehicles yields a speed–flow relationship as shown in Fig. 6.9. For an analysis direction flow rate of 800 veh/h, the resulting average speed (S) is 59.3 mi/h.

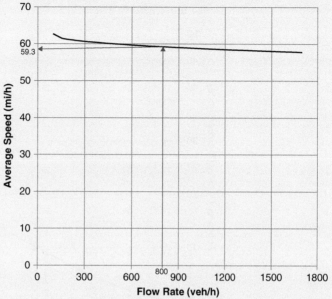

Figure 6.9 Average speed versus directional flow rate for Example 6.6.

Applying the full analysis methodology of the *Highway Capacity Manual* for these roadway characteristics and no heavy vehicles yields a percent followers-flow relationship as shown in Fig. 6.10. For an analysis direction flow rate of 800 veh/h, the resulting percent followers (PF) is 66.2%.

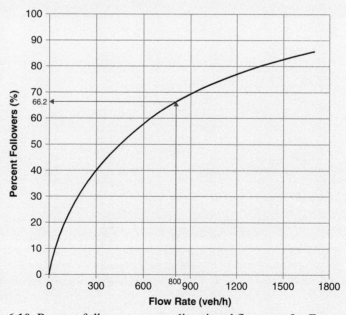

Figure 6.10 Percent followers versus directional flow rate for Example 6.6.

Substituting the average speed and percent followers values into Eq. 6.18 gives a follower density of:

$$FD = \frac{66.2}{100} \times \frac{800}{59.3} = 8.9 \text{ followers/mi}$$

This result is also illustrated in Fig. 6.11.

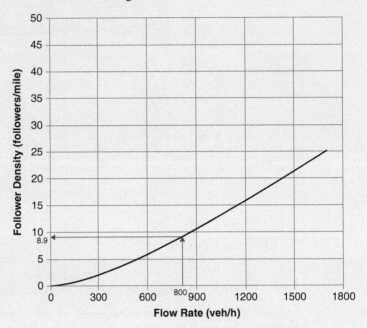

Figure 6.11 Percent follower density versus directional flow rate for Example 6.6.

EXAMPLE 6.7 TWO-LANE HIGHWAY PASSING ZONE SEGMENT

A segment of two-lane highway has the following roadway characteristics:

- Passing zone in the analysis direction
- Segment Length (L) = 5,280 ft
- Vertical Class 2
- FFS = 62.7 mi/h (based on posted speed limit of 55 mi/h)

For these conditions, determine the follower density for an analysis direction flow rate (v_d) of 700 veh/h, opposing direction flow rate (v_o) of 500 veh/h, and 8% heavy vehicles in the analysis direction.

SOLUTION

Applying the full analysis methodology of the *Highway Capacity Manual* for these roadway characteristics, opposing direction traffic flow rate of 500 veh/h, and 8% heavy vehicles in the analysis direction yields a speed–flow relationship as shown in Fig. 6.12. For an analysis direction flow rate of 700 veh/h, the resulting average speed (S) is 58.5 mi/h.

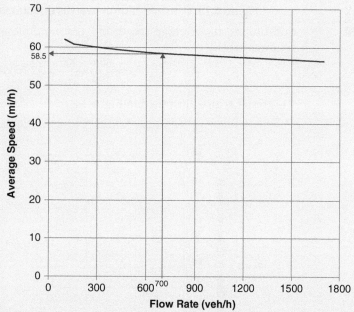

Figure 6.12 Average speed versus directional flow rate for Example 6.7.

Applying the full analysis methodology of the *Highway Capacity Manual* for these roadway characteristics, opposing direction traffic flow rate of 500 veh/h, and 8% heavy vehicles in the analysis direction yields a speed–flow relationship as shown in Fig. 6.13. For an analysis direction flow rate of 700 veh/h, the resulting percent followers (PF) is 60.4%.

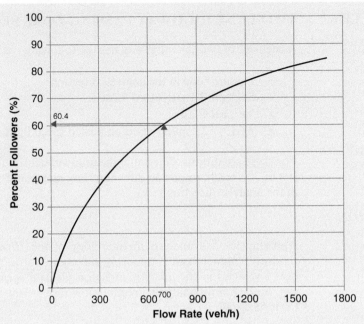

Figure 6.13 Percent followers versus directional flow rate for Example 6.7.

Substituting the average speed and percent followers values into Eq. 6.18 gives a follower density of:

$$FD = \frac{60.4}{100} \times \frac{700}{58.5} = 7.2 \text{ followers/mi}$$

This result is also illustrated in Fig. 6.14.

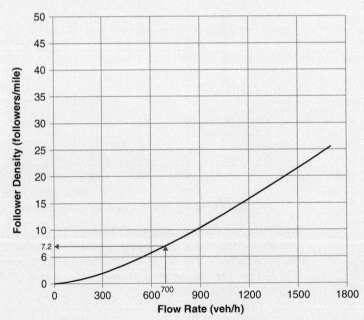

Figure 6.14 Percent follower density versus directional flow rate for Example 6.7.

Calculations for passing lane segments are more complicated than those for passing constrained and passing zone segments, especially in the context of a two-lane highway facility (multiple contiguous segments analyzed together). Performance improvements in roadway operations can persist well downstream of the end of a passing lane segment, particularly with respect to percent followers. Average speed also improves; however, this improvement is relatively minor and persists for a much shorter distance downstream, relative to follower-related improvements.

The distance downstream of the passing lane segment to which the operational improvements resulting from the passing lane applies is referred to as the effective length of the passing lane. Thus, segments downstream of a passing lane, and within this *effective length*, will have an adjustment applied to their base follower density result (assuming no upstream passing lane influence), and referred to as the adjusted follower density. For the passing lane segment itself, additional follower density calculations are performed. This is done to account for operating conditions within the passing lane segment to account for the two lanes, in addition to the results for the end of the passing lane immediately downstream of the two lanes merging together.

6.7   DESIGN TRAFFIC VOLUMES

In the preceding sections of this chapter, consideration was given to the determination of LOS, given some hourly volume. However, a procedure for selecting an appropriate hourly volume is needed to compute the LOS and to determine the number of lanes that need to be provided in a new roadway design to achieve some specified LOS. The selection of an appropriate hourly volume is complicated by two issues. First, there is considerable variability in traffic volume by time of day, day of week, time of year, and type of roadway. Figure 6.15 shows such variations in traffic volumes by hour of day and day of week for typical intra-city and inter-city routes. Figure 6.15 gives variations by time of year by comparing monthly percentages of the annual average daily traffic, AADT (in units of vehicles per day and computed as the total yearly traffic volume divided by the number of days in the year). The second concern is an outgrowth of the first: Given the temporal variability in traffic flow, what hourly volume should be used for design and/or analysis? To answer this question, consider the example diagram shown in Fig. 6.16. This figure plots hourly volume (as a proportion of AADT) against the cumulative number of hours that exceed this volume, per year. For example, the highest traffic flow in the year, on this sample roadway, would have an hourly volume of $0.148 \times \text{AADT}$ (a volume that is exceeded by zero other hours). Sixty hours in the year would have a volume that exceeds $0.11 \times \text{AADT}$.

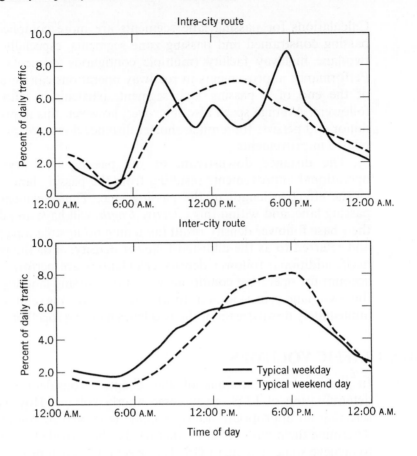

Figure 6.15 Examples of hourly and daily traffic variations for intra-city and inter-city routes.

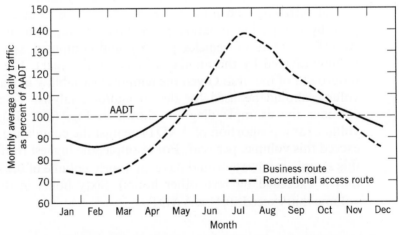

Figure 6.16 Example of monthly traffic volume variations for business and recreational access routes.

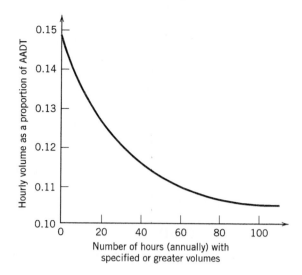

Figure 6.17 Highest 100 hourly volumes over a one-year period for a typical roadway.

In determining the number of lanes that should be provided on a new or redesigned roadway, it is obvious that using the worst single hour in a year (the hour with the highest traffic flow, which would be $0.148 \times$ AADT from Fig. 6.16) would be a wasteful use of resources because additional lanes would be provided for a relatively rare occurrence. In contrast, if the 100th highest volume is used for design, the design LOS will be exceeded 100 times a year, which will result in considerable driver delay. Clearly, some compromise between the expense of providing additional capacity (such as additional lanes) and the expense of incurring additional driver delay must be made.

A common practice in the United States is to use a design hour-volume (DHV) that is between the 10th and 50th highest-volume hours of the year, depending on the type and location of the roadway (urban freeway, rural/suburban multilane highway, etc.), local traffic data, and engineering judgment. Perhaps the most common hourly volume used for roadway design is the 30th highest of the year. In practice, the K-factor is used to convert annual average daily traffic (AADT) to the 30th highest hourly volume. K is defined as

$$K = \frac{\text{DHV}}{\text{AADT}} \tag{6.24}$$

where

> K = factor used to convert annual average daily traffic to a specified annual hourly volume,
>
> DHV = design hour-volume (typically, the 30th highest annual hourly volume), and
>
> AADT = roadway's annual average daily traffic in veh/day.

For example, Fig. 6.16 shows that the K-value corresponding to the 30th highest hourly volume is 0.12. More generally, K_i can be defined as the K-factor corresponding to the ith highest annual hourly volume. Again, for example, the

20th highest annual hourly volume would have a K-value, K_{20}, of 0.126, from Fig. 6.16. If K is not subscripted, the 30th highest annual hourly volume is assumed ($K = K_{30}$).

Finally, in the design and analysis of some highway types (such as freeways and multilane highways), the concern lies with directional traffic flows. Thus, a factor is needed to reflect the proportion of peak-hour traffic volume traveling in the peak direction. This factor is denoted D and is used to arrive at the directional design-hour volume (DDHV) by application of

$$DDHV = K \times D \times AADT \qquad (6.25)$$

where

$\quad$ DDHV = directional design-hour volume,

$\qquad D$ = directional distribution factor to reflect the proportion of peak-hour traffic volume traveling in the peak direction, and

Other terms are as defined previously.

EXAMPLE 6.8 **DETERMINATION OF REQUIRED NUMBER OF FREEWAY LANES**

A freeway is to be designed as a passenger-car–only facility for an AADT of 35,000 vehicles per day. It is estimated that the freeway will have a *FFS* of 70 mi/h. The peak-hour factor is estimated to be 0.85 with 65% of the peak-hour traffic traveling in the peak direction. Assuming that Fig. 6.16 applies, determine the number of lanes required to provide at least LOS C using the highest annual hourly volume and the 30th highest annual hourly volume.

SOLUTION

By inspection of Fig. 6.16, the highest annual hourly volume has $K_1 = 0.148$. Application of Eq. 6.25 gives

$$DDHV = K_1 \times D \times AADT$$
$$= 0.148 \times 0.65 \times 35,000 = 3367 \text{ veh/h}$$

The next step is to determine the maximum service flow rate that can be accommodated at LOS C for *FFS* = 70 mi/h. From Table 6.1, we see that this value is 1735 pc/h/ln. Thus, we must provide enough lanes so that the per-lane traffic flow is less than or equal to this value.

We can use Eq. 6.7 to find v_p, based on an assumed number of lanes. Comparing the calculated value of v_p to the maximum service flow rate of 1735 will determine whether we have an adequate number of lanes. Assuming a four-lane freeway (two lanes in each direction), Eq. 6.7 gives

$$v_p = \frac{3367}{0.85 \times 2 \times 1.0} = 1980.6 \text{ pc/h/ln}$$

with

$V = 3367$ (DDHV from above), and

$f_{HV} = 1.0$ (no heavy vehicles).

This value is higher than 1735, so we need to provide more lanes. The calculation is repeated, this time with an assumed six-lane freeway (three lanes in each direction):

$$v_p = \frac{3367}{0.85 \times 3 \times 1.0} = 1320.4 \text{ pc/h/ln}$$

Since this value is less than 1735, a six-lane freeway is necessary to provide LOS C operation for the design traffic flow rate.

For the 30th highest hourly annual volume, Fig. 6.16 gives $K_{30} = K = 0.12$, which when used in Eq. 6.25 gives

$$DDHV = K \times D \times AADT$$
$$= 0.12 \times 0.65 \times 35{,}000 = 2730 \text{ veh/h}$$

Again applying Eq. 6.7, with an assumed four-lane freeway, yields

$$v_p = \frac{2730}{0.85 \times 2 \times 1.0} = 1605.9 \text{ pc/h/ln}$$

This value is less than 1735, so a <u>four-lane freeway</u> (two lanes each direction) is adequate for this design traffic flow rate.

This example demonstrates the impact of the chosen design traffic flow rate on roadway design. Only a four-lane freeway is necessary to provide LOS C for the 30th highest annual hourly volume, as opposed to a six-lane freeway needed to satisfy the LOS requirement for the highest annual hourly volume.

6.8   PRACTICE PROBLEMS

PRACTICE PROBLEM 6.1

BASIC FREEWAY SEGMENT LOS WITH A SPECIFIC GRADE

A northbound freeway segment is on a 4% upgrade from station 430 + 60 to 397 + 60 and has three 11 ft-wide lanes, a 5 ft right shoulder and has a ramp density of 1 per mile in the 3 miles before and after the mid-point of this freeway segment. The peak-hour factor is 0.9 and northbound traffic during the peak hour is 2400 vehicles, with 20% heavy vehicles (and a 50% SUT/50% TT mix). Determine the density and LOS of the freeway segment.

SOLUTION   **Note: Open boxes in equations "$\boxed{}$" are to be completed by the reader**

We begin by determining the *FFS* according to Eq. 6.6,

$$FFS = 75.4 - f_{LW} - f_{RLC} - 3.22 TRD^{\,0.84}.$$

Here, from Table 6.3 $f_{LW} = \boxed{}$ mi/h, from Table 6.4 $f_{RLC} = \boxed{}$ mi/h (Table 6.4), and
$TRD = 1$ ramp/mi (given), so

$$FFS = 75.4 - \boxed{} - \boxed{} - 3.22(1)^{0.84} = 69.9 \text{ mi/h}.$$

Next, to determine the analysis flow rate v_p, we begin noting that the length of this freeway segment is,

$$430 + 60 \text{ minus } 397 + 60 = \boxed{} \text{ ft.}$$

This is 0.625 mi ($\boxed{}$/5280) and, because it is on an upgrade that exceeds 3% and is longer the 0.25 mi, extended segment analysis cannot be used.

Using values in Table 6.7 (50% SUTs/50% TTs mix) with 20% heavy vehicles and a section length of 0.625 mi, we find $E_T = \boxed{}$ at a 3.5% grade and $E_T = \boxed{}$ at a 4.5% grade. Interpolating, we find $E_T = 2.52$ [($\boxed{} + \boxed{}$)/2] at a grade of 4%. Applying Eq. 6.9 to arrive at the heavy-vehicle adjustment factor, we get,

$$f_{HV} = \frac{1}{1 + 0.20\left(\boxed{} - 1\right)} = \boxed{}$$

So, from Eq. 6.7, the analysis flow rate is,

$$v_p = \frac{\boxed{}}{\boxed{} \times \boxed{} \times \boxed{}} = 1159 \text{ pc/h/ln}.$$

To determine the average speed (*S*), calculate the breakpoint (*BP*) value for the speed–flow relationship, from Eq. 6.4 as,

$$BP = \left[1000 + 40\left(75 - \boxed{}\right)\right] = \boxed{} \text{ pc/h/ln}$$

Since the v_p of 1159 is less than the *BP* value of $\boxed{}$, *S* will be equal to the *FFS*, $\boxed{}$ mi/h. With this, the density can be calculated using Eq. 6.10 as,

$$D = \frac{\boxed{}}{\boxed{}} = 16.6 \text{ pc/mi/ln}.$$

From Table 6.2, it can be seen that this corresponds to LOS B [$\boxed{}$ (max density for LOS B) < 16.6 < $\boxed{}$ (max density for LOS B)]. Thus, this freeway segment operates at <u>LOS B</u>.

PRACTICE
PROBLEM 6.2

BASIC FREEWAY SEGMENT LOS WITH A SPECIFIC GRADE AND INCREASING VOLUME

Consider the freeway and conditions in Practice Problem 6.1. Suppose that traffic increases by 50% and that the peak-hour factor increases from 0.9 to 0.95 (as the increase in traffic makes the traffic flow rate more uniform over the peak hour). The proportion of heavy vehicles remains the same as do all other elements of the problem. Determine the new density and LOS of the freeway segment.

SOLUTION

Note: Open boxes in equations "☐" are to be completed by the reader

To determine the new analysis flow rate (v_p), note that the free-flow speed (*FFS*) and heavy-vehicle adjustment factor (f_{HV}) will remain unchanged from Practice Problem 6.1. The new hourly volume will be $V = 3600$ veh/h (2400×1.5) and, with the new *PHF* = 0.95, the new analysis flow rate from Eq. 6.7 is,

$$v_p = \frac{\boxed{}}{\boxed{} \times \boxed{} \times \boxed{}} = 1647 \text{ pc/h/ln}$$

As in Practice Problem 6.1, to determine the average speed (*S*), the breakpoint (*BP*) value for the speed–flow relationship is calculated from Eq. 6.4 as,

$$BP = \left[1000 + 40\left(75 - \boxed{}\right)\right] = \boxed{} \text{ pc/h/ln}.$$

Unlike Practice Problem 6.1, the v_p of 1647 is greater than the *BP* value of $\boxed{}$ so *S* must be computed from Eq. 6.2. To apply Eq. 6.2, we must compute the capacity *c* (and also check that this value of v_p is less than the capacity to ensure Eq. 6.3 can be applied). For capacity, using Eq. 6.3,

$$c = \text{Min}\left[2200 + 10\left(\boxed{} - 50\right), 2400\right] \qquad 55 \leq FFS \leq 75,$$

which gives $c = \boxed{}$, so $v_p \leq c$. Applying Eq. 6.3 gives,

$$S = \boxed{} - \frac{\left(\boxed{} - \dfrac{\boxed{}}{45}\right)\left(\boxed{} - \boxed{}\right)^{2.0}}{\left(\boxed{} - \boxed{}\right)^{2.0}} \qquad BP < v_p \leq c$$

or $S = 67.6$ mi/h. With this, density is given in Eq. 6.10 as,

$$D = \frac{\boxed{}}{\boxed{}} = 24.4 \text{ pc/mi/ln}$$

From Table 6.2, it can be seen that this density corresponds to LOS C [$\boxed{}$ (max density for LOS B) $< 26.86 < \boxed{}$ (max density for LOS C)]. Thus, this freeway segment now operates at <u>LOS C</u> with the increased traffic.

PRACTICE	**MULTILANE HIGHWAY SEGMENT LOS WITH A SPECIFIC GRADE**
PROBLEM 6.3	

An undivided multilane highway segment has two 11-ft lanes in the eastbound direction with no shoulders and a 55 mi/h speed limit. This highway segment has 40 access points on a 1.25 mile, 2.5% upgrade. During the highest 15 minutes of traffic-flow within the peak hour, there are 755 vehicles and 12% of these are heavy vehicles with a 70%/30% mix of single-unit and tractor-trailer trucks. What are the estimated speed, density, and LOS of upgrade?

SOLUTION

Note: Open boxes in equations " $\boxed{}$ " are to be completed by the reader

We begin by determining the *FFS* according to Eq. 6.14,

$$FFS = BFFS - f_{LW} - f_{TLC} - f_M - f_A$$

In this equation,

$BFFS = \boxed{}$ mi/h (the speed limit pus 5 mi/h),

$f_{LW} = \boxed{}$ mi/h (Table 6.3),

$f_{TLC} = \boxed{}$ mi/h (with $TLC = 6$ since undivided with 0 ft right shoulder, given),

$f_M = \boxed{}$ mi/h (Table 6.12), and

$f_A = \boxed{}$ mi/h ($40/1.25 \times 0.25$, since 0.25 mi/h is lost for every access point per mile), (Table 6.13).

With these values, the *FFS* is computed as,

$$FFS = \boxed{} - \boxed{} - \boxed{} - \boxed{} - \boxed{} = 47.2 \text{ mi/h}.$$

Moving on to the determination of the analysis flow rate v_p, we begin noting that the length of this freeway segment is an upgrade of 1.25 mi in length at 2.5% which is a grade less than or equal to 3% that extends for more than 1 mi, so extended segment analysis cannot be used (see discussion in Section 6.5.5). Thus, using values in Table 6.8 (70% SUTs/30% TTs mix) with 12% heavy vehicles and a section length of 1.25 mi, we find $E_T = \boxed{}$ at a 2.5% grade with 10% heavy vehicles, and $E_T = \boxed{}$ at a 2.5% grade with 15% heavy vehicles. Interpolating, we find $E_T = \boxed{}$ $[\boxed{} - (\boxed{} - \boxed{}) \times \boxed{}/\boxed{}]$. Applying Eq. 6.9 to arrive at the heavy-vehicle adjustment factor we get,

$$f_{HV} = \frac{1}{1 + 0.12\left(\boxed{} - 1\right)} = \boxed{}$$

Because we are given $V_{15} = 755$, Eq. 6.7 can be written as,

$$v_p = \frac{V_{15} \times 4}{N \times f_{HV}}.$$

So, substituting from Eq. 6.7,

$$v_p = \frac{\boxed{} \times 4}{\boxed{} \times \boxed{}} = 1782 \text{ pc/h/ln}.$$

As shown in the text (see Eqs. 6.11 and 6.12) and illustrated in Fig. 6.4, the break point (*BP*) for multilane highways is always 1400 pc/h/ln. Since our $v_p = 1782 > 1400$, Eq. 6.12 needs to be applied to determine the speed, *S*. We begin by determining the capacity (needed to solve for Speed in Eq. 6.12), using Eq. 6.13,

$$c = \text{Min}\left[1900 + 20\left(\boxed{} - 45\right), 2300 \right] \quad 45 \le FFS \le 70,$$

which gives $c = \boxed{}$, so $v_p \le c$. Applying Eq. 6.12 gives,

$$S = \boxed{} - \frac{\left(\boxed{} - \dfrac{\boxed{}}{45}\right)\left(\boxed{} - 1400\right)^{1.31}}{\left(\boxed{} - 1400\right)^{1.31}} \quad 1400 < v_p \le c$$

or $S = 44.7$ mi/h. With this, density is given in Eq. 6.10 as,

$$D = \frac{\boxed{}}{\boxed{}} = 39.9 \text{ pc/mi/ln}$$

From Table 6.2, it can be seen that this density corresponds to LOS E [$\boxed{}$ (max density for LOS D) $< 39.9 < \boxed{}$ (max density for LOS E)]. Thus, this freeway segment now operates at <u>LOS E</u> with the increased traffic.

NOMENCLATURE FOR CHAPTER 6

AADT	annual average daily traffic	f_{LW}	free-flow speed adjustment factor for lane width (freeways and multilane highways)
BFFS	estimated free-flow speed for base conditions	f_M	free-flow speed adjustment factor for median type (multilane highways)
c	roadway capacity		
D	density or factor for directional distribution of traffic	K_i	factor used to convert AADT to ith highest annual hourly volume
DHV	design-hour volume	LC_L	left-side lateral clearance (multilane highways)
DDHV	directional design-hour volume		
E_R	passenger car equivalent value for recreational vehicles	LC_R	right-side lateral clearance (multilane highways)
E_T	passenger car equivalent value for large trucks and buses	N	number of lanes in one direction
		PCE	passenger car equivalent
FD	follower density (two-lane highways)	PHF	peak-hour factor
FFS	measured or estimated free-flow speed	PF	percent followers (two-lane highways)
f_A	free-flow speed adjustment factor for access point frequency (multilane highways)	S	average speed
		TLC	total lateral clearance (multilane highways)
f_{HV}	heavy-vehicle adjustment factor	v	analysis flow rate
f_{ID}	free-flow speed adjustment factor for interchange density (freeways)	V	hourly volume
		V_{15}	highest 15-minute volume
f_{RLC}	free-flow speed adjustment factor for lateral clearance (freeways and multilane highways)	v/c	volume-to-capacity ratio

REFERENCE

Transportation Research Board. *Highway Capacity Manual.* *6th Edition.* National Academy of Sciences, Washington, DC. 2016.

Chapter 7

Traffic Control and Analysis at Signalized Intersections

7.1 INTRODUCTION

Due to conflicting traffic flows, highway intersections are of great concern to traffic engineers. Intersections can be a major source of crashes and vehicle delays (as vehicles yield to avoid conflicts with other vehicles). Most highway intersections are not signalized due to low traffic volumes and adequate sight distances. However, at some point, traffic volumes, accident frequencies and their resulting injury severities, as well as other factors, reach a level that warrants the installation of a traffic signal.

The installation and operation of a traffic signal to control conflicting traffic and pedestrian flows (also referred to as movements) at an intersection has advantages and disadvantages. The advantages include a potential reduction in some types of crashes (particularly angle crashes), provisions for pedestrians to cross the street, provisions for side-street vehicles to enter the traffic stream, provisions for the progressive flow of traffic in a signal-system corridor, and possible improvements in capacity and reductions in delays. However, signals are by no means the perfect solution for delay or potential crash problems at an intersection. A poorly timed signal or one that is not justified can have a negative impact on the operation of the intersection by increasing vehicle delay, increasing the rate of vehicle crashes (particularly rear-end crashes), causing a disruption in traffic progression (adversely impacting the through movement of traffic), and encouraging the use of routes not intended for through traffic (such as routes through residential neighborhoods). Traffic signals are also costly to install, with some basic signal installations costing in excess of $100,000. Therefore, the decision to install a signal must be weighed and studied carefully. To assist transportation engineers in this process, the Federal Highway Administration of the U.S. Department of Transportation publishes the *Manual on Uniform Traffic Control Devices* (MUTCD) [U.S. Federal Highway Administration, 2009], which contains a section on warrants for the installation of a traffic signal. There are a total of nine warrants, which include consideration of vehicle volumes, pedestrian volumes, school crossings, signal coordination, and crash experience. The reader is referred to the MUTCD for details on these warrants.

Unlike uninterrupted flow, in which vehicle flow is affected only by other vehicles and the highway environment, the introduction of a traffic control device such as a signal exerts a significant influence on the flow of vehicles. Thus, the analysis of traffic flow at signalized intersections can become very complex. This chapter will make several simplifying assumptions to keep the material at an accessible level.

The chapter begins by providing an overview of the physical elements of intersection configuration and traffic signal control. A basic understanding of these principles provides the foundation for designing intersection geometry and traffic movement sequence plans. This is followed by a presentation of concepts, definitions, and analytical techniques that are used in the design and analysis of signal timing plans at signalized intersections.

7.2 INTERSECTION AND SIGNAL CONTROL CHARACTERISTICS

An intersection is defined as an at-grade crossing of two or more roadways. For analysis, the roadways entering the intersection are segmented into approaches, which are defined by lane groups (groups of one or more lanes). These lane groups are usually based on the allowed movements (left, through, right) within each lane and the sequencing of allowed movements by the traffic signal. The establishment of lane groups will be discussed in more detail in Section 7.4.2.

To illustrate these concepts, note that approach 1 of the intersection depicted in Fig. 7.1 consists of a lane for the exclusive use of left turns, a lane for the exclusive use of right turns, and two lanes for the exclusive use of through movements. Approach 3 is similar to approach 1 but does not include an exclusive right-turn lane; instead, the right turns share the outside lane with the through movements. Because the lanes for the exclusive use of left and right turns are short, they are usually referred to as bays and are intended to hold a limited number of queued vehicles. Queuing analysis can be used to determine the length of bay necessary to prevent queued turning vehicles from overflowing the bay and blocking the through lanes (known as spillover) and/or the length necessary to prevent queued through vehicles from blocking the entrance of the turn bay (known as spillback). Approach 2 consists of a shared through/right-turn lane and an exclusive left-turn lane (not a bay in this case because it has the same length as the adjacent lane). Approach 4 is similar to approach 2, but the inside lane is a shared through/left-turn lane.

From a driver's perspective, a traffic signal is just a collection of light-emitting devices [usually incandescent bulbs or light-emitting diodes (LEDs)] and lenses that are housed in casings of various configurations (referred to as signal heads) whose purpose is to display red, yellow, and green full circles and/or arrows. Figure 7.2 shows typical configurations of signal heads in the United States. These signal heads are usually mounted to mast arms or wire spanned across the intersection.

The following terminology is commonly used in the design of traffic signal controls.

Indication. The illumination of one or more signal lenses (greens, yellows, reds) indicating an allowed or prohibited traffic movement.

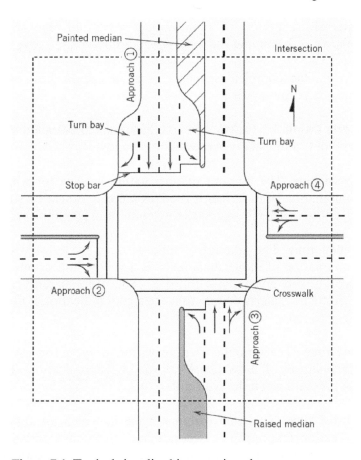

Figure 7.1 Typical signalized intersection elements.

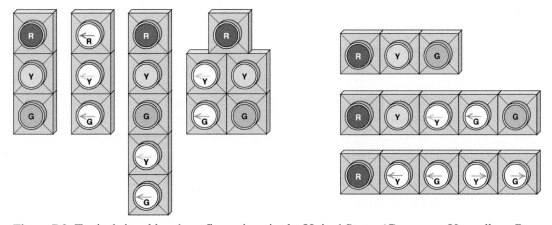

Figure 7.2 Typical signal head configurations in the United States (G = green; Y = yellow; R = red).

Interval. A period of time during which all signal indications (greens, yellows, reds) remain the same for all approaches.

Cycle. One complete sequence (for all approaches) of signal indications (greens, yellows, reds).

Cycle length. The total time for the signal to complete one cycle (given the symbol C and usually expressed in seconds).

Green time. The amount of time within a cycle for which a movement or combination of movements receives a green indication (the illumination of a signal lens). This is expressed in seconds and given the symbol G.

Yellow time. The amount of time within a cycle for which a movement or combination of movements receives a yellow indication. This is expressed in seconds and given the symbol Y. This time is referred to as the change interval, as it alerts drivers that the signal indication is about to change from green to red.

Red time. The amount of time within a cycle for which a movement or combination of movements receives a red indication. This is expressed in seconds and given the symbol R.

All-red time. The time within a cycle in which all approaches have a red indication (expressed in seconds and given the symbol AR). This time is referred to as the clearance interval because it allows vehicles that might have entered at the end of the yellow interval to clear the intersection before the green phase starts for the next conflicting movement(s). This type of interval is becoming increasingly common for safety reasons because the rate of vehicles entering at the end of the yellow and beginning of the red indication has steadily increased in recent years.

Phase. The sum of the displayed green, yellow, and red times for one or more movements that start and stop moving at the same time.

Timing Stage. Consists of the time during which the same combination of movements, across one or more phases, are moving simultaneously. The sum of the timing stage lengths (in seconds) is the cycle length.

The term "movement" is used frequently in the preceding definitions. In addition to a directional descriptor, such as left, through, or right, a distinction is made by categorizing movements as either protected or permitted.

Protected movement. A movement that has the right-of-way and does not need to yield to conflicting movements, such as opposing vehicle traffic or pedestrians. Through movements, which are always protected, are given a green full circle indication (or in some geometric configurations, a green arrow pointing up). Left- or right-turn movements that are protected are given a green arrow indication (pointing either left or right).

Permitted movement. A movement that must yield to opposing traffic flow or a conflicting pedestrian movement. This movement is made during gaps (time headways) in opposing traffic and conflicting pedestrian movements. Left- or right-turn movements with a green full circle indication are permitted movements. Left-turning vehicles in this situation must wait for gaps in the opposing through and right-turning traffic before making their turns. Right-turning vehicles must yield to pedestrians in the adjacent crosswalk before making their turns.

To understand how these control characteristics are implemented, it is useful to analyze the physical implementation of these concepts. The display of the various signal indications (green, yellow, red, protected, permitted) at an intersection is handled by a signal controller (which is typically located in a cabinet next to the intersection). Modern signal controllers are sophisticated pieces of electronic equipment. These controllers, when combined with a method of vehicle detection, offer great flexibility in controlling phase duration and sequence. Traffic signal controllers are designed to operate in one or more of the following modes: pretimed, semi-actuated, or fully actuated.

Pretimed. A signal whose timing (cycle length, green time, etc.) is fixed over specified time periods and does not change in response to changes in traffic flow at the intersection. No vehicle detection is necessary with this mode of operation.

Semi-actuated. A signal whose timing (cycle length, green time, etc.) is affected when vehicles are detected (by video, pavement-embedded inductance loop detectors, etc.) on some, but not all, approaches. This mode of operation is usually found where a low-volume road intersects a high-volume road, often referred to as the minor and major streets, respectively. In such cases, green time is allocated to the major street until vehicles are detected on the minor street; then the green indication is briefly allocated to the minor street and then returned to the major street.

Fully actuated. A signal whose timing (cycle length, green time, etc.) is completely influenced by the traffic volumes, when detected, on all of the approaches. Fully actuated signals are most commonly used at intersections of two major streets and where substantial variations exist in all approach traffic volumes over the course of a day.

7.2.1   Actuated Control

Although pretimed signal control does not require the expense of vehicle detection hardware, it results in signal timing that it is not responsive to real-time traffic demands. The fixed-time values of the pretimed signal are based on expected traffic demands during the time period of interest. However, traffic arrivals can vary significantly from one cycle to the next, as described in Chapter 5. Thus, with fixed timing, a phase may provide excessive green time one cycle (which results in extra delay for the vehicles that move in other phases) and not enough time in another cycle.

Improvements in signal-controller technology over the last 25 years have made possible significant advances in traffic control at signalized intersections. Modern signal controllers are able to accept inputs on traffic demands and utilize this information to adjust the green interval duration and phasing sequence from one cycle to the next.

Actuated control's ability to respond to changes in traffic demands allows the green time to be reduced for a movement when the arrival rate is lower than normal, the green time to be extended when the arrival rate is higher than normal, or a phase to be skipped altogether if no demand is present for that movement. Thus, an intersection operating under actuated control will almost always result in lower overall delays than one operating under pretimed control.

Vehicle Detection

Actuated control requires vehicle detection technology. The most common form of vehicle detection technology is the inductance loop detector (ILD). The ILD, a simple technology that has been in use for several decades now, consists of a loop (or coil) of wire embedded in the pavement through which an electrical current is circulated. This current is monitored by a device that interfaces with the signal controller, and when a vehicle passes over the ILD, the inductance level of the current is altered. When this change in the inductance level is detected by the monitoring device, it sends an input to the signal controller to indicate the presence of a vehicle.

ILDs can take on a variety of shapes, with different shapes having different advantages/disadvantages for detection ability. Furthermore, the sensitivity of the detectors can be tuned. The challenge in tuning the sensitivity is to find the level that allows detection of smaller vehicles (such as motorcycles or bicycles), yet is not so sensitive as to detect objects that are not vehicles.

The limitation of ILD technology is that vehicles can be detected only where the ILDs are placed. This limits the control options because it is prohibitively expensive to implement a large number of ILDs on an intersection approach.

A newer vehicle detection technology that is increasing in popularity at signalized intersections is video imaging processing (VIP). This technology has three main components: video camera, video digitization and processing unit, and signal controller interface. With VIP technology, virtual (software-based) detectors can be placed anywhere within a video camera's field of view. The video processing unit converts the video from the camera into a digital format, and a software algorithm processes the combination of the digitized field of view and virtual detectors to determine vehicle presence, as well as additional measures that cannot be obtained with ILDs (such as queue length). With the additional measures that can be obtained from VIP systems, a greater number of control strategies are possible. Thus, a greater level of traffic responsiveness can be obtained.

Typical Phase Operation

With the most basic configuration for vehicle detection, signal phases operating under actuated control consist of the following phase periods.

Initial Green. This period provides a practical minimum amount of green time to the traffic movement.

Extended Green. After the initial green time expires, the phase enters an extension mode. This extension mode provides a continuation of green time as long as a vehicle arrives (at the detector) within a specified amount of time after the arrival of the previous vehicle.

A maximum green time is also specified for the phase. Thus, even if the arrival rate of vehicles is high enough to continue the extension period indefinitely, the phase will eventually be terminated so that other traffic movements can be served. If the phase terminates in this manner, it is referred to as a max-out condition. If the phase terminates prior to reaching the maximum green time, due to the time gap in successive vehicle arrivals exceeding the extension interval (the time allowed between successive vehicle arrivals before it is assumed no more vehicles are arriving), this is referred to as a gap-out condition.

Initial green periods are intended to serve all, or most, of the vehicle queue that develops during that traffic movement's red period. Vehicles that join the queue at the start of green, or arrive just after queue clearance, are usually handled during the extension period. Therefore, a large enough green time should be specified so that the regularly expected initial queue length can be served, but not too large that excessive green time is wasted during cycles with much lower than average vehicle arrivals. Additionally, the green time should be at least several seconds greater than the lost time for a phase. Typical minimum green times are on the order of 10 seconds.

The time allowed between successive vehicle arrivals at vehicle detectors before it is assumed that no more vehicles are arriving (the extension interval) ranges from 2 to 4 seconds. Overall phase operations are quite sensitive to the selected value of this extension interval (also commonly referred to as the unit extension). A smaller value, such as 2 seconds, will generally result in a "snappy" operation; that is, the phase will tend to gap out quickly after the initial queue is served. In practice, a small extension interval can often result in a phase gapping out prematurely due to a driver in the queue hesitating (from being distracted or inattentive) in starting up from green, thus increasing delays and driver frustration. A larger extension interval, such as 4 seconds, will generally result in "sluggish" operation and a greater likelihood of a max-out. Although the longer extension interval decreases the likelihood of premature gap out, it can also lead to frustration for the drivers of other movements as they may perceive that the green time for the active phase is lasting an unnecessarily long time.

The basic sequence of events for an actuated phase is as follows: The initial green period is provided; the phase enters the extension period; the extension period continues until the phase either gaps out or maxes out; the yellow and all-red intervals commence and then the phase terminates. This sequence is illustrated in Fig. 7.3. This is the most basic operation of an actuated phase. Many more elements can be incorporated, depending on the level of traffic responsiveness desired, the level of coordination with other signalized intersections, and the level of vehicle detection.

For a more complete description of actuated control, the reader is referred to the *Manual of Traffic Signal Design* [Kell and Fullerton, 1998] and the *Highway Capacity Manual* [Transportation Research Board, 2016].

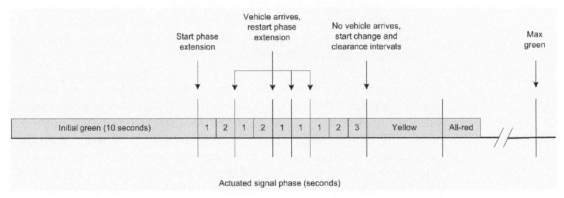

Figure 7.3 Basic operation of an actuated signal phase.

7.2.2   Signal Controller Operation

Most modern signal controllers are designed to operate in what is termed a dual-ring configuration. This configuration allows maximum flexibility for controlling phase duration and sequencing, which is necessary for operating in a fully actuated mode. The dual-ring concept can be best explained with a graphical illustration (see Fig. 7.4a).

In this figure, the three-letter notation in each numbered box refers to direction and movement type. For example, WBL means westbound left and NBT means northbound through. At an intersection with four approaches and separate/exclusive left-turn phases for each approach, a total of eight separate movements are possible. Any right-turn movements are implicitly considered to move with the corresponding through movement. The dual-ring terminology comes from four movements (1–4) being represented on the top ring, and four movements (5–8) being represented on the bottom ring. The movements in each of the numbered boxes must start and end at the same time; thus, the boxes represent phases. The logic behind this configuration is straightforward. Any phase in ring 1 can occur simultaneously with any phase in ring 2 as long as both are on the same side of the barrier (a term used figuratively to separate conflicting traffic movements). For example, the opposing WB and EB left-turn movements (phases 1 and 5) can move simultaneously. However, if no vehicles are detected in the EB left-turn lane, this movement can be skipped and the WB left-turn and WB through movements (phases 1 and 6) can move simultaneously. The same logic applies to movements on the right side of the barrier (phases 3, 4, 7, and 8).

Furthermore, if the volume of a phase movement subsides sooner than the volume of another phase movement currently moving at the same time, the green time for this lower-volume phase can be terminated and another phase can be initiated, according to the previously described logic. For example, suppose the WB and EB left turns (phases 1 and 5) start to move at the same time, but the WB left-turn (phase 1) volume is considerably larger than the EB left-turn (phase 5) volume. If both phases 1 and 5 received enough green time to satisfy the WB left-turn vehicle demand, this would result in wasted green time for phase 5. With the dual-ring configuration, phase 5 can be terminated before phase 1, and phase 6 can be initiated while phase 1 continues. This results in a more efficient allocation of green time and reduced delay. The phase sequence and phase durations can therefore vary from one cycle to the next at a fully actuated intersection, especially with highly variable approach volumes. Consequently, the cycle length can vary from one cycle to the next. Figure 7.4*b* illustrates the typical dual-ring phase sequence options as discussed above. Note that phases can be skipped entirely if no traffic demand is present, and other phase options are possible depending on intersection geometry/lane movement assignment.

It must be pointed out that although no two phases are required to either start or terminate at the same time in a dual-ring configuration, all movements on the left side of the barrier must be terminated before any movement on the right side of the barrier can be initiated, and vice versa. This is a safety feature: no movement on the left side of the barrier can be allowed to move simultaneously with any movement on the right side of the barrier, or else conflicting traffic movements will intersect. Movements on the same side of a barrier are commonly referred to as being in the same concurrency group.

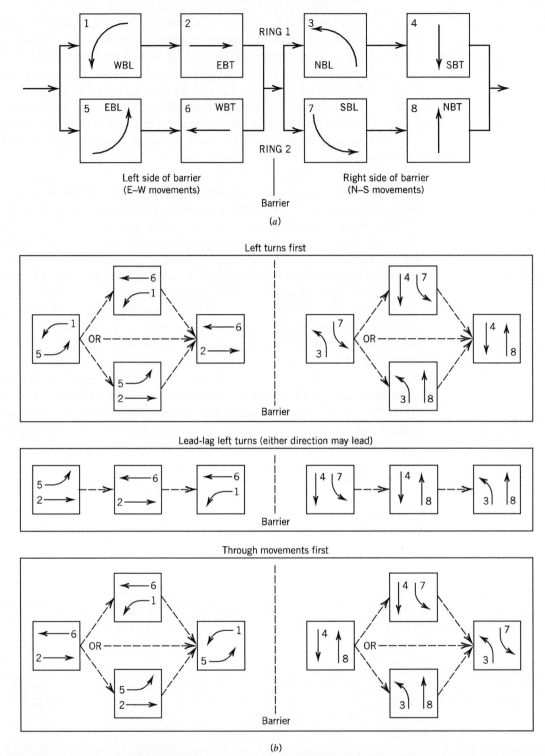

Figure 7.4 Dual-ring signal control. (*a*) Movement/phase-based representation of dual-ring logic. (*b*) Timing-stage–based representation of dual-ring logic.

Part (*a*) Adapted from Exhibit 19-2, p. 19-6, *Highway Capacity Manual*, 6th Edition, Copyright, Transportation Research Board, National Academy of Sciences, Washington, D.C. 2016.

7.3   TRAFFIC FLOW FUNDAMENTALS FOR SIGNALIZED INTERSECTIONS

Before presenting the analytical principles and techniques used for signalized intersections, it is important to introduce several key concepts and definitions used in the development of signal timing plans and the analysis of traffic at signalized intersections.

Saturation Flow Rate

The saturation flow rate is the maximum hourly volume that can pass through an intersection, from a given lane or group of lanes, if that lane (or lanes) were allocated constant green over the course of an hour. Saturation flow rate is given by

$$s = \frac{3600}{h} \tag{7.1}$$

where

$$
\begin{aligned}
s &= \text{saturation flow rate in veh/h,} \\
h &= \text{saturation headway in s/veh, and} \\
3600 &= \text{number of seconds per hour.}
\end{aligned}
$$

Note the similarity between Eq. 7.1 and Eq. 5.4. The difference is that h in Eq. 7.1 is a constant minimum headway value maintained for saturated conditions, as opposed to an average headway value as used in Eq. 5.4 (Eq. 7.1 also directly yields units of veh/h for s because of the numerator). The use of the term "saturation" is an important qualifier in this definition, as it implies the presence of constant vehicle demand in measuring the headway. If the measure of interest is simply the traffic flow through the intersection for some period of time, then the appropriate equation would be 5.1 or 5.4.

Research has found that a typical maximum saturation flow rate of 1900 passenger cars per hour per lane (pc/h/ln) is possible at signalized intersections, and this is referred to as the base saturation flow rate. This corresponds to a saturation headway of about 1.9 seconds. Just as in the analysis of uninterrupted flow, a number of roadway and traffic factors can affect the maximum flow rate through an intersection. These factors include lane widths; grades; curbside parking maneuvers; the distribution of traffic among multiple approach lanes; the level of roadside development; bus stops; and the influence of pedestrians, bicycles, and heavy vehicles (since they occupy more roadway space and have poorer acceleration/deceleration capabilities). Additionally, lanes that allow left or right turns usually have lower saturation flow rates because drivers reduce speed to make a turning maneuver (especially heavy vehicles, with their increased turning radii). Furthermore, if a turning movement is permitted rather than protected, its saturation flow rate will be reduced as a result of the turning vehicles yielding to conflicting through and right-turning vehicles (for left turns only), bicycles, and/or pedestrians. All of these factors are accommodated by applying adjustments to the base saturation flow rate. The end result is usually a value less than 1900 pc/h/ln for each approach lane at an intersection, and is referred to as the adjusted saturation flow rate. Additionally, the units are converted to vehicles per hour per lane (veh/h/ln) due to adjustment of the heavy-vehicle volume with passenger car equivalents (in a manner similar to the procedures of Chapter 6). The process for arriving at an adjusted saturation flow rate by making adjustments to the base saturation flow rate, for the preceding factors, is quite involved and beyond the scope of this book (see the *Highway Capacity Manual* [Transportation Research Board, 2016]). Of course, saturation flow rates can be measured directly in the field, in which case no further adjustments are necessary. For the rest of this chapter, it should be assumed that the provided saturation flow rates have been adjusted for the given conditions; the term "adjusted" has been dropped just for notational convenience.

Lost Time

Due to the traffic signal's function of continuously alternating the right-of-way between conflicting movements, traffic streams are continuously started and stopped. Every time this happens, a portion of the cycle length is not being completely utilized, which translates to lost time (time that is not effectively serving any movement of traffic). Total lost time is a combination of start-up and clearance lost times. Start-up lost times occur because when a signal indication turns from red to green, drivers in the queue do not instantly start moving at the saturation flow rate; there is an initial lag due to drivers reacting to the change of signal indication and the time it takes for vehicles to accelerate to their saturation-flow speed. This start-up delay results in a portion of the green time for that movement not being completely utilized. This start-up lost time has a typical value of around 2 seconds. This concept is illustrated in Fig. 7.5.

In this figure, note that the headway for the first several vehicles is larger than the saturation headway. The saturation headway is typically reached after the fourth vehicle in the queue. The summation of the amount of headway time greater than the saturation headway for each of the first 4 vehicles yields the total start-up lost time for the movement.

The stopping of a traffic movement also results in lost time. When the signal indication turns from green to yellow, the latter portion of time during the yellow interval is generally not utilized by traffic. Additionally, if there is an all-red interval, this time period is generally not utilized by traffic. These periods of time during the change and clearance intervals that are not effectively used by traffic are referred to as clearance lost time. Typically, the last second of the yellow interval and the entire all-red interval are included in the estimate of clearance lost time. However, for intersections with significant red-light running, the clearance lost time may be negligible.

Start-up and clearance lost times are summed to arrive at a total lost time for the phase, given as

$$t_L = t_{sl} + t_{cl} \tag{7.2}$$

where

t_L = total lost time for a movement during a cycle in seconds,

t_{sl} = start-up lost time in seconds (typical value of 2 seconds), and

t_{cl} = clearance lost time in seconds (typically the last second of the yellow interval plus the entire all-red interval).

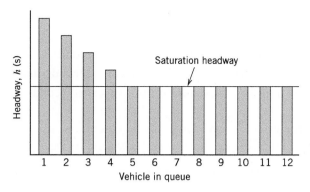

Figure 7.5 Concept of saturation headway and lost time.

This amount of time remains fixed, regardless of phase or cycle length. Thus, for shorter cycle lengths, the lost time will comprise a larger percentage of the cycle length and will result in a larger amount of lost time over the course of a day compared with longer cycle lengths. However, longer cycle lengths usually have more phases than shorter cycle lengths, which may result in similar proportions of lost time.

Effective Green and Red Times

For analysis purposes, the time during a cycle that is effectively (or not effectively) utilized by traffic must be used rather than the time for which green, yellow, and red signal indications are actually displayed, because they are most likely different. This results in two measures of interest: the effective green time and the effective red time. The effective green time is the time during which a traffic movement is effectively utilizing the intersection. The effective green time for a given movement or phase is calculated as

$$g = G + Y + AR - t_L \qquad\qquad (7.3)$$

where

$\quad g$ = effective green time for a traffic movement in seconds,
$\quad G$ = displayed green time for a traffic movement in seconds,
$\quad Y$ = displayed yellow time for a traffic movement in seconds,
AR = displayed all-red time in seconds, and
$\quad t_L$ = total lost time for a movement during a cycle in seconds.

The effective red time is the time during which a traffic movement is not effectively utilizing the intersection. The effective red time for a given movement or phase is calculated as

$$r = R + t_L \qquad\qquad (7.4)$$

where

r = effective red time for a traffic movement in seconds,

R = displayed red time for a traffic movement in seconds, and

t_L = total lost time for a movement during a cycle in seconds.

Alternatively, the effective red time can be calculated as follows, assuming that the cycle length and effective green time have already been determined:

$$r = C - g \qquad (7.5)$$

where

C = cycle length in seconds, and

other terms are as defined previously.

Likewise, the effective green time can be calculated by subtracting the effective red time from the cycle length.

Capacity

Because movements on an intersection approach do not receive a constant green indication (as assumed in the definition for saturation flow rate), another measure must be defined that accounts for the hourly volume that can be accommodated on an intersection approach given that the approach will receive less than 100% green time. This measure is capacity and is given by

$$c = s \times g/C \qquad (7.6)$$

where

c = capacity (the maximum hourly volume that can pass through an intersection from a lane or group of lanes under prevailing roadway, traffic, and control conditions) in veh/h,

s = saturation flow rate in veh/h, and

g/C = ratio of effective green time to cycle length.

7.4   DEVELOPMENT OF A TRAFFIC SIGNAL PHASING AND TIMING PLAN

Assuming the decision to install a traffic signal at an intersection has been made, an appropriate phasing and timing plan must be developed. The development of a traffic signal phasing and timing plan can be complex, particularly if the intersection has multiple-lane approaches and requires protected turning movements (a turn arrow). However, the timing plan analysis can be simplified by dealing with each approach separately. This section provides the basic process and fundamentals needed to develop a phasing and timing plan for an isolated, fixed-time (pretimed) traffic signal. As timing plans become more complex, they simply build on these fundamental principles. The reader is encouraged to review the material in other references to see how actuated and progressive timing plans are developed [Kell and Fullerton, 1998; Transportation Research Board, 2016]. This section describes the basic process that results in the development of a signal phasing and timing plan.

7.4.1   Select Signal Phasing

Recall that a timing stage consists of the time during which the same combination of movements, across one or more phases, are moving simultaneously and that the sum of the timing stage lengths is equal to the cycle length. The most basic traffic signal cycle is made up of two timing stages, as shown in Fig. 7.6. In this case, timing stage 1 accommodates the movement of all the eastbound and westbound vehicles, and timing stage 2 accommodates the movement of all the northbound and southbound vehicles. These timing stages will alternate during the continuous operation of the signal. This timing-stage scheme, however, could prove to be very inefficient if one or more of the approaches includes a high left-turn volume.

Given that each approach consists of one lane, vehicles will be delayed behind a left-turning vehicle waiting for a gap in the opposing traffic stream. If the high volume of left turns is present on both the northbound and southbound approaches, for example, each of these approaches could be given a separate timing stage consisting of the left-turn phases. This would be more efficient because left-turning vehicles on these two approaches would not have to wait for gaps in the opposing traffic stream, thus greatly reducing delays for all vehicles. This would result in a three-timing-stage operation, as shown in Fig. 7.6. When phases for movements on opposing approaches are run in a separate timing stage, as in this case, it is referred to as split phasing.

Some common signal timing-stage configurations, along with the applicable dual-ring phase numbers, are shown in Fig. 7.7. In this figure, note that the dashed lines represent permitted movements and the solid lines represent protected movements. When the left turns precede the through and right-turn movements in the timing-stage sequence for an approach, they are referred to as leading left turns. When the left turns follow the through and right-turn movements, they are referred to as lagging left turns. Although not shown in Fig. 7.7, it is also possible for a movement to be protected for a period of time and then permitted for a period of time, or vice versa. This is most commonly seen with left-turn movements, and is referred to as protected plus permitted or permitted plus protected, depending on the sequence.

It is important to remember that there is lost time (start-up and clearance) associated with each timing stage. Thus, with each timing stage added to a cycle, the lost time increases. Although the lost time may be only 3 to 5 seconds per timing stage, the accumulated lost time throughout the day can be significant.

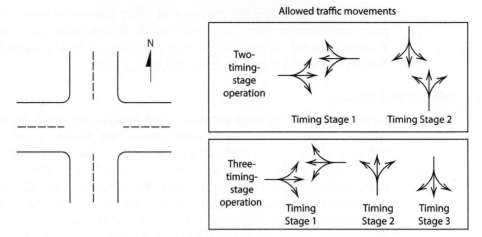

Figure 7.6 Illustration of two-timing and three-timing signal operation.

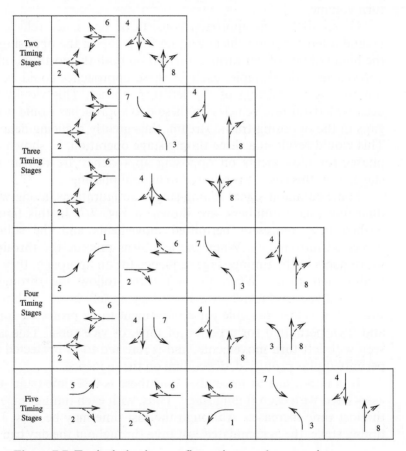

Figure 7.7 Typical phasing configurations and sequencing.

Given this, a point of diminishing returns is reached with the addition of phases as the efficiencies gained by separating traffic movements eventually become outweighed by the inefficiencies of increased lost time. Thus, a primary concern in signal timing is to keep the number of phases to a minimum. Because protected-turn phases add to lost time, they should be used only when warranted. Because of opposing motor vehicle traffic, left-turn movements typically require a protected-turn phase much more often than right turns. There are no nationally established guidelines on when protected left-turn phasing should be used, so local policies and practices should be consulted before a decision is made about whether to provide a protected left-turn phase. In general, decisions on whether to provide a protected left-turn phase are based on one or more of the following factors:

- Volume (just left turn or combination of left turn and opposing volume)
- Delay
- Queuing (spillover)
- Traffic progression
- Opposing traffic speeds
- Geometry (number of left-turn lanes, crossing distance, sight distance)
- Crash experience (which may also be related to any of the above factors)

More specific guidance on this issue can be found in several references, including the *Highway Capacity Manual* [Transportation Research Board, 2016], the *Traffic Control Devices Handbook* [ITE, 2001], and the *Manual of Traffic Signal Design* [Kell and Fullerton, 1998].

One of the more common guidelines is the use of the cross product of left-turn volume and opposing through and right-turn volumes. The *Highway Capacity Manual* offers the following criteria for this guideline: The use of a protected left-turn phase should be considered when the product of left-turning vehicles and opposing traffic volume exceeds 50,000 during the peak hour for one opposing lane, 90,000 for two opposing lanes, or 110,000 for three or more opposing lanes.

EXAMPLE 7.1   DETERMINE LEFT-TURN PHASING

Refer to the intersection shown in Fig. 7.8. Use the cross-product guideline to determine if protected left-turn phases should be provided for any of the approaches.

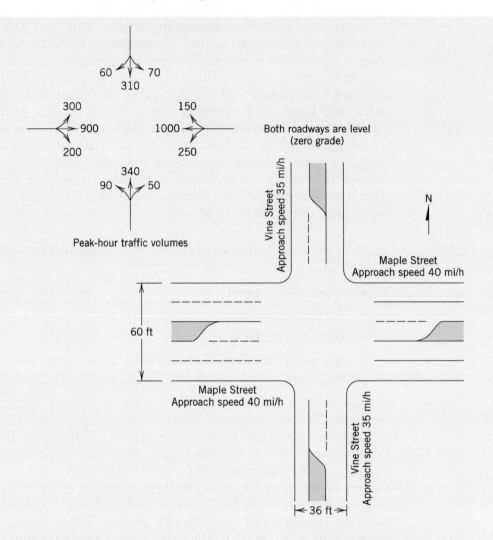

Figure 7.8 Intersection geometry and peak traffic volumes for example problems.

SOLUTION

There are 250 westbound vehicles that turn left during the peak hour. The product of the westbound left-turning vehicles and the opposing eastbound traffic (right-turn and straight-through vehicles) is 275,000 [250 × (900 + 200)]. There are 300 eastbound vehicles that turn left during the peak hour. The product of the eastbound left-turning vehicles and the opposing westbound traffic (right-turn and straight-through vehicles) is 345,000 [300 × (1000 + 150)].

Because the cross product for each of these approaches is greater than 90,000 (the requirement for two opposing lanes), a protected left-turn phase is suggested for the WB and EB left-turn movements. The NB and SB approaches do not require a separate timing stage for the left-turn movements using this criterion because the cross products for these approaches are less than 50,000 (for one opposing lane). Therefore, a three-timing-stage plan is recommended, as shown in Fig. 7.9.

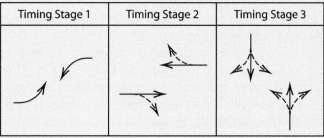

Figure 7.9 Recommended signal phasing plan for the intersection in Example 7.1.

7.4.2   Establish Analysis Lane Groups

Each intersection approach is initially treated separately, and the results are later aggregated. Thus, each approach must be subdivided into logical groupings of traffic movements for analysis purposes.

The *Highway Capacity Manual* [Transportation Research Board, 2016] provides detailed guidelines on this process. Generally, the process consists of placing like movements into *movement groups* (left-turn, right-turn, and through movements would be identified) and then translating *movement groups* to *lane groups* based on the allowable movements from each lane. The only time these two group designations differ is when a shared lane (left-turn and through movements allowed from the same lane) is present on an approach with two or more lanes. When shared lanes are present on an approach with two or more lanes, the *Highway Capacity Manual* [Transportation Research Board, 2016] employs an iterative procedure to identify the expected distribution of each movement type in each lane, based on the principle that drivers will choose the lane that they perceive will minimize their travel time through the intersection (delay). This procedure is beyond the scope of this book and, as a result, the subsequent example problems and end-of-chapter problems will provide specific lane distributions of traffic as necessary. Consequently, it is only necessary to make reference to lane groups for the remainder of the chapter.

Based on the lane and traffic movement distribution on an approach, lane groups can be readily determined. The following general guidelines are offered for establishing lane groups [Transportation Research Board, 2016]:

- If an exclusive turn lane (or lanes) is present, it should be treated as a separate lane group.
- Each shared lane on an approach should be treated as a separate lane group.
- Any remaining lanes, which would be exclusive through lanes, should be treated as a separate lane group.

Figure 7.10 shows some typical lane groupings for analysis purposes. Note that when multiple lanes are combined into a lane group, the subsequent analysis calculations for this lane group should treat these lanes as a single unit. Because of the aforementioned complexity of dealing with shared lanes, the subsequent example and end-of-chapter problems in this book assume that adjacent exclusive through lanes and shared through/right-turn lanes will be combined into one lane group (such as is done in Example 7.2). This assumption will often introduce only very minimal error when compared to treating the exclusive through lane and shared through/right-turn lane as separate lane groups and does not affect learning the basic concept of aggregating lane group results into overall approach results (this concept is covered later in this chapter).

Number of Approach Lanes	Movements by Lane and Corresponding Lane Groups	Number of Approach Lanes	Movements by Lane and Corresponding Lane Groups
1	LT + TH + RT	3	EXC LT EXC TH TH + RT
2	EXC LT TH + RT	3	EXC LT EXC TH EXC RT
2	LT + TH TH + RT		

Figure 7.10 Example lane groupings for analysis (LT = left turn; TH = through; RT = right turn; EXC = exclusive).

Adapted from Exhibit 19-19, p. 19-43, *Highway Capacity Manual*, 6th Edition. Copyright, Transportation Research Board, National Academy of Sciences, Washington, D.C. 2016.

EXAMPLE 7.2   DETERMINE LANE GROUPS

Determine the lane groups to use for analysis of the Maple and Vine Streets intersection (Fig. 7.8).

SOLUTION

The EB and WB left-turn movements will each be a lane group because they have a separate lane and move in a separate phase from the through/right-turn movements. Likewise, the EB and WB through/right-turn movements proceed together in a separate phase and will therefore be separate lane groups. Although the right turns use only the outside lane, this movement's impact on the saturation flow rate for the two lanes combined will be determined. The NB and SB left turns will also each be a separate lane group. Even though they move during the same phase as the adjacent through and right-turn movements, these left turns are permitted and will have very different operating characteristics from the through and right-turn movements.

Because the through and right-turn movements use the same lane, they will be an individual lane group for both the NB and SB approaches. The recommended lane groups for analysis for each of the approaches are shown in Fig. 7.11.

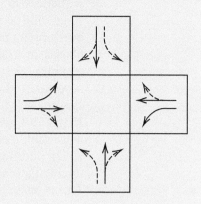

Figure 7.11 Analysis lane groups for the three-timing-stage design at the intersection of Maple Street and Vine Street.

7.4.3   Calculate Analysis Flow Rates and Adjusted Saturation Flow Rates

Just as for the analysis of uninterrupted flow, the hourly traffic volume arriving on each intersection approach must be converted to an analysis flow rate that accounts for the peak 15-minute flow within that hour (typically the peak hour). This is accomplished by calculating the peak-hour factor (*PHF*) and dividing this into the hourly volume (as shown in Chapter 6), which yields the analysis flow rate.

One note about adjusting for the *PHF*. With the multiple traffic streams entering an intersection, a separate *PHF* can be calculated for each approach's traffic stream. However, adjusting each approach volume by its specific *PHF* can yield unrealistically high combined analysis volumes, because the different approach volumes usually do not peak during the same 15-minute period. Applying a single *PHF* determined for the intersection as a whole will result in more reasonable analysis volumes.

The adjustment of the saturation flow rate was discussed in Section 7.3. Again, it is assumed that the approach volumes and saturation flow rates provided in this chapter have already been adjusted.

7.4.4   Determine Critical Lane Groups and Total Cycle Lost Time

For any combination of lane group movements during a particular timing stage, one of these lane groups will control the necessary green time for that timing stage. This lane group is referred to as the critical lane group. When the traffic movements of each lane group occur during only one timing stage of the signal cycle, the determination of the critical lane group for each timing stage is straightforward. In this case, the critical lane group for each phase is simply the lane group with the highest ratio of vehicle arrival rate to vehicle departure rate (λ/μ). This quantity is referred to as the flow ratio (which was called the traffic intensity, ρ, in Chapter 5) and is designated v/s (arrival flow rate divided by saturation flow rate). If the allocation of green time for each timing stage is based on the flow ratio of the critical lane group, then the noncritical lane group movements will be accommodated as well.

As previously discussed, with dual-ring controllers, a wide variety of phasing sequences is possible. The situation where a phase starts in one signal timing stage and continues in the next signal timing stage is referred to as an overlapping phase (see the last two rows of Fig. 7.7). For the case of overlapping phases in a signal cycle, the identification of the critical lane groups is more complex, as the lane group with the highest flow ratio in each timing stage is not necessarily the critical lane group for that timing stage. The remainder of this chapter will focus only on non-overlapping phases. The reader is referred to the *Highway Capacity Manual* [Transportation Research Board 2016] for details on determining critical lane groups for overlapping phases.

In addition, the sum of the flow ratios for the critical lane groups can be used to calculate a suitable cycle length, which will be discussed in the next section. This is given by

$$Y_c = \sum_{i=1}^{n} \left(\frac{v}{s}\right)_{ci} \tag{7.7}$$

where

Y_c = sum of flow ratios for critical lane groups,
$(v/s)_{ci}$ = flow ratio for critical lane group i, and
n = number of critical lane groups.

The total lost time for the cycle will also be used in the calculation of cycle length. In determining the total lost time for the cycle, the general rule is to apply the lost time for a critical lane group when its movements are initiated (the start of its green interval). The total cycle lost time is given as

$$L = \sum_{i=1}^{n} \left(t_L\right)_{ci} \tag{7.8}$$

where

L = total lost time for cycle in seconds,
$(t_L)_{ci}$ = total lost time for critical lane group i in seconds, and
n = number of critical lane groups.

EXAMPLE 7.3   CALCULATE SUM OF FLOW RATIOS

Calculate the sum of the flow ratios for the critical lane groups for the three-timing-stage timing plan determined in Example 7.1 given the saturation flow rates in Table 7.1.

Table 7.1 Saturation Flow Rates for Three-Timing-Stage Design at Intersection of Maple Street and Vine Street

Timing Stage 1	Timing Stage 2	Timing Stage 3
EB L: 1750 veh/h	EB T/R: 3400 veh/h	SB L: 450 veh/h
		NB L: 475 veh/h
WB L: 1750 veh/h	WB T/R: 3400 veh/h	SB T/R: 1800 veh/h
		NB T/R: 1800 veh/h

SOLUTION

Note that the saturation flow rates are relatively low for the SB and NB left (L) turns because they are permitted only, and the opposing through and right-turn (T/R) vehicles limit the number of usable gaps for these vehicles. The saturation flow rates for the WB and EB through and right-turn (T/R) movements account for both through lanes.

The flow ratios will now be calculated, with the critical lane group for each timing stage indicated with a check mark in Table 7.2. As indicated in the table, the critical lane group for timing stages 1, 2, and 3, respectively, are the EB left turn, the WB through and right turn, and the NB through and right turn.

Table 7.2 Flow Ratios and Critical Lane Groups for Three-Timing-Stage Design at Intersection of Maple Street and Vine Street

Timing Stage 1	Timing Stage 2	Timing Stage 3
EB L: $\dfrac{300}{1750} = 0.171$ ✓	EB T/R: $\dfrac{1100}{3400} = 0.324$	SB L: $\dfrac{70}{450} = 0.156$
		NB L: $\dfrac{90}{475} = 0.189$
WB L: $\dfrac{250}{1750} = 0.143$	WB T/R: $\dfrac{1150}{3400} = 0.338$ ✓	SB T/R: $\dfrac{370}{1800} = 0.206$
		NB T/R: $\dfrac{390}{1800} = 0.217$ ✓

The sum of the flow ratios for the critical lane groups for this timing plan will be needed for the next section. Since this timing plan does not include any overlapping phases, this value is simply the sum of the highest lane group v/s ratios for the three timing stages, as follows:

$$Y_c = \sum_{i=1}^{n}\left(\frac{v}{s}\right)_{ci}$$
$$= 0.171 + 0.338 + 0.217 = \underline{\underline{0.726}}$$

Assuming 2 seconds of start-up lost time and 2 seconds of clearance lost time (1 second of yellow time plus 1 second of all-red time), for each critical lane group, gives a lost time of 4 s/timing stage. The total lost time for the cycle is then 12 seconds (3 timing stages × 4 s/ timing stage).

EXAMPLE 7.4 CALCULATE SUM OF FLOW RATIOS AND TOTAL LOST TIME

Suppose it is necessary to run the NB and SB movements in a split-phase configuration (with timing stage 3 for SB movements and a new timing stage 4 for NB movements). Calculate the sum of the flow ratios for the critical lane groups and total cycle lost time for this situation, assuming that the EB and WB movement phasing remains the same. Table 7.3 summarizes the calculation of the flow ratios and the identification of the critical lane groups.

SOLUTION

The sum of the flow ratios for the critical lane groups for this phasing plan is

$$\sum_{i=1}^{n} \left(\frac{v}{s}\right)_{ci} = 0.171 + 0.338 + 0.206 + 0.217 = \underline{\underline{0.932}}$$

The total lost time for the cycle is $\underline{\underline{16 \text{ seconds}}}$ (4 timing stages × 4 s/timing stage).

Table 7.3 Flow Ratios and Critical Lane Groups for Four-Timing-Stage Design (Split Phasing for N-S Movements) at Intersection of Maple Street and Vine Street

Timing Stage 1	Timing Stage 2	Timing Stage 3	Timing Stage 4
EB L: $\frac{300}{1750} = 0.171$ ✓	EB T/R: $\frac{1100}{3400} = 0.324$	SB L: $\frac{70}{1750} = 0.040$	NB L: $\frac{90}{1750} = 0.051$
WB L: $\frac{250}{1750} = 0.143$	WB T/R: $\frac{1150}{3400} = 0.338$ ✓	SB T/R: $\frac{370}{1800} = 0.206$ ✓	NB T/R: $\frac{390}{1800} = 0.217$ ✓

7.4.5 Calculate Cycle Length

In practice, cycle lengths are generally kept as short as possible, typically between 60 and 75 seconds. However, complex intersections with five or more timing stages can have cycle lengths of 120 seconds or more. The minimum cycle length necessary for the lane group volumes and phasing plan of an intersection is given by

$$C_{\min} = \frac{L \times X_c}{X_c - \sum_{i=1}^{n} \left(\frac{v}{s}\right)_{ci}} \tag{7.9}$$

where

$\quad C_{\min}$ = minimum necessary cycle length in seconds (typically rounded up to the nearest 5-second increment in practice),

$\qquad L$ = total lost time for cycle in seconds,

$\qquad X_c$ = critical *v/c* ratio for the intersection,

$\quad (v/s)_{ci}$ = flow ratio for critical lane group i, and

$\qquad n$ = number of critical lane groups.

In this equation, the total lost time for the cycle and the sum of the flow ratios for the critical lane groups are predetermined. However, a critical intersection volume/capacity ratio, X_c, must be chosen for the desired degree of utilization. In other words, if it is desired that the intersection operate at its full capacity, a value of 1.0 is used for X_c. A value of 1.0 is not generally recommended, however, due to the randomness of vehicle arrivals, which can result in occasional cycle failures. Note that this equation gives the minimum cycle length necessary for the intersection to operate at a specified degree of capacity utilization. This cycle length does not necessarily minimize the average vehicle delay experienced by motorists at the intersection.

A practical equation for the calculation of the cycle length that seeks to minimize vehicle delay was developed by Webster [1958]. Webster's optimum cycle length formula is

$$C_{opt} = \frac{1.5 \times L + 5}{1.0 - \sum_{i=1}^{n} \left(\frac{v}{s}\right)_{ci}} \tag{7.10}$$

where

      C_{opt} = cycle length to minimize delay in seconds, and
      other terms are as defined previously.

The cycle length determined from this calculation is only approximate. Webster noted that values between $0.75C_{opt}$ and $1.5C_{opt}$ will likely give similar values of delay. Calculating an accurate optimal cycle length (and phase length) can be a very computationally intensive exercise for all but the simplest signalized intersections, especially if coordination among multiple signals is involved.

It should be noted that regardless of the minimum or optimal cycle length calculated, practical maximum cycle lengths must generally be observed. Public acceptance or tolerance of large cycle lengths will vary by location (urban vs. rural), but as a rule, cycle lengths in excess of 3 minutes (180 seconds) should be used only in exceptional circumstances.

EXAMPLE 7.5   **CALCULATE MINIMUM AND OPTIMAL CYCLE LENGTHS**

Calculate the minimum and optimal cycle lengths for the intersection of Maple and Vine Streets, using the information provided in the preceding examples, for both the three-timing-stage and four-timing-stage design.

SOLUTION

For the three-timing-stage design (Example 7.3), the sum of the flow ratios for the critical lane groups and the total cycle lost time were determined to be 0.726 and 12 seconds, respectively. For the minimum cycle length, a somewhat conservative value of 0.9 will be used for the critical intersection *v/c* ratio to minimize the potential of cycle failures due to occasionally high arrival volumes. Using these values in Eq. 7.9 gives

$$C_{min} = \frac{12 \times 0.9}{0.9 - 0.726} = 62.1 \rightarrow \underline{\underline{65}} \text{ s (rounding up to nearest 5 seconds)}$$

Using Eq. 7.10 for the optimal cycle length gives

$$C_{opt} = \frac{1.5 \times 12 + 5}{1.0 - 0.726} = 83.9 \rightarrow \underline{\underline{85}} \text{ s (rounding up to nearest 5 seconds)}$$

For the four-timing-stage design (Example 7.4), the sum of the critical flow ratios and the total cycle lost time were determined to be 0.932 and 16 seconds, respectively. The first issue with this design is that a higher X_c will need to be used because the sum of flow ratios for critical lane groups is higher than the 0.90 used for the three-phase design (otherwise the denominator of Eq. 7.9 will be negative). To minimize the cycle length, the maximum value of 1.0 will be used for X_c in Eq. 7.9, as follows:

$$C_{min} = \frac{16 \times 1.0}{1.0 - 0.932} = \underline{\underline{235.3 \text{ s}}}$$

The second issue is that despite the use of an X_c value of 1.0 (the intersection operating at capacity) to minimize the cycle length, an unreasonably high cycle length is still required for this design. Thus, this design is not nearly as desirable as the three-timing-stage design.

Generally a split-phase design is recommended only under one or more of the following conditions:

- The left turns are the dominant movement.
- The left turns share a lane with the through movement.
- There is a large difference in the total approach volumes.
- There are unusual opposing approach geometrics.

It should also be noted that serving pedestrians in an efficient manner on split-phase approaches can be difficult.

7.4.6   Allocate Green Time

After a cycle length has been calculated, the next step in the traffic signal timing process is to determine how much green time should be allocated to each timing stage. The cycle length is the sum of all effective green times plus the total lost time. Thus, after subtracting the total lost time from the cycle length, the remaining time can be distributed as green time among the timing stages of the cycle.

There are several strategies for allocating the green time to the various timing stages. One of the most popular and simplest is to distribute the green time so that the v/c ratios are equalized for the critical lane groups, as by the following equation:

$$g_i = \left(\frac{v}{s}\right)_{ci} \left(\frac{C}{X_i}\right) \tag{7.11}$$

where

$$g_i = \text{effective green time for timing stage } i,$$
$$(v/s)_{ci} = \text{flow ratio for critical lane group } i,$$
$$C = \text{cycle length in seconds, and}$$
$$X_i = v/c \text{ ratio for lane group } i.$$

EXAMPLE 7.6   DETERMINE GREEN TIMES

Determine the green-time allocations for the 65-second cycle length found in Example 7.5, using the method of v/c ratio equalization.

SOLUTION

Because the calculated cycle length was rounded up a few seconds, the critical intersection v/c ratio for this rounded cycle length will be calculated for use in the green-time allocation calculations. Equation 7.9 can be rearranged to solve for X_c as follows:

$$X_c = \frac{\sum_{i=1}^{n}\left(\frac{v}{s}\right)_i \times C}{C - L}$$

Using this equation with

$$\sum (v/s)_{ci} = 0.726 \text{ (Example 7.3)}$$
$$C = 65 \text{ s (Example 7.5)}$$
$$L = 12 \text{ s (Example 7.4)}$$

gives

$$X_c = \frac{0.726 \times 65}{65 - 12} = 0.890$$

Therefore, the cycle length of 65 seconds and X_c of 0.890 are used to calculate the effective green times, from Eq. 7.11, for the three timing stages, as follows:

$$g_1 = \left(\frac{v}{s}\right)_{c1}\left(\frac{C}{X_1}\right)$$

(EB and WB left-turn movements)

$$= 0.171 \times \frac{65}{0.890} = \underline{\underline{12.5 \text{ s}}}$$

$$g_2 = \left(\frac{v}{s}\right)_{c2}\left(\frac{C}{X_2}\right)$$

(EB and WB through and right-turn movements)

$$= 0.338 \times \frac{65}{0.890} = \underline{\underline{24.7 \text{ s}}}$$

$$g_3 = \left(\frac{v}{s}\right)_{c3}\left(\frac{C}{X_3}\right)$$

(NB and SB left-turn, through, and right-turn movements)

$$= 0.217 \times \frac{65}{0.890} = \underline{\underline{15.8\,\text{s}}}$$

The cycle length is checked by summing these effective green times and the lost time, giving

$$C = g_1 + g_2 + g_3 + L$$
$$= 12.5 + 24.7 + 15.8 + 12 = 65.0$$

Therefore, all calculations are correct.

7.4.7   Calculate Change and Clearance Intervals

Recall that the change interval corresponds to the yellow time and the clearance interval corresponds to the all-red time. If an all-red interval does not exist, then the yellow time is considered as both the change and clearance intervals. The change interval alerts drivers that the green interval is about to end and that they should come to stop before entering the intersection, or continue through the intersection if they are too close to come to a safe stop. The clearance interval allows those vehicles that might have entered the intersection at the end of the yellow to clear the intersection before conflicting traffic movements are given a green signal indication. In the past, the yellow indication was intended to also allow for clearance time. Today, however, there is routine red-indication abuse and frequent running of red indications after the yellow time. As a result, the all-red indication is often implemented.

Typically, the yellow time is in the range of 3 to 5 seconds. Warning times that are shorter than 3 seconds and longer than 5 seconds are not practical because long warning times encourage motorists to continue to enter the intersection whereas short times can place the driver in a dilemma zone. A dilemma zone is created for the driver if a safe stop before the intersection cannot be accomplished, and continuing through the intersection at a constant speed (without accelerating) will result in the vehicle entering the intersection during a red indication. If a dilemma zone exists, drivers always make the wrong decision, whether they decide to stop or to continue through the intersection. Figure 7.12 illustrates the dilemma zone. Referring to this figure, suppose a vehicle traveling at a constant speed requires distance x_s to stop. If the vehicle is closer to the intersection than distance d_d, then it can enter before the all-red indication. If the vehicle is in the shaded area ($x_s - d_d$ from the intersection) when the yellow light is displayed, the driver is in the dilemma zone and can neither stop in time nor continue through the intersection at a constant speed without passing through a red indication.

Formulas and policies for calculating yellow (Y) and all-red (AR) times vary by agency, but one set of commonly accepted formulas is provided in the *Traffic Engineering Handbook* [ITE 1999] and are as follows:

$$Y = t_r + \frac{V}{2a + 2gG} \qquad\qquad (7.12)$$

where

Y = yellow time (usually rounded up to the nearest 0.5 seconds),
t_r = driver perception/reaction time, usually taken as 1.0 second,
V = speed of approaching traffic in ft/s,
a = deceleration rate for the vehicle, usually taken as 10.0 ft/s^2,
g = acceleration due to gravity [32.2 ft/s^2], and
G = percent grade divided by 100.

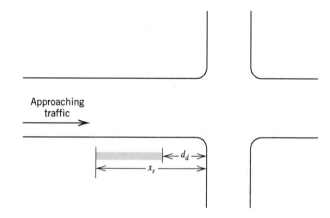

Figure 7.12 The dilemma zone for traffic approaching a signalized intersection.

and

$$AR = \frac{w+l}{V}$$ (7.13)

where

AR = all-red time (usually rounded up to the nearest 0.5 seconds),
w = width of the cross street in ft,
l = length of the vehicle, usually taken as 20 ft, and
V = speed of approaching traffic in ft/s.

To avoid a dilemma zone and the possibility of a vehicle being in the intersection when a conflicting movement receives a green-signal indication, the total of the change and clearance intervals (yellow plus all-red times) should always be equal to or greater than the sum of Eqs. 7.12 and 7.13.

EXAMPLE 7.7   DETERMINE YELLOW AND ALL-RED TIMES

Determine the yellow and all-red times for vehicles traveling on Vine and Maple Streets as shown in Fig 7.8.

SOLUTION

For the Vine Street timing stages (applying Eqs. 7.12 and 7.13),

$$Y = 1.0 + \frac{(35 \times 5280/3600)}{2(10)}$$

$$= 3.6 \rightarrow \underline{\underline{4.0\,\text{s}}}\,(\text{rounding up to the nearest } 0.5\,\text{s})$$

$$AR = \frac{60 + 20}{35 \times 5280/3600}$$

$$= 1.6 \rightarrow \underline{\underline{2.0\,\text{s}}}\,(\text{rounding up to the nearest } 0.5\,\text{s})$$

For the Maple Street timing stages (applying Eqs. 7.12 and 7.13),

$$Y = 1.0 + \frac{(40 \times 5280/3600)}{2(10)}$$

$$= 3.9 \rightarrow \underline{\underline{4.0\,\text{s}}}\,(\text{rounding up to the nearest } 0.5\,\text{s})$$

$$AR = \frac{36 + 20}{40 \times 5280/3600}$$

$$= \underline{\underline{1.0\,\text{s}}}$$

Note that separate calculations are usually required for exclusive left-turn phases, as vehicle approach speeds are often lower than for through vehicles and intersection crossing distances may be longer (due to the width of the opposing direction and the circular travel path).

7.4.8   Check Pedestrian Crossing Time

In urban areas and other locations where pedestrians are present, the signal-timing plan should be checked for its ability to provide adequate pedestrian crossing time. At locations where streets are wide and green times are short, it is possible that pedestrians can be caught in the middle of the intersection when the phase changes. To avoid this problem, the minimum green time required for pedestrian crossing should be checked against the apportioned green time for the phase. If there is not enough green time for a pedestrian to safely cross the street, the apportioned green time should be increased to meet the pedestrian needs. If pedestrian pushbuttons are provided at an intersection (for actuated control), the green time can be increased to meet pedestrian crossing needs only when the pushbuttons are activated.

The minimum pedestrian green time is given by

$$G_p = 3.2 + \frac{L}{S_p} + \left(0.27 N_{ped}\right) \quad \text{for } W_E \leq 10\,\text{ft} \tag{7.14}$$

$$G_p = 3.2 + \frac{L}{S_p} + \left(2.7 \frac{N_{ped}}{W_E}\right) \quad \text{for } W_E > 10\,\text{ft} \tag{7.15}$$

where

G_p = minimum pedestrian green time in seconds,

3.2 = pedestrian start-up time in seconds,

L = crosswalk length in ft,

S_p = walking speed of pedestrians, usually taken as 3.5 ft/s,

N_{ped} = number of pedestrians crossing during an interval, and

W_E = effective crosswalk width in ft.

The generally recommended walking speed of 3.5 ft/s [U.S. Federal Highway Administration 2009] represents a slower-than-average speed. However, at intersections where a significant number of slower pedestrians (elderly, vision impaired, etc.) are served, the use of a slower walking speed may be warranted.

EXAMPLE 7.8   DETERMINE PEDESTRIAN GREEN TIME

Determine the minimum amount of pedestrian green time required for the intersection of Vine and Maple Streets. Assume a maximum of 15 pedestrians crossing either street during any one phase and a crosswalk width of 8 ft.

SOLUTION

A pedestrian who crosses Maple Street will cross while Vine Street has a green interval. The minimum pedestrian green time needed on Vine Street is (using Eq. 7.14, as the effective crosswalk width is less than or equal to 10 ft)

$$G_p = 3.2 + \frac{60}{3.5} + (0.27 \times 15) = \underline{\underline{24.4 \text{ s}}}$$

In Example 7.6, Vine Street was assigned 15.8 seconds of effective green time [13.8 seconds of displayed green time (from Eq. 7.3)]. This amount of time is insufficient for pedestrians crossing Maple Street. Therefore, the green time for this phase will have to be increased to accommodate crossing pedestrians, and the overall signal timing plan adjusted accordingly (although we will continue to use the previously computed green time in subsequent examples). The minimum pedestrian green time needed on Maple Street (for the through/right-turn phase, when pedestrian movement would be permitted) is

$$G_p = 3.2 + \frac{36}{3.5} + (0.27 \times 15) = \underline{17.5 \text{ s}}$$

In Example 7.6, Maple Street was assigned 24.7 seconds of effective green time (23.7 seconds of displayed green time) for this phase, so this green time is adequate for pedestrians crossing Vine Street.

7.5   ANALYSIS OF TRAFFIC AT SIGNALIZED INTERSECTIONS

This section will utilize and build upon the elements of traffic flow theory introduced in Chapter 5 and Section 7.3 to make possible the basic analysis of traffic flow at signalized intersections.

7.5.1 Signalized Intersection Analysis with D/D/1 Queuing

The assumption of *D/D/1* queuing (as discussed in Chapter 5) provides a strong intuitive appeal that helps in understanding the analytical fundamentals underlying traffic analysis at signalized intersections. To begin applying *D/D/1* queuing to signalized intersections, we consider the case where the approach capacity exceeds the approach arrivals. Under these conditions, and the assumption of uniform arrivals throughout the cycle and uniform departures during green, a *D/D/1* queuing system as shown in Fig. 7.13 will result.

Note that this chapter will use the variables *v* (for arrival rate) and *s* (for departure/saturation flow rate), rather than the variables λ and μ used in Chapter 5, as these variables are more commonly used in signalized intersection analyses.

The "Arrivals *v×t*" line gives the total number of vehicle arrivals at time *t*, and the "Departures *s×t*" line gives the slope of vehicle departures (number of vehicles that depart) during the effective greens. Note that the per-cycle approach arrivals will be *vC* and the corresponding approach capacity (maximum departures) per cycle will be *sg*. Figure 7.13 is predicated on the assumption that *sg* exceeds *vC* for all cycles (no queues exist at the beginning or end of a cycle).

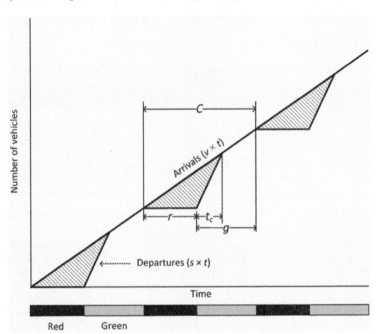

Figure 7.13 *D/D/1* signalized intersection queuing with approach capacity (*sg*) exceeding arrivals (*vC*) for all cycles.

v = arrival rate, typically in veh/s,

s = departure rate, typically in veh/s,

t = elapsed time since a reference time, typically the start of green or red, in seconds,

t_c = time from the start of the effective green until queue clearance in seconds,

r = effective red time in seconds,

g = effective green time in seconds, and

C = cycle length in seconds.

Given the properties of *D/D/1* queues presented in Chapter 5, a number of general equations can be derived by simple inspection of Fig. 7.13:

1. The time to queue clearance after the start of the effective green, t_c (note that $v(r + t_c) = st_c$),

$$t_c = \frac{vr}{s - v} \tag{7.16}$$

2. The proportion of the cycle with a queue, P_q,

$$P_q = \frac{r + t_c}{C} \tag{7.17}$$

3. The proportion of vehicles stopped, P_s,

$$P_s = \frac{v(r + t_c)}{v(r + g)} = \frac{r + t_c}{C} = P_q$$

$$\text{also, } P_s = \frac{v(r + t_c)}{v(r + g)} = \frac{st_c}{vC} = \frac{t_c}{\dfrac{v}{s}C} \tag{7.18}$$

4. The maximum number of vehicles in the queue, Q_{max},

$$Q_{max} = vr \tag{7.19}$$

5. The total vehicle delay per cycle, D_t,

$$D_t = \frac{vr^2}{2(1 - v/s)} \tag{7.20}$$

6. The average delay per vehicle, d_{avg},

$$d_{avg} = \frac{vr^2}{2(1 - v/s)} \times \frac{1}{vC} = \frac{r^2}{2C(1 - v/s)}$$

$$\text{also, } d_{avg} = \frac{0.5C\left(1 - \dfrac{g}{C}\right)^2}{1 - \left(\dfrac{v}{c} \times \dfrac{g}{C}\right)} \tag{7.21}$$

7. The maximum delay of any vehicle, assuming a FIFO queuing discipline, d_{max},

$$d_{max} = r \tag{7.22}$$

EXAMPLE 7.9   SIGNALIZED INTERSECTION ANALYSIS USING EQS. 7.16–7.22

An approach at a pretimed signalized intersection has a constant saturation flow rate of 2400 veh/h and is allocated 24 seconds of effective green time in an 80-second signal cycle. If the total approach flow rate is 500 veh/h and arrivals are uniform throughout the cycle, provide an analysis of the approach assuming $D/D/1$ queuing.

SOLUTION

Putting arrival and departure rates into common units of vehicles per second,

$$v = \frac{500 \text{ veh/h}}{3600 \text{ s/h}} = 0.139 \text{ veh/s}$$

$$s = \frac{2400 \text{ veh/h}}{3600 \text{ s/h}} = 0.667 \text{ veh/s}$$

Checking to make certain that capacity exceeds arrivals, note that the capacity (sg) is 16 veh/cycle (0.667×24), which is greater than (permitting fractions of vehicles for the sake of clarity) the 11.12 arrivals ($vC = 0.139 \times 80$). Therefore, Eqs. 7.16 to 7.22 are valid. By definition,

$$r = C - g$$
$$= 80 - 24 = 56 \text{ s}$$

This leads to the following values:

1. Time to queue clearance after the start of the effective green (Eq. 7.16),

$$t_c = \frac{0.139(56)}{(0.667 - 0.139)}$$
$$= 14.74 \text{ s}$$

2. Proportion of the cycle with a queue (Eq. 7.17),

$$P_q = \frac{56 + 14.74}{80}$$
$$= 0.884$$

3. Proportion of vehicles stopped (Eq. 7.18),

$$P_s = \frac{14.74}{0.139/0.667(80)}$$
$$= 0.884$$

4. Maximum number of vehicles in the queue (Eq. 7.19),

$$Q_{max} = 0.139(56)$$
$$= 7.78 \text{ veh}$$

5. Total vehicle delay per cycle (Eq. 7.20),

$$D_t = \frac{0.139(56)^2}{2(1 - 0.139/0.667)}$$
$$= 275.33 \text{ veh-s}$$

6. Average delay per vehicle (Eq. 7.21),

$$d_{avg} = \frac{56^2}{2(80)(1 - 0.139/0.667)}$$
$$= \underline{\underline{24.76 \text{ s/veh}}}$$

7. Maximum delay of any vehicle (Eq. 7.22),

$$d_{max} = r$$
$$= \underline{\underline{56 \text{ s}}}$$

EXAMPLE 7.10   SIGNALIZED INTERSECTION ANALYSIS WITH *D/D/*1 QUEUING

Confirm the average delay result from Example 7.9 using a *D/D/*1 queuing diagram.

SOLUTION

Again, with arrival and departure rates in units of vehicles per second,

$$v = 0.139 \text{ veh/s}$$
$$s = 0.667 \text{ veh/s}$$

and 24 seconds of effective green time in an 80-second signal cycle yields the following graph (Fig. 7.14) of cumulative arrivals and cumulative departures for one cycle.

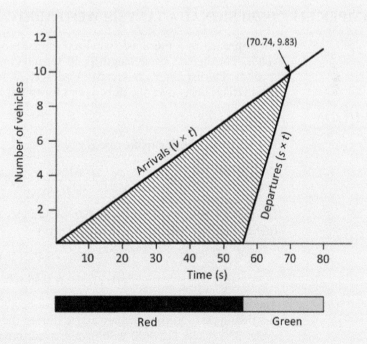

Figure 7.14 *D/D/*1 queuing diagram for Example 7.10.

This leads to the following calculations for average delay:

1. Time for queue clearance after the start of the effective green remains unchanged at 14.74 s,

2. Total vehicle delay per cycle,

$$D_t = 0.5 \times r \times \left[v \times (r + t_c) \right]$$
$$= 0.5 \times 56 \times \left[0.139 \times (56 + 14.74) \right]$$
$$= \underline{\underline{275.32 \text{ veh-s}}}$$

3. Average delay per vehicle,

$$d_{avg} = \frac{D_t}{vC} = \frac{275.32}{0.139 \times 80}$$
$$= \underline{\underline{24.76 \text{ s/veh}}}$$

This average delay value matches with that from Example 7.9.

Recall that Eqs. 7.16 through 7.22 are valid only when the arrivals are uniform throughout the cycle, the saturation flow rate is constant during the effective green period, and the approach capacity exceeds approach arrivals. For the case when approach arrivals exceed capacity for some signal cycles, $D/D/1$ queuing can again be used, as illustrated in the following example.

EXAMPLE 7.11 *$D/D/1$ SIGNAL ANALYSIS WITH ARRIVALS EXCEEDING CAPACITY*

An approach to a pretimed signalized intersection has a saturation flow rate of 1700 veh/h. The signal's cycle length is 60 seconds and the approach's effective red is 40 seconds. During three consecutive cycles 15, 8, and 4 vehicles arrive. Determine the total vehicle delay over the three cycles for this approach assuming $D/D/1$ queuing.

SOLUTION

For all cycles, the departure rate is

$$s = \frac{1700 \text{ veh/h}}{3600 \text{ s/h}} = 0.472 \text{ veh/s}$$

During the first cycle, the number of vehicles that will depart from the signal is (permitting fractions for the sake of clarity)

$$sg = 0.472(20)$$
$$= 9.44 \text{ veh}$$

Therefore, 5.56 vehicles (15 – 9.44) will not be able to pass through the intersection on the first cycle even though they arrive during the first cycle. At the end of the second cycle, 23 vehicles (15 + 8) will have arrived, but only 18.88 (2sg) will have departed, leaving 4.12 vehicles waiting at the beginning of the third cycle. At the end of the third cycle, a total of 27 vehicles will have arrived and as many as 28.32 (3sg) could have departed, so the queue that began to form during the first cycle will dissipate at some time during the third cycle. This process is shown graphically in Fig. 7.15.

From this figure, the total vehicle delay of the first cycle is (the area between arrival and departure curves) is

$$D_1 = 0.5(60)(15) - 0.5(20)(9.44)$$
$$= 355.6 \text{ veh-s}$$

Similarly, the delay in the second cycle is

$$D_2 = 0.5(60)(15+23) - (40)(9.44) - 0.5(20)(9.44+18.88)$$
$$= 479.2 \text{ veh-s}$$

To determine the delay in the third cycle, it is necessary to first know exactly when, in this cycle, the queue dissipates. The time to queue clearance after the start of the effective green, t_c, is (with v_3 being the arrival rate during the third cycle and n_3 being the number of vehicles in the queue at the start of the third cycle)

$$n_3 + v_3(r + t_c) = st_c$$

where

$$v_3 = \frac{4 \text{ veh}}{60 \text{ sec}} = 0.067 \text{ veh/s}$$

Therefore,

$$(23 - 18.88) + 0.067(40 + t_c) = 0.472t_c$$

which gives t_c = 16.8 seconds. Thus the queue will clear 56.8 seconds (40 + 16.8) after the start of the third cycle, at which time a total of 26.8 vehicles ($0.067 \times 56.8 + 15 + 8$) will have arrived at, and departed from, the intersection. The vehicle delay for the third cycle is

$$D_3 = 0.5(56.8)(23+26.8) - (40)(18.88) - 0.5(16.8)(18.88+26.8)$$
$$= 275.4 \text{ veh-s}$$

giving the total delay over all three cycles as

$$D_t = D_1 + D_2 + D_3$$
$$= 355.6 + 479.2 + 275.4$$
$$= \underline{\underline{1110.2 \text{ veh-s}}}$$

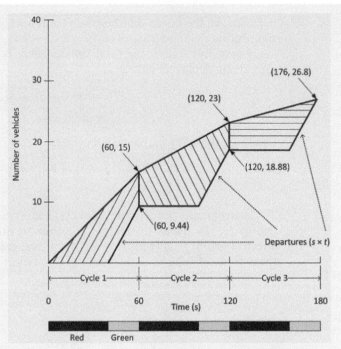

Figure 7.15 *D/D/*1 queuing diagram for Example 7.11.

It is also possible to handle the case where intersection arrivals and/or departures are deterministic but time-varying in a fashion similar to that shown in Chapter 5. An example of time-varying arrivals is presented next.

EXAMPLE 7.12   *D/D/*1 SIGNAL ANALYSIS WITH TIME-VARYING ARRIVALS

The saturation flow rate of an approach to a pretimed signal is 6000 veh/h. The signal has a 60-second cycle with 20 seconds of effective red allocated to the approach. At the beginning of an effective red (with no vehicles remaining in the queue from a previous cycle), vehicles start arriving at a rate $v(t) = 0.4 + 0.01t + 0.00057t^2$ [where $v(t)$ is in vehicles per second and t is the number of seconds from the beginning of the cycle]. Thirty seconds into the cycle the arrival rate remains constant at its 30-second level and stays at that rate until the end of the cycle. What is the total vehicle delay for this approach over the cycle (in vehicle seconds), assuming *D/D/*1 queuing?

SOLUTION

Vehicle arrivals for the first 30 seconds (again allowing fractions of vehicles for the sake of clarity) are

$$\int_0^{30} 0.4 + 0.01t + 0.00057t^2 = 0.4t + 0.005t^2 + 0.00019t^3 \Big|_0^{30}$$

$$= 21.63 \text{ veh}$$

The arrival rate after 30 seconds is 1.213 veh/s [$0.4 + 0.01(30) + 0.00057(30)^2$] and is no longer time-varying. During the effective green, the departure rate is 1.667 veh/s (6000/3600). To determine when the queue will clear, let t' be the time after 30 seconds (the time after which the arrival rate is no longer time-varying), so

$$21.63 + 1.213t' = 1.667(t' + 10)$$

which gives $t' = 10.93$ seconds. Thus the queue clears at 40.93 seconds after the beginning of the cycle. For delay, the area under the arrival curve is

$$\int_0^{30} 0.4t + 0.005t^2 + 0.00019t^3 + \frac{1}{2}\Big[21.63 + \big(1.213(10.93) + 21.63\big)\Big](10.93)$$

$$= 0.2t^2 + 0.00167t^3 + 0.0000475t^4 \Big|_0^{30} + 308.87$$

$$= 263.48 + 308.87$$

$$= 572.35 \text{ veh-s}$$

The area under the departure curve is

$$\frac{1}{2}[34.90 \times 20.94] = 365.42 \text{ veh-s}$$

So total vehicle delay over the cycle is <u><u>206.93 veh-s</u></u> (572.35 − 365.42).

7.5.2   Signal Coordination

The main limitation with Eqs. 7.16–7.22 is that their derivations are based on the assumption of uniform traffic arrivals throughout the cycle and a constant saturation flow rate during the effective green period, as well as no queue at the start of the effective red period or at the end of the effective green period. As discussed in Chapter 5, nonuniform arrivals are very likely in most traffic flow situations, and are frequently the case for signalized arterials. Thus, the assumption of uniform arrivals throughout the cycle is generally unrealistic and is likely to yield considerable error compared with values measured in the field.

One of the biggest influences on the arrival pattern of vehicles at each approach to the intersection is the level of signal coordination along the roadway. The term signal coordination generally refers to the level of timing coordination, or synchronization, between adjacent signals on the roadway.

Ideally, traffic signals should be timed so that as many vehicles as possible arrive at the signalized intersection when the signal indication is green, which would indicate good signal coordination. If a large proportion of vehicles arrive at an intersection during the red interval, the signal coordination would be considered poor. From a delay perspective, the larger the proportion of vehicles arriving during green, the lower the signal delay, and the larger the proportion of vehicles arriving during red, the higher the signal delay.

The effect of signal coordination on traffic arrival patterns is referred to as progression quality. Quantitatively, progression quality is expressed as the number of vehicles that arrive at an intersection approach while the signal indication is green for that approach, relative to all vehicles that arrive at that intersection approach during the entire signal cycle. This value is denoted as *PVG*, for Proportion of Vehicles arriving on Green. A general overview of signal coordination is provided in the rest of this section, as well as an example delay calculation for different arrivals rates during the green and red periods.

Fundamental Relationships

The three most significant factors affecting progression quality are signal spacing, vehicle speed, and cycle length. The relationship between signal spacing and vehicle speed is most easily illustrated by considering a one-way arterial. Consider two intersections on a street running east-west separated by some distance d_o. With traffic traveling westbound, the time at which the signal phase of the westernmost signal (downstream signal) turns green after the easternmost (upstream) signal phase turns green should be equal to the travel time between the two intersections. The time difference between the start of the green between corresponding phases at adjacent signalized intersections is referred to as the offset, and is calculated as

$$offset = \frac{d_o}{V} \tag{7.23}$$

where

$offset$ = start of green phase for downstream intersection relative to upstream intersection, for the same traffic movement, in seconds,

d_o = distance between upstream and downstream intersection for offset calculation, in feet, and

V = travel speed between upstream and downstream intersection, in ft/s.

In practice, to take driver perception-reaction times into account, the downstream signal should turn green a few seconds before the lead vehicles reach the stop bar. For simplification, this time is not considered in the forthcoming example.

EXAMPLE 7.13   CALCULATE OFFSET FOR IDEAL ONE-WAY PROGRESSION

Two intersections on a one-way arterial are separated by 2640 ft. The average running speed of vehicles along this segment is 40 mi/h. Calculate the offset necessary for ideal progression between the two intersections.

SOLUTION

Applying Eq. 7.23,

$$offset = \frac{2640 \text{ ft}}{40 \text{ mi/h} \times \dfrac{5280 \text{ ft/mi}}{3600 \text{ s/h}}}$$

$$= \underline{\underline{45 \text{ s}}}$$

It is worth noting that coordinating signal timing between adjacent intersections for good progression in only one direction is very straightforward. However, for an arterial with traffic in both directions, the setting of the offset for ideal progression in one direction may lead to poor progression in the other direction.

EXAMPLE 7.14   DRAW A TIME-SPACE DIAGRAM

For Example 7.13, draw the time-space diagram, assuming that the cycle length is 60 seconds and the g/C ratio is 0.5, for both intersections.

SOLUTION

For a 60-second cycle length and a g/C of 0.5, the major street effective green and effective red times will each be 30 seconds. The time-space diagram for this situation is shown in Fig. 7.16. In this figure, the solid diagonal lines represent vehicles traveling from intersection 1 to intersection 2, and the dashed diagonal lines represent vehicles traveling from intersection 2 to intersection 1. The slope of these lines represents the vehicle speed.

In Fig. 7.16, the traffic flowing from intersection 1 to intersection 2 has ideal progression. However, also notice that the traffic flowing from intersection 2 to intersection 1 has the worst-case progression; that is, all traffic arriving on green at intersection 2 arrives on the red at intersection 1.

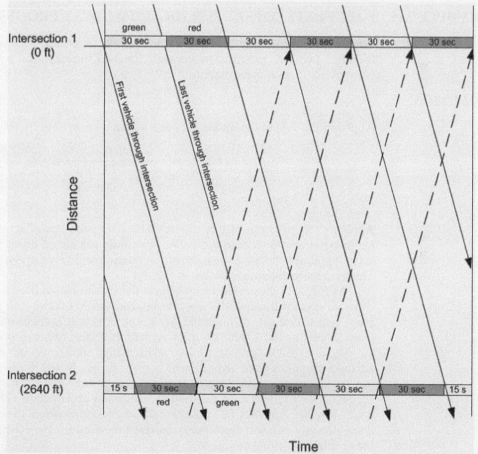

Figure 7.16. Time-space diagram illustrating good coordination for only one direction, for Example 7.14.

To obtain good progression for both directions of travel in the previous example problem, the cycle length must be considered. For good progression in both directions, the cycle length (for both intersections) needs to be twice the travel time from Intersection 1 to Intersection 2. That is, Eq. 7.23 is multiplied by 2, as shown in Eq. 7.24 (assuming that the travel speed is the same in both directions).

$$C_{prog} = \frac{d_o}{V} \times 2 \tag{7.24}$$

where

C_{prog} = cycle length necessary for ideal two-way progression, in seconds, and other terms are as defined previously.

When planning new construction, this relationship can help guide the location of new signalized intersections. For an analysis of an existing roadway, however, the objective is to usually find a common cycle length and corresponding offset values that produce the best progression through the existing intersections.

EXAMPLE 7.15   CALCULATE OFFSET FOR IDEAL TWO-WAY PROGRESSION

For Example 7.13, assume a two-way arterial and determine the cycle length necessary for ideal two-way progression between the two intersections, and then sketch the corresponding time-space diagram.

SOLUTION

Applying Eq. 7.24 to determine the cycle length,

$$C_{prog} = \frac{2640 \text{ ft}}{40 \text{ mi/h} \times \dfrac{5280 \text{ ft/mi}}{3600 \text{ s/h}}} \times 2$$

$$= 90 \text{ s}$$

Applying the g/C ratio of 0.5, the new effective green and effective red times for the major street are 45 seconds each. The previously calculated offset value of 45 seconds is still applicable because the intersection spacing and travel speed remain unchanged. The time-space diagram is shown in Fig. 7.17.

As Fig. 7.17 illustrates, every vehicle that gets through the green at intersection 1 will also make it through the green at intersection 2. Likewise, in the other direction, every vehicle that gets through the green at intersection 2 will also make it through the green at intersection 1. This situation represents perfect two-way progression.

Because of the precise timing relationships that must be maintained between adjacent signals, a basic requirement is that all signals within the coordinated system run on a common cycle length. It is possible, however, to run one or more of the signals at a cycle length of one-half the common cycle length. This is referred to as double-cycling, and may be applicable for intersections where the cross-street demand is significantly different from the cross-street demand for the other intersections along the arterial.

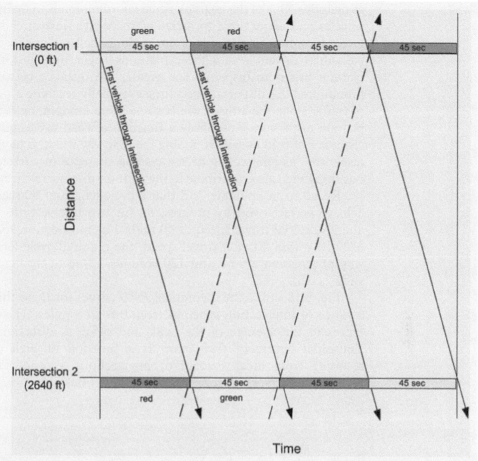

Figure 7.17. Time-space diagram illustrating good coordination for both directions, for Example 7.15.

Two other factors that can significantly affect progression quality are g/C ratio and platoon dispersion, as described in the following sections.

Effective Green to Cycle Length Ratio (g/C)

To illustrate the influence of the g/C ratio, first consider an intersection approach that has 100% g/C ratio (constant green). For this situation, the PVG would be 100% because there would never be a red indication. Next, consider the opposite case where an approach has a g/C ratio of 0% (constant red). In this case the PVG would be 0%, because there would never be a green indication for vehicle arrivals. Thus, for the unlikely case of uniform vehicle arrivals, the PVG will be equal to the g/C ratio. In more realistic vehicle distributions, the PVG will not equal the g/C ratio, but the g/C ratio still serves as a limiting condition, as shown later in Fig. 7.18.

Platoon Dispersion

When queued vehicles depart an intersection after the start of a green phase, they are usually closely spaced. These closely spaced groupings of vehicles are referred to as platoons. One of the goals of signal coordination is to maintain these platoons of vehicles and allow them to arrive at successive downstream intersections on the green. However, as platoons progress along the length of roadway between signals, individual drivers within these platoons begin to adjust their speeds, and the platoon begins to disperse. The greater the distance between signals, the more pronounced this dispersion becomes, eventually reaching a point at which the flow of traffic along the arterial will become more random, or even uniform. Although platoon dispersion is primarily a function of roadway length between signals, the character of land uses surrounding the roadway will also have an effect on platoon dispersion, as intersecting driveways and presence of curbside activities (parking, bus stops, etc.) also contribute to the platoon-dispersing effect.

Recall from Example 7.15 that a cycle length of 90 seconds was required to achieve perfect two-way progression for two intersections separated by 2640 ft and an arterial travel speed of 40 mi/h. For intersection spacings of 1760 ft and 3520 ft, with a 40 mi/h travel speed, the required cycle lengths for perfect two-way progression are 60 and 120 seconds, respectively (again, the assumed g/C value is 0.5).

Fig. 7.18 shows the theoretical PVG curves for these three cycle lengths over a range of intersection spacings from 0 ft to 3 miles. The graphed PVG values represent the average of the peak and off-peak directional PVG values. The sinusoidal nature of the curves is a function of high average PVG values resulting from ideal spacing of intersections and low average PVG values resulting from worst-case spacing of intersections.

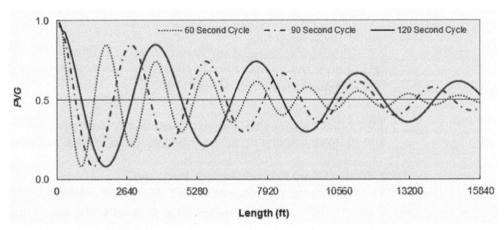

Figure 7.18 Proportion of vehicles arriving on green (PVG) for different cycle lengths over a range of intersection spacings.

The curve peaks and valleys occur at different distances for each of the cycle lengths; however, the magnitude of the peaks and valleys is consistent across cycle lengths. For example, the second peak for the 120-second cycle (which occurs at ≈ 7400 ft) has a PVG value of about 70%, and the second peaks for the 90-second (at ≈ 5500 ft) and 60-second (at ≈ 3600 ft) cycles also have PVG values of about 70%. Except for the zero-length case, the 60-second cycle has the best PVG when intersection spacing is 1760 ft, the 90-second cycle has the best PVG

when spacing is 2640 ft, and the 120-second cycle has the best *PVG* when spacing is 3520 ft.

Note that *PVG* values eventually "dampen out" as the length between signals gets large, reflecting the effect of platoon dispersion. Also note that the *PVG* values converge on the *g/C* ratio of 0.5.

The following example demonstrates how *D/D/*1 queuing can be used to estimate signal delay for non-uniform arrivals.

EXAMPLE 7.16   CALCULATE SIGNAL DELAY CONSIDERING PROGRESSION

Consider Example 7.10, but instead of uniform arrivals throughout the cycle, 60% of the traffic volume arrives during the green indication. Determine the average delay for this condition.

SOLUTION

In this case, the arrivals are not uniform throughout the cycle; therefore, the arrival rate during the effective green and effective red periods will be different. These rates are readily calculated by using the overall arrival rate, the effective green time, the cycle length, and the proportion of vehicles arriving on green, as follows:

$$v_{green} = \frac{v \times PVG}{g/C} = \frac{0.139 \times 0.6}{24/80}$$

$$= 0.278 \text{ veh/s}$$

$$v_{red} = \frac{v \times (1 - PVG)}{1 - g/C} = \frac{0.139 \times (1 - 0.6)}{1 - 24/80}$$

$$= 0.0794 \text{ veh/s}$$

With these arrival rates and the other values for Example 7.10, the graph in Fig. 7.19 of cumulative arrivals and cumulative departures for one cycle results.

This leads to the following calculations for average delay:

1. Time to queue clearance after the start of the effective green,

$$t_c = \frac{v_{red} \times r}{(s - v_{green})} = \frac{0.0794(56)}{(0.667 - 0.278)}$$

$$= \underline{11.43 \text{ s}}$$

2. Total vehicle delay per cycle,

$$D_t = \left[0.5 \times (v_{red} \times r) \times r \right] + \left[0.5 \times (v_{red} \times r) \times t_c \right]$$

$$= \left[0.5 \times (0.0794 \times 56) \times 56 \right] + \left[0.5 \times (0.0794 \times 56) \times 11.43 \right]$$

$$= 124.50 + 25.41$$

$$= \underline{149.91 \text{ veh-s}}$$

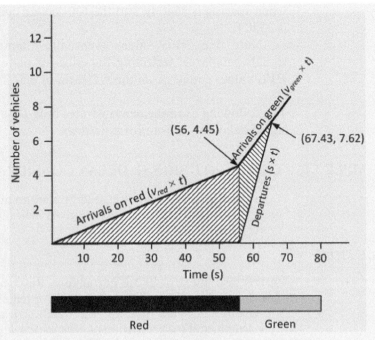

Figure 7.19 $D/D/1$ queuing diagram for Example 7.16.

3. Average delay per vehicle,

$$d_{avg} = \frac{D_t}{vC} = \frac{149.91}{0.139 \times 80}$$
$$= \underline{\underline{13.48 \text{ s/veh}}}$$

By timing the signal system so that 60% of the vehicles arrive on green at this approach, the average delay would be reduced by 11.28 s (24.76 − 13.48). Note that the PVG value for the uniform arrival case would be 0.3 (i.e., the same as the g/C ratio).

State of the Practice

The examples presented in the previous section were highly simplified and idealized. Real signalized arterials rarely consist of only two intersections, and the spacing between intersections is seldom consistent. Determining the appropriate offset values, green times, and common cycle length to minimize signal delay, for a two-way arterial with several intersections, is a complex problem. A closed-form analytical solution is usually not possible; thus, an iterative solution technique must be used.

The number of combinations for the above variable values for two directions of traffic flow through several intersections quickly reaches an extremely large number. When considering a network of streets, where several major arterials may intersect at several points, finding optimal signal timing settings (to support coordination and minimize delay) for the entire network cannot be done from a practical perspective. For these network-wide optimization efforts, the number of iterations required to find an optimal solution becomes so extreme that the solution algorithm must be run on a supercomputer for several days. Practicing engineers performing these kinds of studies rarely have access to these resources, or the available time.

Accordingly, modern software packages that perform signal timing "optimization" for arterial or network coordination employ a search algorithm that attempts to find the best solution amongst a greatly reduced number of variable-value combinations. This allows a solution to be found within a reasonable amount of time, using commonly available computing hardware. However, because these algorithms do not perform an exhaustive search of all of the variable-value combinations (as is generally required to find a truly optimal solution for this type of problem), the identified solution is considered to be only a "local" optimum rather than a "global" optimum. So while the local optimum solution may provide reasonable results, it is unlikely that this particular set of variable settings will provide the best results. Nonetheless, with the tremendous increase in desktop computing power over recent decades, the differences between a local optimum and global optimum solution have decreased, as the search space (the set of variable values considered by the solution algorithm) has accordingly been expanded.

While computer software can identify optimal, or nearly optimal, signal timing settings, there are practical situations that can still prevent a signal system from running at its highest possible efficiency, such as accommodating emergency and transit vehicle preemption (priority). Additionally, recall that signals operating under actuated control do not have a fixed cycle length. To include such signals in a coordinated system, they must be reprogrammed to run on a fixed cycle length, which could possibly result in a net loss of efficiency for those signals. One alternative that is commonly employed in this situation is to convert the signal operation from fully actuated to semi-actuated, where any green time not used for the minor street phases is allocated to the major street phases to keep the cycle length constant.

It should also be pointed out that sometimes signal timing may be optimized for measures other than signal delay, such as number of stopped vehicles, emissions output, and so on. Because different optimization criteria almost always provide different results, many signal timing strategies seek to balance the values of more than one factor (such as vehicle stops and vehicle delays). Needless to say, the optimization-measurement problem makes this aspect of traffic analysis a fruitful area for future research.

7.5.3   Control Delay Calculation for Level of Service Analysis

The level-of-service concept was introduced in Chapter 6 for uninterrupted-flow facilities. This same concept also applies to interrupted-flow facilities. Although a variety of performance measures can be calculated for signalized intersections, as demonstrated previously, only one measure has been chosen as the service measure in the *Highway Capacity Manual* [Transportation Research Board 2016]. This measure is control delay, and it applies to both signalized and unsignalized intersections. Control delay (i.e., signal delay for the case of a signalized intersection) represents the total delay experienced by the driver as a result of the control, which includes delay due to deceleration time, queue move-up time, stop time, and acceleration time, as illustrated in Fig. 7.20.

Analytic methods for estimating delay, such as the $D/D/1$ queuing approach described previously, are generally not able to capture the delay due to deceleration and acceleration and thus usually underestimate the actual delay. Furthermore, while the assumption of uniform arrivals leads to the intuitive and straightforward $D/D/1$ queuing analysis approach, it has been found to underestimate delay when the v/c ratio for an approach exceeds 0.5. This is because as the traffic intensity increases from moderate to a level nearing the capacity of the intersection, the probability of having cycle failures, where not all queued vehicles get through during a particular cycle, increases substantially.

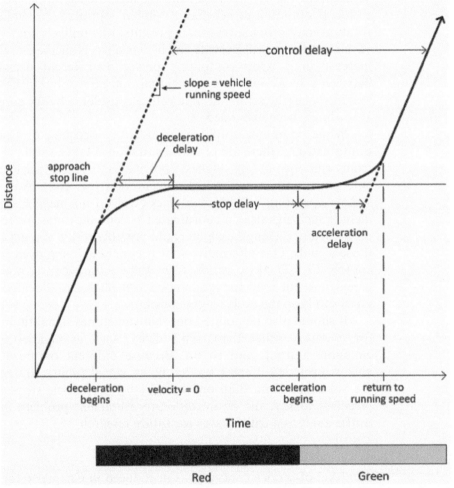

Figure 7.20 Illustration of control delay for a single vehicle traveling through a signalized intersection.

Adapted from Exhibit 19-5, p. 19-10, *Highway Capacity Manual*, 6th Edition. Copyright, Transportation Research Board, National Academy of Sciences, Washington, D.C. 2016.

These cycle failures are random occurrences for the most part, but must be accounted for in the estimation of overall delay to achieve reasonably accurate results under higher flow conditions. Whereas Eq. 7.21 is based on a purely theoretical derivation, the need to account for stochastic vehicle arrivals which lead to occasional oversaturation adds considerable complexity to the analysis of delay at signalized intersections. Numerous researchers over the last several decades have proposed delay formulas and refinements to meet this need, based on combinations of analytical, empirical, and simulation-based methods. The following formulation, denoted as d_2, is one that has been adopted by the *Highway Capacity Manual* [Transportation Research Board 2016], given as

$$d_2 = 900T \left[(X-1) + \sqrt{(X-1)^2 + \frac{8kIX}{cT}} \right] \qquad (7.25)$$

where

d_2 = average incremental delay per vehicle due to random arrivals and occasional oversaturation in seconds,

T = duration of analysis period in h,

X = v/c ratio for lane group,

k = delay adjustment factor that is dependent on signal controller mode,

I = upstream filtering/metering adjustment factor, and

c = lane group capacity in veh/h.

Assuming the analysis flow rate is based on the peak 15-minute traffic flow within the analysis hour, T is equal to 0.25.

The signal-controller–mode delay adjustment factor, k, accounts for whether the intersection is operating in an actuated or pretimed mode. A value of 0.5 is used for intersections under pretimed control. Given that actuated intersection control usually results in more efficient handling of traffic volumes, k can take on values less than 0.5 to account for this efficiency and resultant reduced delay. For actuated control, k depends upon the v/c ratio and the passage time (an actuated controller setting related to the maximum amount of time the green display can be extended due to a vehicle detector actuation). For the purposes of the example and end-of-chapter problems, pretimed signal control is assumed.

The upstream filtering/metering factor is used to adjust for the effect that an upstream signal has on the randomness of the arrival pattern at a downstream intersection. An upstream signal will typically have the effect of reducing the variance of the number of arrivals at the downstream intersection. I is defined as

$$I = \frac{\text{variance of the number of arrivals per cycle}}{\text{mean number of arrivals per cycle}} \qquad (7.26)$$

Recall from Chapter 5 that for a Poisson distribution (which is used to represent random arrivals), the variance is equal to the mean. Thus, if the downstream arrival pattern of vehicles conforms to the Poisson distribution, this factor will be equal to 1.0. This is considered to be the case for signalized intersections operating in an isolated mode (intersections sufficiently distant from adjacent signalized intersections). For nonisolated signalized intersections, the I value will be less than 1.0 due to the reduced variance of vehicle arrivals. The I value is dependent on the v/c ratio of the upstream movements that contribute to the downstream intersection volume. Again, for the purposes of this chapter, it will be assumed that a signal is operating in isolated mode, and thus the I value will be equal to 1.0.

The *Highway Capacity Manual* [Transportation Research Board 2016] also includes a formula for estimating the delay caused by an initial queue of vehicles at the beginning of the analysis time period, denoted as d_3. This equation and its inputs are beyond the scope if this book; however, a simple illustration of this issue was included in Example 7.11. If no initial queue is present at the start of the analysis period, the d_3 term is simply set to zero.

Thus, estimation of signal delay as prescribed in the *Highway Capacity Manual* [Transportation Research Board 2016] includes two terms in addition to a term for calculating delay due to uniform arrivals, as follows:

$$d = d_1 + d_2 + d_3 \tag{7.27}$$

where

d = average signal delay per vehicle in seconds,

d_1 = average delay per vehicle due to uniform arrivals in seconds,

d_2 = average delay per vehicle due to random arrivals in seconds, and

d_3 = average delay per vehicle due to initial queue at start of analysis time period, in seconds.

For signalized intersections where vehicle arrivals are not influenced by coordination, isolated signalized intersections for one, the PVG is considered to be equal to the g/C ratio. In this case, Eq. 7.21 can be applied to determine d_1. If signal coordination results in different arrival rates for the green and red intervals, a calculation process similar to that used in Example 7.16 must be applied to determine d_1.

Note that Eq. 7.27 is applied to each established lane group in the intersection. Individual lane group delays must be aggregated to arrive at an overall intersection delay, which will be discussed in the following section.

EXAMPLE 7.17   CALCULATE SIGNAL DELAY USING EQ. 7.27

Compute the average approach delay per cycle using Eq. 7.27, given the conditions described in Example 7.9. Assume that the intersection is isolated, the traffic flow accounts for the peak 15-min period, and that there is no initial queue at the start of the analysis period.

SOLUTION

In this example, the uniform delay is computed as (using the alternative version of Eq. 7.21)

$$d_1 = \frac{0.5C\left(1 - \frac{g}{C}\right)^2}{1 - \left(\frac{v}{c} \times \frac{g}{C}\right)}$$

with

$C = 80$ s

$g = 24$ s

The v/c ratio is calculated as

$$\frac{v}{c} = \frac{v}{s \times g/C}$$

with

$v = 500$ veh/h

$s = 2400$ veh/h

giving

$$\frac{v}{c} = \frac{500}{2400 \times 24/80} = \frac{500}{720} = 0.694$$

Uniform delay is calculated as

$$d_1 = \frac{0.5(80)\left(1 - \frac{24}{80}\right)^2}{1 - \left[0.694 \times \frac{24}{80}\right]} = 24.75 \text{ s}$$

This value matches the delay value computed in Example 7.9 (the small difference is due to rounding of arrival and departure rates in Example 7.9). Random delay will be computed as (using Eq. 7.25)

$$d_2 = 900T\left[(X-1) + \sqrt{(X-1)^2 + \frac{8kIX}{cT}}\right]$$

with

$T = 0.25$ (15 min, from problem statement)

$X = 0.694$ (from above)

$k = 0.5$ (pretimed control)

$I = 1.0$ (isolated mode)

$c = 720$ veh/h (from above)

$$d_2 = 900(0.25)\left[(0.694-1)+\sqrt{(0.694-1)^2+\frac{8(0.5)(1.0)0.694}{(720)0.25}}\right]$$

$$= 5.45 \text{ s}$$

Now the total signal delay is computed using Eq. 7.27:

$$d = d_1 + d_2 + d_3$$

with

d_1 = 24.75 s

d_2 = 5.45 s

d_3 = 0 s (given)

$$d = 24.75 + 5.45 + 0 = \underline{\underline{30.20}} \text{ s/veh}$$

As this example illustrates, ignoring the d_2 component of delay when the v/c ratio is above 0.5 results in a significant underestimation of average delay.

7.5.4 Level-of-Service Determination

Before the implementation of any developed signal phasing and timing plan, the level of service should be determined to assess whether the intersection will operate at an acceptable level under this plan. As previously mentioned, the service measure (the performance measure by which level of service is assessed) for signalized intersections is delay. In previous examples, we calculated the delay for a specific lane group, but now we must do this for all lane groups and then aggregate the delay values to arrive at an overall intersection delay measure and corresponding level of service.

The first step is to aggregate the delays of all lane groups for an approach, and then repeat the procedure for each approach of the intersection. This will result in approach-specific delays and levels of service. The aggregated lane group delay for each approach is given by

$$d_A = \frac{\sum_i d_i v_i}{\sum_i v_i} \tag{7.28}$$

where

d_A = average delay per vehicle for approach A in seconds,

d_i = average delay per vehicle for lane group i (on approach A) in seconds, and

v_i = analysis flow rate for lane group i in veh/h.

Once all the approach delays have been calculated, they can be aggregated to arrive at the overall intersection delay. The aggregated approach delay for the intersection is given by

$$d_I = \frac{\sum\limits_{A} d_A v_A}{\sum\limits_{A} v_A} \qquad (7.29)$$

where

d_I = average delay per vehicle for the intersection in seconds,
d_A = average delay per vehicle for approach A in seconds, and
v_A = analysis flow rate for approach A in veh/h.

The delay level-of-service criteria for signalized intersections are specified in the *Highway Capacity Manual* [Transportation Research Board 2016] and are given in Table 7.4. These delay criteria can be used to determine the level of service for a lane group, an approach, and the intersection.

LOS	Control delay per vehicle (s/veh)
A	≤ 10
B	$> 10\text{--}20$
C	$> 20\text{--}35$
D	$> 35\text{--}55$
E	$> 55\text{--}80$
F	> 80

Table 7.4 Level-of-Service Criteria for Signalized Intersections

Reproduced with permission of the Transportation Research Board, *Highway Capacity Manual*, 6th Edition, Copyright, National Academy of Sciences, Washington, D.C. 2016. Exhibit 19-8, p. 19-16.

EXAMPLE 7.18   DETERMINE LEVEL OF SERVICE FOR APPROACH

Determine the level of service for the eastbound approach of Maple Street assuming no initial queuing at the start of the analysis period.

SOLUTION

Lane groups for this intersection were established in Example 7.2. Two lane groups were established for the eastbound (EB) approach, one for the left-turn movement and the other for the combined through and right-turn movements. The delay will be calculated for the left-turn lane group first, followed by the through/right-turn lane group.

For the left-turn lane group, the uniform delay is computed using Eq. 7.21 with

$$C = 65 \text{ s}$$
$$g = 12.5 \text{ s}$$
$$\frac{v}{c} = \frac{v}{s \times g/C} = \frac{300}{1750 \times 12.5/65} = \frac{300}{337} = 0.891$$

giving

$$d_1 = \frac{0.5(65)\left(1 - \frac{12.5}{65}\right)^2}{1 - \left[0.891 \times \frac{12.5}{65}\right]} = 25.6 \text{ s}$$

Random delay is computed using Eq. 7.25 with

T = 0.25 (15 min)
X = 0.891 (from above)
k = 0.5 (pretimed control)
I = 1.0 (isolated mode)
c = 337 veh/h (from above)

giving

$$d_2 = 900(0.25)\left[(0.891-1) + \sqrt{(0.891-1)^2 + \frac{8(0.5)(1.0)0.891}{(337)0.25}}\right]$$

$$= 27.8 \text{ s}$$

Now, the total signal delay is computed using Eq. 7.27 with

d_1 = 25.6 s
d_2 = 27.8 s
d_3 = 0 s (given)

$$d = 25.6 + 27.8 + 0 = 53.4 \text{ s}$$

From Table 7.4, this lane group delay corresponds to a level of service of D.

For the through/right-turn lane group, the uniform delay is computed using Eq. 7.21 with

$$C = 65 \text{ s}$$
$$g = 24.7 \text{ s}$$
$$\frac{v}{c} = \frac{v}{s \times g/C} = \frac{1100}{3400 \times 24.7/65} = \frac{1100}{1292} = 0.851$$

giving

$$d_1 = \frac{0.5(65)\left(1 - \frac{24.7}{65}\right)^2}{1 - \left[0.851 \times \frac{24.7}{65}\right]} = 18.5 \text{ s}$$

Random delay is computed using Eq. 7.25 with

T = 0.25 (15 min)
X = 0.851 (from above)

$$k = 0.5 \text{ (pretimed control)}$$
$$I = 1.0 \text{ (isolated mode)}$$
$$c = 1292 \text{ veh/h (from above)}$$

giving

$$d_2 = 900(0.25)\left[(0.851-1) + \sqrt{(0.851-1)^2 + \frac{8(0.5)(1.0)0.851}{(1292)0.25}} \right]$$

$$= 7.2 \text{ s}$$

Now, the average signal delay for this lane group is computed using Eq. 7.27 with

$$d_1 = 18.5 \text{ s}$$
$$d_2 = 7.2 \text{ s}$$
$$d_3 = 0 \text{ s (given)}$$

$$d = 18.5 + 7.2 + 0 = 25.7 \text{ s}$$

From Table 7.4, this lane group delay corresponds to a level of service of C.

Now, to compute the volume-weighted aggregate delay for the approach, we use Eq. 7.28:

$$d_A = \frac{\sum d_i v_i}{\sum v_i}$$

with

$$d_{LT} = \text{delay for left-turn lane group}$$
$$d_{T/R} = \text{delay for through/right-turn lane group}$$
$$v_{LT} = \text{analysis flow rate for left-turn lane group}$$
$$v_{T/R} = \text{analysis flow rate for through/right-turn lane group}$$

giving

$$d_{EB} = \frac{v_{LT} \times d_{LT} + v_{T/R} \times d_{T/R}}{v_{LT} + v_{T/R}}$$

$$= \frac{300 \times 53.4 + 1100 \times 25.7}{300 + 1100}$$

$$= \frac{44,290}{1400}$$

$$= \underline{31.6 \text{ s}}$$

From Table 7.4, this approach delay corresponds to a level of service of C.

EXAMPLE 7.19   DETERMINE LEVEL OF SERVICE FOR INTERSECTION

Determine the level of service for the intersection of Maple and Vine Streets.

SOLUTION

The delay for each of the other three approaches (WB, NB, SB) can be determined by the exact same process used in Example 7.18 for the EB approach. Due to the length of the calculations involved for the remaining three approaches, the results are summarized in Table 7.5.

Table 7.5 Summary of Delay and Level-of-Service Calculations for the Intersection of Maple and Vine Streets

Approach	EB		WB		NB		SB	
Lane group	L	T/R	L	T/R	L	T/R	L	T/R
Analysis flow rate (v)	300	1100	250	1150	90	390	70	370
Saturation flow rate (s)	1750	3400	1750	3400	475	1800	450	1800
Flow ratio (v/s)	0.171	0.324	0.143	0.338	0.189	0.217	0.156	0.206
Critical lane group (✓)	✓			✓		✓		
Y_c				0.726				
Lost time/phase				4				
Total lost time				12				
Cycle length (C_{min})				65				
X_c				0.891				
Eff. green time (g)	12.5	24.7	12.5	24.7	15.8	15.8	15.8	15.8
g/C	0.192	0.380	0.192	0.380	0.243	0.243	0.243	0.243
Lane group capacity (c)	337	1292	337	1292	115	438	109	438
v/c (X)	0.891	0.851	0.743	0.890	0.779	0.891	0.640	0.846
d_1	25.6	18.5	24.7	18.9	23.0	23.8	22.1	23.4
k				0.5				
I				1.0				
T				0.25				
d_2	27.8	7.2	13.8	9.5	39.4	23.0	25.3	17.9
Lane group delay (d)	53.4	25.7	38.5	28.3	62.4	46.7	47.3	41.4
Lane group LOS	D	C	D	C	E	D	D	D
Approach delay	31.6		30.2		49.7		42.3	
Approach LOS	C		C		D		D	
Intersection delay				34.7				
Intersection LOS				C				

In this table, note that the v/c ratios for the critical lane groups match (rounding differences aside) the calculated critical intersection v/c ratio, X_c, as they should, because green time was allocated based on the strategy of equalizing v/c ratios for the critical lane groups in each phase (using Eq. 7.11).

The overall intersection delay calculation will be shown for the sake of clarity. Using Eq. 7.29, the intersection delay is given by

$$d_I = \frac{31.6 \times 1400 + 30.2 \times 1400 + 49.7 \times 480 + 42.3 \times 440}{1400 + 1400 + 480 + 440}$$

$$= \frac{128{,}988}{3{,}720}$$

$$= \underline{34.7 \text{ s}}$$

It is worth pointing out that, although all but two lane groups (EB T/R and WB T/R) have a level of service of D or worse, the much higher volumes for the EB T/R and WB T/R lane groups relative to the other lane groups keeps the level of service at C (albeit barely) for the intersection due to the volume weighting in the delay aggregation.

7.6   PRACTICE PROBLEMS

PRACTICE PROBLEM 7.1

TIMING-PLAN DEVELOPMENT

An isolated, pretimed signalized intersection with 1 left-turn lane and 1 through/right-turn lane on each of four approaches has the following timing-stage scheme.

Timing Stage 1	Timing Stage 2	Timing Stage 3

Arrival flow rates and saturation flow rates are given in the following table.

	Timing Stage 1	**Timing Stage 2**	**Timing Stage 3**
Arrival Flow Rate (veh/h)	WB L: 325 EB L: 350	WB T/R: 510 EB T/R: 490	SB L: 150; SB T/R: 530 NB L: 130; NB T/R: 580
Saturation Flow Rate (veh/h)	WB L: 1600 EB L: 1650	WB T/R: 1725 EB T/R: 1700	SB L: 450; SB T/R: 1750 NB L: 425; NB T/R: 1775

The startup lost time is 2.5 s/timing stage, the clearance lost time is 1.5 s/timing stage, the traffic speed is 40 mi/h on all approaches, the design vehicle length is 20 ft, all lanes are 12 ft wide, and all approaches are on level grade.

Determine the following:
- Minimum cycle length assuming 95% capacity utilization
- Effective green time for each timing stage assuming equal critical v/c ratios
- Yellow and all-red times for timing stage 2

SOLUTION

Note: Open boxes in equations " $\square$ " are to be completed by the reader

First we calculate the flow ratios for each movement in each timing stage and then identify the critical lane groups with a check mark.

Timing Stage 1	Timing Stage 2	Timing Stage 3
WB L: $\dfrac{\square}{\square}=\square$	WB T/R: $\dfrac{\square}{\square}=\square$ ✓	SB L: $\dfrac{\square}{\square}=\square$ ✓
		SB T/R: $\dfrac{\square}{\square}=\square$
EB L: $\dfrac{\square}{\square}=\square$ ✓	EB T/R: $\dfrac{\square}{\square}=\square$	NB L: $\dfrac{\square}{\square}=\square$
		NB T/R: $\dfrac{\square}{\square}=\square$

The sum of the flow ratios for the critical lane groups for this timing-stage plan is calculated from Eq. 7.7, as follows:

$$Y_c = \sum_{i=1}^{n}\left(\frac{v}{s}\right)_{ci} = \square + \square + \square = 0.841$$

The total lost time for the cycle is:

$$(\square \text{ s} + \square \text{ s})/\text{timing stage} \times \square \text{ timing stages} = 12 \text{ s}.$$

The minimum cycle length can now be calculated from Eq. 7.9, substituting the total cycle lost time and X_c of 0.95 (for 95% capacity utilization), giving,

$$C_{min} = \frac{\square \times \square}{\square - \square} = \underline{\underline{104.6}} \rightarrow \underline{\underline{105 \text{ s}}}.$$

Now, the cycle length of 105 seconds and X_c of 0.95 are used to calculate the effective green times, from Eq. 7.11, for the three timing stages, as follows:

$$g_1 = \square_{c1}\left(\frac{\square}{\square}\right) = \underline{\underline{23.4 \text{ s}}} \text{ (WB and EB left-turn movements)},$$

$$g_2 = \square_{c2}\left(\frac{\square}{\square}\right) = \underline{\underline{32.7 \text{ s}}} \text{ (WB and EB through and right-turn movements), and}$$

$$g_3 = \square_{c3}\left(\frac{\square}{\square}\right) = \underline{\underline{36.8 \text{ s}}} \text{ (SB and NB left-turn, through, and right-turn movements)}.$$

To determine the yellow and all-red times for timing stage 2, we first identify which movements are served in this timing stage. Timing stage 2 serves the WB and EB through and right-turn movements.

The yellow time is calculated from Eq. 7.12, using the traffic speed of 40 mi/h, which gives,

$$Y = 1.0 + \frac{\left(\boxed{} \times 5280/3600\right)}{2(10)}$$

$$= 3.9 \rightarrow \underline{\underline{4.0\,\text{s}}} \left(\text{rounding up to the nearest } 0.5\,\text{s}\right)$$

To calculate the all-red time, from Eq. 7.13, we need the clearance distance in addition to the traffic speed. The clearance distance is equal to the design vehicle length, 20 ft, plus the sum of the lane widths for the number of lanes being crossed in the intersection area. The WB and EB through movements cross a longer distance than the right-turn movements; thus, the through movement distance is the critical distance to use for design. The WB and EB through movements will cross three lanes (each 12 ft wide) along the NB/SB street. The three lanes consist of one through/right-turn lane in each direction and a left-turn lane (the left-turn lanes in each direction are directly opposing one another). Thus, the clearance distance is

$$3 \text{ lanes} \times 12 \text{ ft/lane} + \boxed{} \text{ ft} = 56 \text{ ft}.$$

Substituting the clearance distance into the numerator of Eq. 7.13 and the vehicle speed into the denominator gives an all-red time of

$$AR = \frac{56}{\boxed{} \times 5280/3600} = \underline{\underline{1.0\,\text{s}}}.$$

PRACTICE PROBLEM 7.2

***D/D/*1 SIGNAL ANALYSIS WITH TIME-VARYING ARRIVALS AND MINIMUM GREEN TIME**

At the beginning of an approach's effective red there are 2 cars left in a queue, cars continue to arrive at the approach at a rate of $v(t) = 0.055 - 0.0002t$ [with $v(t)$ in veh/s and t in seconds after the beginning of the effective red] for the entire cycle. The saturation flow of this approach is 1200 veh/h, the yellow time is 4 s, all-red is 2 s, and the approach has typical lost-time values. The cycle length is 90 s. What minimum displayed green time must be provided to ensure that the queue clears before the end of the effective green, and what is the total delay per cycle for this green time on this approach? (Assume *D/D/*1 queuing.)

SOLUTION

Note: Open boxes in equations "$\boxed{}$" are to be completed by the reader

This problem demonstrates the fundamentals of *D/D/*1 intersection queuing and also illustrates how to determine the minimum green time required to dissipate queued traffic on an approach. To begin, for the queue to clear by the end of the effect green, the total arrivals during the 90 s cycle, plus the 2 vehicles remaining from the previous effective red, must equal the total departures (saturation flow times the effective greed). With the arrival rate given as $v(t) = 0.055 - 0.0002t$, the number of vehicles arriving in the 90 s cycle is,

$$\text{arrivals during the cycle} = \int_0^t 0.055 - 0.0002t \; dt \;\; = \Box \times \Box - \Box \times \Box^{\Box} \Big|_0^{90}.$$

and for departures we have (with saturation flow and effective green),

$$\text{departures during the cycle} \; = s \times g = \frac{\Box}{3600} \times g.$$

So, with arrival and departure equations, and the 2 vehicles already in the queue at the start of the effective red, equating arrivals (plus 2 vehicles) with departures and solving for the necessary effective green (g) gives,

$$\text{arrivals the during cycle} + 2 = \text{departures the during cycle}$$

$$\Box \times \Box - \Box \times \Box^{\Box} \Big|_0^{90} + 2 \;\; = \frac{\Box}{3600} \times g$$

$$\Box \times 90 - \Box \times 90^{\Box} + 2 \;\; = \frac{\Box}{3600} \times g,$$

or an effective of $g = 18.42$ s.

To continue, the problem states that typical lost-time values should be used. The typical start-up lost time (t_{sl}) is 2 s (see text). The typical clearance lost time (t_{cl}) is the last second of the yellow plus the all-red (see text). With 2 s of all-red, $t_{cl} = 1 + 2 = 3$. So, from Eq. 7.2 $t_L = t_{sl} + t_{cl} = 2 + 3 = 5$ s.

To arrive at the minimum displayed green time G, Eq. 7.3 is rearranged and applied giving,

$$G = g - Y - AR + t_L$$

$$= \Box - \Box - \Box + \Box$$

$$= \underline{\underline{17.42 \text{ s}}}$$

Now moving on to compute the total delay during the 90 s cycle, the delay will be the area under the arrival curve minus the area under the departure curve. The area under the arrival curve can be obtained by integrating the expression for arrivals during the cycle,

$$= \int_0^t \Box \times t - \Box \times t^{\Box} + 2 \; dt$$

$$= \Box \times t^{\Box} - \Box \times t^{\Box} + 2 \times t \Big|_0^{90}$$

$$= 378.45 \text{ veh-s}$$

For the area under the departure curve, it is noted that during the effective green a total of 6.14 veh will have departed [$s \times g = \Box /3600 \times \Box$]. Because the departure rate is temporally constant (does not vary as time passes, unlike the case of the arrival rate), the area under the departure curve is a triangle with the height equal to the number of vehicles that have departed and the base equal to the effective green time. So this area is,

$$= \frac{1}{2}\boxed{}\times\boxed{} = 56.55 \text{ veh-s}$$

So the area under the arrival curve minus the area under the departure curve gives the total vehicle delay of <u>321.90 veh-s</u> (378.45 − 56.44).

<table>
<tr><td>**PRACTICE
PROBLEM 7.3**</td><td>**THE POTENTIAL EFFECT OF AUTONOMOUS VEHICLES
ON TRAFFIC-SIGNAL DELAY**</td></tr>
</table>

A left-turn movement at a signalized intersection has a saturation flow of 900 veh/h per lane. The signal is allocated 19 s of green time in a 120 s cycle with 4 s of yellow and 1 s of all-red. The signal has typical start-up and clearance lost times. At the beginning of a cycle's effective red there are no vehicles in the queue and vehicles arrive throughout the cycle at a rate of $v(t) = 0.10 + 0.000278t$ [with $v(t)$ in veh/s and t in seconds after the beginning of the effective red]. An analyst wants to consider the effects of autonomous vehicles (driverless cars controlled by computers and appropriate sensor technology) on this left-turn movement. With autonomous vehicles it is estimated that the saturation flow rate will increase from 900 veh/h to 2800 veh/h (since vehicles can follow more closely and maintain a high safety level because drivers' reaction time is no longer a factor). In addition, there is no longer the need for yellow and all-red times since vehicle-to-infrastructure communication tells vehicles in advance of a change from displayed green to displayed red. So the green time for this left-turn movement is increased from 19 to 24 seconds. With autonomous vehicles, start-up lost time is reduced to just 1 second (it still exists to account for the time for the vehicles to accelerate to their saturation-flow speeds). How much shorter would the total delay for this movement be in this cycle (in veh-s, assuming the cycle length and green time do not change) if autonomous vehicles were present instead of conventional vehicles? (Assume $D/D/1$ queuing.)

SOLUTION

Note: Open boxes in equations "$\boxed{}$" are to be completed by the reader

This problem demonstrates the potential benefits of autonomous vehicles at signalized intersections. To begin, the delay for the conventional-vehicle case needs to be determined. We will start by calculating the number of vehicles that arrive and depart at the left-turn movement during the cycle. The total arrivals during the 120 s cycle [with the arrival rate given as $v(t) = 0.10 + 0.000278t$] is,

$$\text{arrivals during the cycle} = \int_0^t 0.10 + 0.0002t \; dt \;=\; \boxed{}\times\boxed{} + \boxed{}\times\boxed{}^{\boxed{}}\Big|_0^{120}$$

$$= 14 \text{ veh.}$$

To determine the departures during the cycle, we must first determine the effective green using Eq. 7.3 ($g = G + Y + AR - t_L$). In this equation all terms are given except the total loss time t_L. The problem states that typical lost-time values should be used. The typical start-up lost time (t_{sl}) is 2 s (see text). The typical clearance lost time (t_{cl}) is the last second of the yellow plus the all-red (see text). With 1 s of all-red, $t_{cl} = 1 + 1 = 2$. So, from Eq. 7.2 $t_L = t_{sl} + t_{cl} = 2 + 2 = 4$ s. With this, and given values from the problem, the effective green for the left-turn movement is (from Eq. 7.3),

$$g = G + Y + AR - t_L$$

$$= \boxed{} + \boxed{} + \boxed{} - \boxed{}$$

$$= \boxed{}.$$

With this, the departures on the left-turn movement are,

$$\text{departures during the cycle} = s \times g = \frac{\boxed{}}{3600} \times \boxed{} = 5 \text{ veh.}$$

With 14 vehicles arriving and only 5 departing there will be 9 vehicles remaining the queue at the end of the effective green. The total delay during the 120 s cycle will be the area under the arrival curve minus the area under the departure curve. The area under the arrival curve can be obtained by integrating the expression for arrivals during the cycle,

$$= \int_0^t \boxed{} \times t + \boxed{} \times t^{\boxed{}}$$

$$= \boxed{} \times t^{\boxed{}} + \boxed{} \times t^{\boxed{}} \Big|_0^{120}$$

$$= 800.06 \text{ veh-s.}$$

For the area under the departure curve, with only 5 vehicles having departed during the effective green, the area under the departure curve is a triangle with the height equal to the number of vehicles that have departed and the base equal to the effective green time. So this area is,

$$= \frac{1}{2} \boxed{} \times \boxed{} = 50.00 \text{ veh-s.}$$

So the area under the arrival curve minus the area under the departure curve gives the total vehicle delay of 750.06 veh-s (800.06 – 50.00).

With the information given in the problem, the arrival pattern will not change in the case of autonomous vehicles. However, the departure patter will change due to the increase in saturation flow and the change in effective green. The problem states that $G = 24$ s, $Y = 0$ s, $AR = 0$ s, $t_{cl} = 0$ s, and $t_{sl} = 1$ s. So, from Eq. 7.2 $t_L = t_{sl} + t_{cl} = 0 + 1 = 1$ s and effective green from Eq. 7.3 is,

$$g = G + Y + AR - t_L$$

$$= \boxed{} + \boxed{} + \boxed{} - \boxed{}$$

$$= \boxed{}.$$

With this, the departures on the left-turn movement can be as high as (using the entire effective green),

$$\text{departures during the cycle} = s \times g = \frac{\boxed{}}{3600} \times \boxed{} = 17.89 \text{ veh.}$$

With as many as 16 vehicles arriving during the 120-s cycle and as many as 17.89 vehicles departing, the queue will clear at some point during the effective green. To determine this point, arrivals and departures are equated to solve for time t as,

$$\text{arrivals} = \text{departures}$$

$$\int_0^t 0.10 + 0.0002t \ dt = s \times \left[t - (c - g) \right]$$

$$\boxed{} \times t + \boxed{} \times t^{\boxed{}} = \boxed{} \times \left[t - \left(\boxed{} - \boxed{} \right) \right],$$

which solving for t gives t = 113.97 s, so the queue will clear at 113.97 s into the cycle or, since the effective green of the left-turn movement starts at 97 s into the cycle (120 – 23), the queue will clear 16.97 s into the effective green (113.97 – 97).

To determine delay, with this value of t, the area under the arrival curve can be obtained by integrating the expression for arrivals during the cycle,

$$= \int_0^t \boxed{} \times t + \boxed{} \times t^{\boxed{}}$$

$$= \boxed{} \times t^{\boxed{}} + \boxed{} \times t^{\boxed{}} \ \Big|_0^{113.97}$$

$$= 718.05 \text{ veh-s}.$$

The area under the departure curve is again going to be a triangle with a height of 13.20 vehicles [vehicles departing before the queue is cleared ($s \times 16.97$)] and a base of 16.97 s (the time before the queue clears in the effective green), so this area is,

$$= \frac{1}{2} \boxed{} \times \boxed{} = 112.00 \text{ veh-s}.$$

So the area under the arrival curve minus the area under the departure curve gives the total vehicle delay of 606.05 veh-s (718.05 – 112.00).

Thus the delay in the autonomous vehicle case would be <u>144.01 veh-s</u> shorter (750.06-606.05) or a decrease in delay of 19.20%.

PRACTICE PROBLEM 7.4	**DETERMINE LEVEL OF SERVICE FOR APPROACH**
	For Practice Problem 1, determine the level of service for the northbound approach, assuming no initial queuing at the start of the analysis period.
SOLUTION	**Note: Open boxes in equations "$\boxed{}$" are to be completed by the reader**
	The northbound (NB) approach consists of two lane groups, one for the left-turn movement (because it is a permitted movement) and one for the combined through and right-turn movements. The delay will be calculated for the left-turn lane group first, followed by the through/right-turn lane group.

For the left-turn lane group, the uniform delay is computed using Eq. 7.21 with

$$C = 105 \text{ s}$$
$$g = 36.8 \text{ s}$$

$$\frac{v}{c} = \frac{v}{s \times g/C} = \frac{\Box}{\Box \times 36.8/105} = 0.873$$

giving

$$d_1 = \frac{0.5(105)\left(1 - \dfrac{36.8}{105}\right)^2}{1 - \left[\Box \times \dfrac{36.8}{105}\right]} = 31.9 \text{ s}$$

Random delay is computed using Eq. 7.25 with

$T = 0.25$ (15 min)
$X = 0.873$ (from above)
$k = 0.5$ (pretimed control)
$I = 1.0$ (isolated mode)
$c = 149$ veh/h (from above)

giving

$$d_2 = 900(0.25)\left[(\Box - 1) + \sqrt{(\Box - 1)^2 + \frac{8(0.5)(1.0)\Box}{(\Box)0.25}}\right]$$

$$= 45.8 \text{ s}$$

Now, the total signal delay is computed using Eq. 7.27 with

$d_1 = 31.9$ s
$d_2 = 45.8$ s
$d_3 = 0$ s (given)

$$d = 31.9 + 45.8 + 0 = 77.7 \text{ s}$$

From Table 7.4, this lane group delay corresponds to a level of service of E.

For the through/right-turn lane group, the uniform delay is computed using Eq. 7.21 with

$$\frac{v}{c} = \frac{v}{s \times g/C} = \frac{\Box}{\Box \times 36.8/105} = 0.931$$

giving

$$d_1 = \frac{0.5(105)\left(1 - \frac{36.8}{105}\right)^2}{1 - \left[\boxed{} \times \frac{36.8}{105}\right]} = 32.9 \text{ s}$$

Random delay is computed using Eq. 7.25 with

$T = 0.25$ (15 min)
$X = 0.931$ (from above)
$k = 0.5$ (pretimed control)
$I = 1.0$ (isolated mode)
$c = 623$ veh/h (from above)

giving

$$d_2 = 900(0.25)\left[\left(\boxed{} - 1\right) + \sqrt{\left(\boxed{} - 1\right)^2 + \frac{8(0.5)(1.0)\boxed{}}{\left(\boxed{}\right)0.25}}\right]$$

$$= 22.6 \text{ s}$$

Now, the average signal delay for this lane group is computed using Eq. 7.27 with

$d_1 = 32.9$ s
$d_2 = 22.6$ s
$d_3 = 0$ s (given)

$$d = 32.9 + 22.6 + 0 = 55.5 \text{ s}$$

From Table 7.4, this lane group delay corresponds to a level of service of E.

Now, to compute the volume-weighted aggregate delay for the approach, we use Eq. 7.28:

$$d_A = \frac{\sum d_i v_i}{\sum v_i}$$

with

d_{LT} = delay for left-turn lane group
$d_{T/R}$ = delay for through/right-turn lane group
v_{LT} = analysis flow rate for left-turn lane group
$v_{T/R}$ = analysis flow rate for through/right-turn lane group

giving

$$d_{EB} = \frac{v_{LT} \times d_{LT} + v_{T/R} \times d_{T/R}}{v_{LT} + v_{T/R}}$$

$$= \frac{\boxed{} \times 77.7 + \boxed{} \times 55.5}{\boxed{} + \boxed{}}$$

$$= \underline{59.6 \text{ s}}$$

From Table 7.4, this approach delay corresponds to a level of service of E.

NOMENCLATURE FOR CHAPTER 7

a	deceleration rate for vehicle at an intersection
AR	all-red time
C	cycle length
C_{min}	minimum cycle length
C_{opt}	optimum cycle length
c	capacity
d_{avg}	average vehicle delay per cycle ($D/D/1$ queuing)
d_1	average vehicle delay per cycle assuming uniform arrivals
d_2	average vehicle delay per cycle due to random arrivals and occasional oversaturation
d_3	average vehicle delay per cycle due to initial queue at start of analysis time period
d_d	distance from the intersection for which the dilemma zone is avoided
d_{max}	maximum delay of any vehicle ($D/D/1$ queuing)
D	deterministic arrivals or departures
D_t	total vehicle delay ($D/D/1$ queuing)
g	effective green time or acceleration due to gravity
G	displayed green time or grade of roadway
G_p	pedestrian green time
I	upstream filtering/metering adjustment factor
k	delay adjustment factor dependent on signal controller mode
l	vehicle length
L	total cycle lost time or crosswalk length
n	number of phases or number of vehicles or number of critical lane groups

N_{ped}	number of crossing pedestrians per timing stage
P_q	proportion of the signal cycle with a queue ($D/D/1$ queuing)
P_s	proportion of stopped vehicles ($D/D/1$ queuing)
PVG	proportion of vehicles arriving on green
Q	number of vehicles in the queue
Q_{max}	maximum number of vehicles in queue ($D/D/1$ queuing)
r	effective red time
R	displayed red time
s	saturation flow rate
S_p	pedestrian walking speed
t	time
t_c	time after the start of effective green until queue clearance ($D/D/1$ queuing)
t_L	total lost time for a movement during a cycle
t_r	driver perception/reaction time
v	analysis flow rate (also referred to as volume, arrival rate)
V	travel speed of vehicle
w	width of street
W_E	effective crosswalk width
X_i	volume-to-capacity ratio for lane group i
X_c	critical volume-to-capacity ratio for the intersection
x_s	distance required to stop
Y	displayed yellow time
Y_c	sum of flow ratios for critical lane groups
λ	arrival rate
μ	departure rate
ρ	traffic intensity

REFERENCES

Institute of Transportation Engineers (ITE). *Traffic Engineering* Handbook, 5th ed. Washington, DC, 1999.

Institute of Transportation Engineers (ITE). *Traffic Control Devices*, 2nd ed. Washington, DC, 2001.

Kell, James H., and Iris J. Fullerton. *Manual of Traffic Signal Design*, 2nd ed. Washington, DC: Institute of Transportation Engineers, 1998.

Transportation Research Board. *Traffic Flow Theory: A Monograph*. Special Report 165. Washington, DC: National Research Council, 1975.

Transportation Research Board. *Highway Capacity Manual*, Washington, DC: National Research Council, 2016.

U.S. Federal Highway Administration. *Manual on Uniform Traffic Control Devices for Streets and Highways*. Washington, DC: U.S. Government Printing Office, 2009.

Webster, F. V. *Traffic Signal Settings*, Road Research Technical Paper No. 39. London: Great Britain Road Research Laboratory, 1958.

Chapter 8

Travel Demand and Traffic Forecasting

8.1 INTRODUCTION

Traffic volumes change over time in response to changes in economic activity, individual travel patterns and preferences, and travelers' social/recreational activities. In addition, traffic volumes are affected by any significant modification of a highway network, which includes items such as new road construction or operational changes on existing roads (use or retiming of traffic signals). Analysts therefore must develop methodological approaches for forecasting changes in traffic volumes. For new road construction, traffic forecasts are needed to determine an appropriate pavement design (number of equivalent axle loads, as discussed in Chapter 4) and geometric design (number of lanes, shoulder widths, and so on) that will provide an acceptable level of service. For operational improvements, traffic forecasts are needed to estimate the effectiveness of alternate improvement options.

In forecasting vehicle traffic, two interrelated elements must be considered: the overall regional traffic growth/decline and possible traffic diversion. Overall traffic growth/decline is clearly an important concern, because projects such as highway construction and operational improvements will be undertaken in a dynamic environment with continual change in economic activity and individual traveler activities and preferences. In the design of highway projects, engineers must seek to provide a sufficient highway level of service and an acceptable pavement ride quality for future traffic volumes. One would expect that factors affecting long-term regional traffic growth/decline trends are primarily economic and, to a historically lesser extent, social in nature. The economics of the region in which a highway project is being undertaken determine the amount of traffic-generating activities (work, social/recreational, and shopping) and the spatial distribution of residential, industrial, and commercial areas. The social aspects of the population determine attitudes and behavioral tendencies with regard to possible traffic-generating decisions. For example, some regional populations may have social characteristics that make them more likely than other regional populations to make fewer trips, to carpool, to vanpool, or to take public transportation (buses or subways), all of which significantly impact the amount of highway traffic.

In addition to overall regional traffic growth/decline, there is the more microscopic, short-term phenomenon of traffic diversion. As new roads are constructed, as operational improvements are made, and/or as roads gradually become more congested, traffic will divert as drivers change routes or trip-departure times in an effort to avoid congestion and improve the level of service that they experience. Thus the highway network must be viewed as a system with the realization that a capacity or level of service change on any one segment of the highway network will impact traffic flows on many of the surrounding highway segments.

Travel demand and traffic forecasting is a formidable problem because it requires accurate regional economic forecasts as well as accurate forecasts of highway users' social and behavioral attitudes regarding trip-oriented decisions, in order to predict growth/decline trends and traffic diversion. Virtually everyone is aware how inaccurate economic forecasts can be, which is testament to the complexity and uncertainty associated with such forecasts. Similarly, one can readily imagine the difficulty associated with forecasting individuals' travel decisions.

Despite the difficulties involved in accurately forecasting traffic, over the years analysts have persisted in the development and refinement of a wide variety of travel demand and traffic forecasting techniques. An overriding consideration has always been the ease with which such techniques can be implemented in terms of data requirements and the ability of users to comprehend the underlying methodological approach. The field has evolved such that many traffic analysts can legitimately argue that the more recent developments in travel demand and traffic forecasting are largely beyond the reach of practice-oriented implementation. In many respects, this is an expected evolution because, due to the complexity of the problem, there will always be a tendency for theoretical work to exceed the limits of practical implementability. Unfortunately, the outgrowth of the methodological gap between theory and practice has resulted in the use of a wide variety of travel demand and traffic forecasting techniques, the selection of which is a function of the technical expertise of forecasting agencies' personnel as well as time and financial concerns.

In the past, textbooks have attempted to cover the full range of travel demand and traffic forecasting techniques from the readily implementable, simplistic approaches to the more theoretically refined methods. In so doing, such textbooks have often sacrificed depth of coverage, and as a consequence, travel demand and traffic forecasting frequently had the appearance of being confusing and disjointed. This chapter attempts to convey the basic principles underlying travel demand and traffic forecasting as opposed to reviewing the many techniques available to forecast traffic. This is achieved by focusing on an approach that is fairly advanced technically and effectively and efficiently conveys the fundamental concepts of travel demand and traffic forecasting. For more information on travel demand and traffic forecasting techniques that are more implementable or more theoretically advanced than the concepts provided in this chapter, the reader is referred to other sources [Meyer and Miller, 2001; Sheffi, 1985; Washington et al., 2011].

8.2   TRAVELER DECISIONS

Forecasts of highway traffic should, at least in theory, be predicated on some understanding of traveler decisions, because the various decisions that travelers make regarding trips will ultimately determine the quantity, spatial distribution (by route), and temporal distribution of vehicles on a highway network. Within this context, travelers can be viewed as making four distinct but interrelated decisions regarding trips: temporal decisions, destination decisions, modal decisions, and spatial or route decisions. The temporal decision includes the decision to travel and, more importantly, when to travel. The destination decision is concerned with the selection of a specific destination (shopping center, recreational facility, etc.), and the modal decision relates to how the trip is to be made (by automobile, bus, walking, or bicycling). Finally, spatial decisions focus on which route is to be taken from the traveler's origin (the traveler's initial location) to the desired destination. Being able to understand, let alone predict, such decisions is a monumental task. The remaining sections of this chapter seek to define the dimensions of this decision-prediction task and, through examples and illustrations, to demonstrate methods of forecasting traveler decisions and, ultimately, traffic volumes.

8.3   SCOPE OF THE TRAVEL DEMAND AND TRAFFIC FORECASTING PROBLEM

Because travel demand and traffic forecasting are predicated on the accurate forecasting of traveler decisions, two factors must be addressed in the development of an effective travel demand and traffic forecasting methodology: the complexity of the traveler decision-making process and system equilibration.

To begin the development of a fuller understanding of the complexity of traveler decisions, consider the schematic presented in Fig. 8.1. This figure indicates that traveler socioeconomics and activity patterns constitute a major driving force in the decision-making process. Socioeconomics, including factors such as household income, number of household members, and traveler age, affect the types of activities that the traveler is likely to be involved in (work, yoga classes, shopping, children's day care, dancing lessons, community meetings, etc.), which in turn are primary factors in many travel decisions. Socioeconomics can also have a direct effect on travel-related decisions by, for example, limiting modal availability (e.g., travelers in low-income households may be forced to take a bus due to the unavailability of a household automobile).

If we look more directly at the decision to travel, mode/destination choice, and highway route choice, Fig. 8.1 indicates that both long-term and short-term factors affect these decisions. For the decision to travel as well as mode/destination choice, the long-term factors of modal availability, residential and commercial distributions, and modal infrastructure play a significant role. These factors are considered long term because they change relatively slowly over time. For example, the development and/or relocation of residential neighborhoods and commercial centers is a process that may take years. Changes in modal infrastructure (construction/relocation of highways, subways, commuter rail systems) and modal availability (changes in automobile ownership, bus routing/scheduling) are also factors that evolve over relatively long periods of time. In contrast, a short-term factor, such as modal traffic, is one that can vary within a short period of time, as discussed in Chapter 6.

Moving down the illustration presented in Fig. 8.1, we see that the choice of a traveler's highway route is also determined by both long-term (highway infrastructure) and short-term (highway traffic) factors.

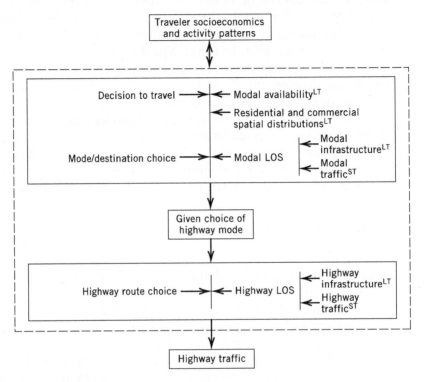

LT: Long-term factors.
ST: Short-term factors.

Figure 8.1 Overview of the process by which highway traffic is determined.

The outcome of the combination of these traveler decisions is, of course, highway traffic, the prediction of which is the objective of travel demand and traffic forecasting.

Aside from the complexities involved in the traveler decision-making process, the issue of system equilibration (mentioned at the beginning of this section) must also be considered. Note that Fig. 8.1 indicates not only that long- and short-term factors affect traveler decisions and choices but also that these decisions and choices in turn affect the long- and short-term factors. Such a reciprocal relationship is most apparent in considering the relationship between traveler choices and short-term factors. For example, consider a traveler's choice of highway route. One would expect that the traveler would be more likely to select a route between origin and destination that provides a shorter travel time. The travel time on various routes will be a function of route distance (highway infrastructure, long-term) and route traffic (higher traffic volumes reduce travel speed and increase travel time, as discussed in Chapter 5). But travelers' decisions to take specific routes ultimately determine the route traffic on which their route decisions are based. This interdependence between traveler decisions and modal traffic is schematically presented in Fig. 8.2. In addition to these short-term effects, persistently high traffic volumes may lead to a change in the highway infrastructure (construction of additional lanes and/or new highways to reduce congestion), again resulting in an interdependence. This interdependence creates the problem of equilibration, which is common to many modeling applications.

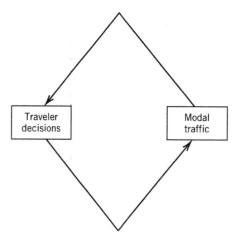

Figure 8.2 Interdependence of traveler decisions and traffic flow.

Perhaps the most recognizable equilibration problem is determination of price in a classic model of economic supply and demand for a product. From a modeling perspective, as will be shown, equilibration adds yet another dimension of difficulty to an already complex travel demand and traffic forecasting problem. It is safe to say that no existing methodology has come close to accurately capturing the complexities involved in traveler decisions or fully addressing the issue of equilibration. However, within rather obvious limitations, the field of travel demand and traffic forecasting has, over the years, made progress toward more accurately modeling traveler decision complexities and equilibration concerns. This evolution of travel demand and traffic forecasting methodology has led to the popular approach of viewing traveler decisions as a sequence of three distinct decisions, as shown in Fig. 8.3, the result of which is forecasted traffic flow (a direct outgrowth of the highway route choice decision). Clearly, the sequential structure of traveler decisions is a considerable simplification of the actual decision-making process in which all trip-related decisions are considered simultaneously by the traveler. However, this sequential simplification permits the development of a sequence of mathematical models of traveler behavior that can be applied to forecast traffic flow. The following sections of this chapter present and discuss typical functional forms of the mathematical models used to forecast the three sequential traveler decisions shown in Fig. 8.3.

8.4 TRIP GENERATION

The first traveler decision to be modeled in the sequential approach to travel demand and traffic forecasting is trip generation. The objective of trip generation modeling is to develop an expression that predicts exactly when a trip is to be made. This is an inherently difficult task due to the wide variety of trip types (working, social/recreational, shopping, etc.) and activities (eating lunch, exercising, visiting friends, etc.) undertaken by a traveler in a sample day, as is schematically shown in Fig. 8.4. To address the complexity of the trip generation decision, the following approach is typically taken:

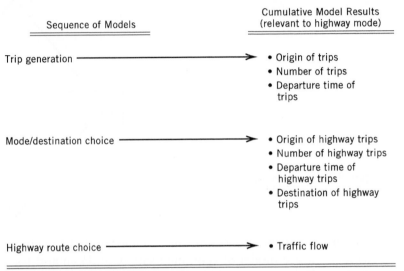

Figure 8.3 Overview of the sequential approach to traffic estimation.

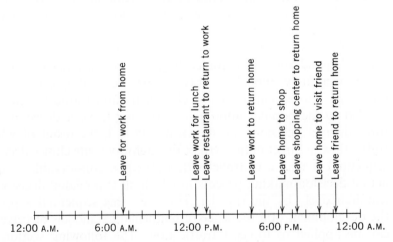

Figure 8.4 Weekday trip generation for a typical traveler.

1. *Aggregation of decision-making units.* Predicting trip generation behavior is simplified by considering the trip generation behavior of a household (a group of travelers sharing the same domicile) as opposed to the behavior of individual travelers. Such an aggregation of traveler

decisions is justified on the basis of the comparatively homogeneous nature of household members (socially and economically) and household members' often intertwined trip-generating activities (joint shopping trips, etc.).

2. *Segmentation of trips by type.* Different types of trips have different characteristics that make them more or less likely to be taken at various times of the day. For example, work trips are more likely to be taken in the morning hours than are shopping trips, which are more likely to be taken during the evening hours. Also, it is more likely that a traveler will take multiple shopping trips during the course of a day as opposed to multiple work trips. To account for this, three distinct trip types are used: (1) work trips, including trips to and from work; (2) shopping trips; and (3) social/recreational trips, which include vacations, visiting friends, church meetings, sporting events, and so on.

3. *Temporal aggregation.* Although research has been undertaken to develop mathematical expressions that predict when a traveler is likely to make a trip (Hamed and Mannering 1993), trip generation more often focuses on the number of trips made over some period of time. Thus trips are aggregated temporally, and trip generation models seek to predict the number of trips per hour or per day.

8.4.1   Typical Trip Generation Models

Trip generation models generally assume a linear form in which the number of vehicle-based (automobile, bus, or subway) trips is a function of various socioeconomic and/or distributional (residential and commercial) characteristics. An example of such a model, for a given trip type, is

$$T_i = b_0 + b_1 z_{1i} + b_2 z_{2i} + \ldots + b_k z_{ki} \tag{8.1}$$

where

T_i = number of vehicle-based trips of a given type (shopping or social/recreational) in some specified time period made by household i,

b_k = coefficient estimated from traveler survey data and corresponding to characteristic k, and

z_{ki} = characteristic k (income, employment in neighborhood, number of household members) of household i.

The estimated coefficients (b's) are usually estimated by the method of least squares regression (linear regression) using data collected from traveler surveys. A brief description and example of this method are presented in Appendix 8A.

EXAMPLE 8.1   SHOPPING-TRIP GENERATION

A simple linear regression model is estimated for shopping-trip generation during a shopping-trip peak hour. The model is

Number of peak-hour vehicle-based shopping trips per household
$$= 0.12 + 0.09(\text{household size})$$
$$+ 0.011(\text{annual household income in thousands of dollars})$$
$$- 0.15(\text{employment in the household's neighborhood, in hundreds})$$

A particular household has six members and an annual income of $50,000. They currently live in a neighborhood with 450 retail employees, but are moving to a new home in a neighborhood with 150 retail employees. Calculate the predicted number of vehicle-based peak-hour shopping trips the household makes before and after the move.

SOLUTION

Note that the signs of the model coefficients (b's, +0.09, and +0.011) indicate that as household size and income increase, the number of shopping trips also increases. This is reasonable because wealthier, larger households can be expected to make more vehicle-based shopping trips. The negative sign of the employment coefficient (−0.15) indicates that as retail employment in a household's neighborhood increases, fewer vehicle-based shopping trips will be generated. This reflects the fact that larger retail employment in a neighborhood implies more shopping opportunities nearer to the household, thereby increasing the possibility that a shopping trip can be conducted without the use of a vehicle (a non–vehicle-based trip, such as walking).

Turning to the problem solution, before the household moves,

$$\text{Number of vehicle trips} = 0.12 + 0.09(6) + 0.011(50) - 0.15(4.5) = \underline{\underline{0.535}}$$

After the household moves,

$$\text{Number of vehicle trips} = 0.12 + 0.09(6) + 0.011(50) - 0.15(1.5) = \underline{\underline{0.985}}$$

Thus the model predicts that the move will result in 0.45 additional peak-hour vehicle-based shopping trips due to the decline in neighborhood shopping opportunities as reflected by the decline in neighborhood retail employment.

EXAMPLE 8.2   SOCIAL/RECREATIONAL TRIP GENERATION

A model for social/recreational trip generation is estimated, with data collected during a major holiday, as

Number of peak-hour vehicle-based social/recreational trips per household
$$= 0.04 + 0.018(\text{household size})$$
$$+ 0.009(\text{annual household income in thousands of dollars})$$
$$+ 0.16(\text{number of nonworking household members})$$

If the household described in Example 8.1 has one working member, how many peak-hour social/recreational trips are predicted?

SOLUTION

The positive signs of the model coefficients indicate that increasing household size, income, and number of nonworking household members result in more social/recreational trips. Again, wealthier and larger households can be expected to be involved in more vehicle-based trip-generating activities, and the larger the number of nonworking household members, the larger the number of people available at home to make peak-hour social/recreational trips.

The solution to this problem is

$$\text{Number of vehicle trips} = 0.04 + 0.018(6) + 0.009(50) + 0.16(5) = \underline{\underline{1.398}}$$

EXAMPLE 8.3   TOTAL TRIP GENERATION

A neighborhood has 205 retail employees and 700 households that can be categorized into four types, with each type having characteristics as follows:

Type	Household size	Annual income	Number of nonworkers in the peak hour	Workers departing
1	2	$40,000	1	1
2	3	$50,000	2	1
3	3	$55,000	1	2
4	4	$40,000	3	1

There are 100 type 1, 200 type 2, 350 type 3, and 50 type 4 households. Assuming that shopping, social/recreational, and work vehicle-based trips all peak at the same time (for exposition purposes), determine the total number of peak-hour trips (work, shopping, social/recreational) using the generation models described in Examples 8.1 and 8.2.

SOLUTION

For vehicle-based shopping trips,

Type 1: $0.12 + 0.09(2) + 0.011(40) - 0.15(2.05)$ $= 0.4325$ trips/household
                                                                                             $\times 100$ households
                                                                                             $= 43.25$ trips

Type 2: $0.12 + 0.09(3) + 0.011(50) - 0.15(2.05)$ $= 0.6325$ trips/household
                                                                                             $\times 200$ households
                                                                                             $= 126.5$ trips

Type 3: $0.12 + 0.09(3) + 0.011(55) - 0.15(2.05)$ $= 0.6875$ trips/household
                                                                                             $\times 350$ households
                                                                                             $= 240.625$ trips

Type 4: $0.12 + 0.09(4) + 0.011(40) - 0.15(2.05)$ $= 0.6125$ trips/household
$\times 50$ households
$= 30.625$ trips

Therefore, there will be a total of 441 vehicle-based shopping trips.
     For vehicle-based social/recreational trips,

Type 1: $0.04 + 0.018(2) + 0.009(40) + 0.16(1)$ $= 0.596$ trips/household
$\times 100$ households
$= 59.6$ trips

Type 2: $0.04 + 0.018(3) + 0.009(50) + 0.16(2)$ $= 0.864$ trips/household
$\times 200$ households
$= 172.8$ trips

Type 3: $0.04 + 0.018(3) + 0.009(55) + 0.16(1)$ $= 0.749$ trips/household
$\times 350$ households
$= 262.15$ trips

Type 4: $0.04 + 0.018(4) + 0.009(40) + 0.16(3)$ $= 0.952$ trips/household
$\times 50$ households
$= 47.6$ trips

Therefore, there will be a total of 542.15 vehicle-based social/recreational trips.
     For vehicle-based work trips, there will be 100 generated from type 1 households (1×100), 200 from type 2 (1×200), 700 from type 3 (2×350), and 50 from type 4 (1×50), for a total of 1050 vehicle-based work trips. Summing the totals for the three trip types gives 2033 peak-hour vehicle-based trips.

It should be noted that the trip generation models used in Examples 8.1, 8.2, and 8.3 are simplified representations of the actual trip generation decision-making process. First, there are many more traveler and household characteristics that affect trip-generating behavior (age, lifestyles, etc.), and second, the models have no variables to capture the equilibration concept discussed earlier. The equilibration concern is important, because if the highway system is heavily congested, travelers are likely to make fewer peak-hour trips as a result of either canceling trips or postponing them until a less congested time period. Unfortunately, such obvious model defects must often be accepted due to data and resource limitations.

8.4.2 Trip Generation with Count Data Models

Although linear regression has been a popular method for estimating trip generation models, there is a problem in that the estimated linear regression models can produce fractions of trips for a given time period. As an example, the model presented in Example 8.2 predicted that the household presented in Example 8.1 with one working member would produce 1.398 peak-hour social/recreational trips during the major holiday. Because fractions of trips are not realistic, a modeling approach that gives the probability of making a nonnegative-integer number of trips (0, 1, 2, 3, . . .) may be more appropriate [Washington et al., 2011]. One such model is the Poisson regression, which can be formulated for trip generation (for a given trip type) as

$$P(T_i) = \frac{e^{-\lambda_i}\lambda_i^{T_i}}{T_i!} \tag{8.2}$$

where

T_i = number of vehicle-based trips of a given type (shopping or social/recreational) made in some specified time period by household i,

$P(T_i)$ = probability of household i making exactly T_i trips (where T_i is a nonnegative integer),

e = base of the natural logarithm ($e = 2.718$), and

λ_i = Poisson parameter for household i, which is equal to household i's expected number of vehicle-based trips in some specified time period, $E[T_i]$.

Poisson regressions are estimated by specifying the Poisson parameter λ_i (the expected number of trips of a specific type made by household i over some time period). The most common relationship between explanatory variables (variables that determine the Poisson parameter) and the Poisson parameter is the log-linear relationship

$$\lambda_i = e^{BZ_i} \tag{8.3}$$

where

B = vector of estimable coefficients,

Z_i = vector of household characteristics determining trip generation, and

other terms are as defined previously.

Note that the Poisson parameter λ_i (the expected number of trips of a specific type made by household i over some time period) is a real number (with fractions of trips) but when applied in Eq. 8.2 gives the probability of making a specified nonnegative-integer number of trips (T_i).

In Poisson regressions, the coefficient vector B is estimated by maximum-likelihood procedures. A brief description and example of this estimation procedure are presented in Appendix 8B.

EXAMPLE 8.4   SHOPPING-TRIP GENERATION WITH THE POISSON MODEL

Following Example 8.1, a Poisson regression is estimated for shopping-trip generation during a shopping-trip peak hour. The estimated coefficients are

BZ_i = −0.75 + 0.03(household size)
+ 0.004(annual household income in thousands of dollars)
+ 0.01(retail employment in the household's neighborhood in hundreds)

Given that the household has six members, has an annual income of $50,000, and lives in their new neighborhood with a retail employment of 150, what is the expected number of peak-hour shopping trips and what is the probability that the household will not make a peak-hour shopping trip?

SOLUTION

For the expected number of peak-hour shopping trips (a real number),

$$E[T_i] = \lambda_i = e^{BZ_i} = e^{-0.75+0.03(6)+0.04(50)+0.01(1.5)} = \underline{\underline{0.701}} \text{ vehicle trips}$$

For the probability of making zero peak-hour shopping trips (a nonnegative integer), Eq. 8.2 is used to give

$$P(0) = \frac{e^{-0.701} 0.701^0}{0!} = \underline{\underline{0.496}}$$

8.5 MODE AND DESTINATION CHOICE

Once the number of trips generated per unit time is known, the next step in the sequential approach to travel demand and traffic forecasting is to determine traveler mode and destination. As was the case with trip generation, trips are classified as work, shopping, and social/recreational. For both shopping and social/recreational trips, a traveler will have the option to choose a mode of travel (automobile, vanpool, or bus) as well as a destination (different shopping centers). In contrast, work trips offer only the mode option, because the choice of work location (destination) is usually a long-term decision that is beyond the time range of most traffic forecasts.

8.5.1 Methodological Approach

Following recent advances in the travel demand and traffic forecasting field, development of a model for mode/destination choice necessitates the use of some consistent theory of travelers' decision-making processes. Of the decision-making theories available, one that is based on the microeconomic concept of utility maximization has enjoyed widespread acceptance in mode/destination choice modeling. The basic assumption is that a traveler will select the combination of mode and destination that provides the most utility. The problem then becomes one of developing an expression for the utility provided by various mode and destination alternatives. Because it is unlikely that individual travelers' utility functions can ever be specified with certainty, the unspecifiable portion is assumed to be random. To illustrate this approach, consider a utility function of the following form:

$$V_{im} = \sum_k b_{mk} z_{imk} + \varepsilon_{im} \tag{8.4}$$

where

V_{im} = total utility (specifiable and unspecifiable) provided by mode/destination alternative m to a traveler i,

b_{mk} = coefficient estimated from traveler survey data for mode/destination alternative m corresponding to mode/destination or traveler characteristic k,

z_{imk} = traveler or mode/destination characteristic k (income, travel time of mode, commercial floor space at destination, etc.) for mode/destination alternative m for traveler i, and

ε_{im} = unspecifiable portion of the utility of mode/destination alternative m for traveler i, which will be assumed to be random.

For notational convenience, define the specifiable nonrandom portion of utility V_{im} as

$$U_{im} = \sum_k b_{mk} z_{imk} \tag{8.5}$$

With these definitions of utility, the probability that a traveler will choose some alternative, say m, is equal to the probability that the given alternative's utility is greater than the utility of all other possible alternatives. The probabilistic component arises from the fact that the unspecifiable portion of the utility expression is not known and is assumed to be a random variable. The basic probability statement is

$$P_{im} = \text{prob}\left[U_{im} + \varepsilon_{im} > U_{is} + \varepsilon_{is}\right] \text{ for all } s \neq m \tag{8.6}$$

where

P_{im} = probability that traveler i will select alternative m,

prob[·] = notation for probability,

s = notation for available alternatives, and

other terms are as defined previously.

With this basic probability and utility expression and an assumed random distribution of the unspecifiable components of utility (ε_{im}), a probabilistic choice model can be derived and the coefficients in the utility function (b_{mk}'s in Eqs. 8.4 and 8.5) can be estimated with data collected from traveler surveys, along the same lines as for the coefficients in the trip generation models. A popular approach to deriving such a probabilistic choice model is to assume that the random, unspecifiable component of utility (ε_{im} in Eq. 8.4) is generalized extreme value distributed. With this assumption, a rather lengthy and involved derivation gives rise to the logit model formulation [Washington et al., 2011],

$$P_{im} = \frac{e^{U_{im}}}{\sum_s e^{U_{is}}} \tag{8.7}$$

where e is the base of the natural logarithm ($e = 2.718$).

The coefficients that comprise the specifiable portion of utility (b_{mk}'s in Eq. 8.5) are estimated by the method of maximum likelihood (see Appendix 8B). For further information on logit model coefficient estimation and maximum-likelihood estimation techniques, refer to more specialized references [Washington et al., 2011].

8.5.2 Logit Model Applications

With the total number of vehicle-based trips made in specific time periods known (from trip generation models), the allocation of trips to vehicle-based modes and likely destinations can be undertaken by applying appropriate logit models. This process is best demonstrated by example.

EXAMPLE 8.5   LOGIT MODEL OF WORK-MODE–CHOICE

A simple work-mode–choice model is estimated from data in a small urban area to determine the probabilities of individual travelers selecting various modes. The mode choices include automobile drive-alone (*DL*), automobile shared-ride (*SR*), and bus (*B*), and the utility functions are estimated as

$$U_{DL} = 2.2 - 0.2(\text{cost}_{DL}) - 0.03(\text{travel time}_{DL})$$

$$U_{SR} = 0.8 - 0.2(\text{cost}_{SR}) - 0.03(\text{travel time}_{SR})$$

$$U_{B} = -0.2(\text{cost}_{B}) - 0.01(\text{travel time}_{B})$$

where cost is in dollars and travel time is in minutes. Between a residential area and an industrial complex, 4000 workers (generating vehicle-based trips) depart for work during the peak hour. For all workers, the cost of driving an automobile is $6.00 with a travel time of 20 minutes, and the bus fare is $1.00 with a travel time of 25 minutes. If the shared-ride option always consists of two travelers sharing costs equally, how many workers will take each mode?

SOLUTION

Note that the utility function coefficients logically indicate that as modal costs and travel times increase, modal utilities decline and, consequently, so do modal selection probabilities (see Eq. 8.7). Substitution of cost and travel time values into the utility expressions gives

$$U_{DL} = 2.2 - 0.2(6) - 0.03(20) = 0.4$$

$$U_{SR} = 0.8 - 0.2(3) - 0.03(20) = -0.4$$

$$U_{B} = -0.2(1.0) - 0.01(25) = -0.45$$

Substituting these values into Eq. 8.7 yields

$$P_{DL} = \frac{e^{0.4}}{e^{0.4} + e^{-0.4} + e^{-0.45}} = \frac{1.492}{1.492 + 0.670 + 0.638} = \frac{1.492}{2.80} = 0.533$$

$$P_{SR} = \frac{0.670}{2.80} = 0.239$$

$$P_{B} = \frac{0.638}{2.80} = 0.228$$

Multiplying these probabilities by 4000 (the total number of workers departing in the peak hour) gives 2132 workers driving alone, 956 sharing a ride, and 912 using a bus.

EXAMPLE 8.6   FORECASTING MODE CHOICE WITH THE LOGIT MODEL

A bus company is making costly efforts in an attempt to increase work-trip bus usage for the travel conditions described in Example 8.5. An exclusive bus lane is constructed that reduces bus travel time to 10 minutes and the bus fare is reduced to $0.50.

   a. Determine the modal distribution of trips after the lane is constructed and the bus fare is lowered.

   b. If shared-ride vehicles are also permitted to use the facility, and travel time for bus and shared-ride modes is 10 min, determine the modal distribution.

   c. Given the conditions described in part (b), determine the modal distribution if the bus company offers free bus service.

SOLUTION

a. After the bus lane construction, the modal utilities of drive-alone and shared-ride are unchanged from those in Example 8.5. However, the bus modal utility becomes

$$U_B = -0.2(0.5) - 0.01(10) = -0.2$$

From Eq. 8.7 with 4000 work trips,

$$P_{DL} = \frac{e^{0.4}}{e^{0.4} + e^{-0.4} + e^{-0.2}} = \frac{1.492}{2.981} = 0.500 \quad \text{and} \quad 0.500(4000) = 2000 \text{ trips}$$

$$P_{SR} = \frac{0.670}{2.981} = 0.225 \quad \text{and} \quad 0.225(4000) = 900 \text{ trips}$$

$$P_B = \frac{0.819}{2.981} = 0.275 \quad \text{and} \quad 0.275(4000) = 1100 \text{ trips}$$

or an increase of 188 bus patrons from the prediction of Example 8.5.

b. With the bus lane opened to shared-ride vehicles, only the modal utility of shared rides will change from those in part (a) to:

$$U_{SR} = 0.8 - 0.2(3) - 0.03(10) = -0.1$$

From Eq. 8.7 with 4000 work trips,

$$P_{DL} = \frac{e^{0.4}}{e^{0.4} + e^{-0.1} + e^{-0.2}} = \frac{1.492}{3.216} = 0.464 \quad \text{and} \quad 0.464(4000) = 1856 \text{ trips}$$

$$P_{SR} = \frac{0.905}{3.216} = 0.281 \quad \text{and} \quad 0.281(4000) = 1124 \text{ trips}$$

$$P_B = \frac{0.819}{3.216} = 0.255 \quad \text{and} \quad 0.255(4000) = 1020 \text{ trips}$$

or a loss of 80 bus patrons and a gain of 224 shared-ride users relative to part (a).

c. With free bus fare, the bus modal utility becomes [with other utilities unchanged from part (b)],

$$U_B = -0.2(0) - 0.01(10) = -0.1$$

From Eq. 8.7 with 4000 work trips,

$$P_{DL} = \frac{e^{0.4}}{e^{0.4} + e^{-0.1} + e^{-0.1}} = \frac{1.492}{3.301} = 0.452 \quad \text{and} \quad 0.452(4000) = 1808 \text{ trips}$$

$$P_{SR} = \frac{0.905}{3.301} = 0.274 \quad \text{and} \quad 0.274(4000) = 1096 \text{ trips}$$

$$P_B = \frac{0.905}{3.301} = 0.274 \quad \text{and} \quad 0.274(4000) = 1096 \text{ trips}$$

or 76 more bus patrons compared with part (b).

EXAMPLE 8.7   **LOGIT MODEL OF SHOPPING MODE/DESTINATION CHOICE**

Consider a residential area and two shopping centers that are possible destinations. From 7:00 to 8:00 P.M. on Friday night, 900 vehicle-based shopping trips leave the residential area for the two shopping centers. A joint shopping-trip mode-destination choice logit model (choice of either auto or bus) is estimated, giving the following coefficients:

Variable	Auto coefficient	Bus coefficient
Auto constant	0.6	0.0
Travel time in minutes	−0.3	−0.3
Commercial floor space (in thousands of ft^2)	0.012	0.012

Initial travel times to shopping centers 1 and 2 are as follows:

	By auto	By bus
Travel time to shopping center 1 (in minutes)	8	14
Travel time to shopping center 2 (in minutes)	15	22

If shopping center 2 has 400,000 ft^2 of commercial floor space and shopping center 1 has 250,000 ft^2, determine the distribution of Friday night shopping trips by destination and mode.

SOLUTION

The utility function coefficients indicate that as modal travel times increase, the likelihood of selecting the mode-destination combination declines. Also, as the destination's floor space increases, the probability of selecting that destination will increase, as suggested by the positive coefficient (+0.012). This reflects the fact that bigger shopping centers tend to have a greater variety of merchandise and hence are more attractive shopping destinations. Note that because this is a joint mode-destination choice model, there are four mode-destination combinations and four corresponding utility functions. Let U_{A1} be the utility of the auto mode to shopping center 1, U_{A2} the utility of the auto mode to shopping center 2, and U_{B1} and U_{B2} the utility of the bus mode to shopping centers 1 and 2, respectively. The utilities are

$$U_{A1} = 0.6 - 0.3(8) + 0.012(250) = 1.2$$
$$U_{B1} = -0.3(14) + 0.012(250) = -1.2$$
$$U_{A2} = 0.6 - 0.3(15) + 0.012(400) = 0.9$$
$$U_{B2} = -0.3(22) + 0.012(400) = -1.8$$

Substituting these values into Eq. 8.7 gives

$$P_{A1} = \frac{3.32}{6.246} = 0.532$$

$$P_{B1} = \frac{0.301}{6.246} = 0.048$$

$$P_{A2} = \frac{2.46}{6.246} = 0.394$$

$$P_{B2} = \frac{0.165}{6.246} = 0.026$$

Multiplying these probabilities by the 900 trips gives 479 trips by auto to shopping center 1, 43 trips by bus to shopping center 1, 355 trips by auto to shopping center 2, and 23 trips by bus to shopping center 2.

EXAMPLE 8.8   LOGIT MODEL OF SOCIAL/RECREATIONAL MODE/DESTINATION CHOICE

A joint mode-destination vehicle-based social/recreational trip logit model is estimated with the following coefficients:

Variable	Auto coefficient	Bus coefficient
Auto constant	0.9	0.0
Travel time in minutes	−0.22	−0.22
Population in thousands	0.16	0.16
Amusement floor space (in thousands of ft²)	0.11	0.11

It is known that 500 social/recreational trips will depart from a residential area during the peak hour. There are three possible trip destinations with the following characteristics:

	Travel time (in minutes)		Population (in thousands)	Amusement floor space (in thousands of ft²)
	Auto	Bus		
Destination 1	14	17	12.4	13.0
Destination 2	5	8	8.2	9.2
Destination 3	18	24	5.8	21.0

Determine the distribution of trips by mode and destination.

SOLUTION

As was the case for the shopping mode-destination model presented in Example 8.7, the signs of the coefficient estimates indicate that increasing travel time decreases an alternative's selection probability. Also, increasing population (reflecting an increase in social opportunities) and increasing amusement floor space (reflecting more recreational opportunities) both increase the probability of an alternative being selected. With two modes and three destinations, there are six alternatives, providing the following utilities (using the same subscripting notation as in Example 8.7):

$$U_{A1} = 0.9 - 0.22(14) + 0.16(12.4) + 0.11(13) = 1.234$$
$$U_{B1} = -0.22(17) + 0.16(12.4) + 0.11(13) = -0.326$$
$$U_{A2} = 0.9 - 0.22(5) + 0.16(8.2) + 0.11(9.2) = 2.124$$
$$U_{B2} = -0.22(8) + 0.16(8.2) + 0.11(9.2) = 0.564$$
$$U_{A3} = 0.9 - 0.22(18) + 0.16(5.8) + 0.11(21) = 0.178$$
$$U_{B3} = -0.22(24) + 0.16(5.8) + 0.11(21) = -2.042$$

Using Eq. 8.7 with 500 trips, the total number of trips to the six mode-destination alternatives are

$$P_{A1} = \frac{3.435}{15.607} = 0.220 \quad \text{and} \quad 0.220 \times 500 = 110 \text{ trips}$$

$$P_{B1} = \frac{0.722}{15.607} = 0.046 \quad \text{and} \quad 0.046 \times 500 = 23 \text{ trips}$$

$$P_{A2} = \frac{8.365}{15.607} = 0.536 \quad \text{and} \quad 0.536 \times 500 = 268 \text{ trips}$$

$$P_{B2} = \frac{1.76}{15.607} = 0.113 \quad \text{and} \quad 0.113 \times 500 = 57 \text{ trips}$$

$$P_{A3} = \frac{1.195}{15.607} = 0.077 \quad \text{and} \quad 0.077 \times 500 = 38 \text{ trips}$$

$$P_{B3} = \frac{0.13}{15.607} = 0.008 \quad \text{and} \quad 0.008 \times 500 = 4 \text{ trips}$$

EXAMPLE 8.9   FORECASTING SOCIAL/RECREATIONAL MODE/DESTINATION CHOICE

Consider the situation described in Example 8.8. A labor dispute results in a bus union slowdown that increases travel times from the origin by 4, 2, and 8 minutes to destinations 1, 2, and 3, respectively. If the total number of trips remains constant, determine the resulting distribution of trips by mode and destination.

SOLUTION

The mode-destination utilities are computed as

$$U_{A1} = 1.234 \text{ (as in Example 8.8)}$$
$$U_{B1} = -0.22(21) + 0.16(12.4) + 0.11(13) = -1.206$$
$$U_{A2} = 2.124 \text{ (as in Example 8.8)}$$
$$U_{B2} = -0.22(10) + 0.16(8.2) + 0.11(9.2) = 0.124$$
$$U_{A3} = 0.178 \text{ (as in Example 8.8)}$$
$$U_{B3} = -0.22(32) + 0.16(5.8) + 0.11(21) = -3.802$$

Applying Eq. 8.7 with 500 trips gives the following distribution of trips among mode-destination alternatives:

$$P_{A1} = \frac{3.435}{14.45} = 0.238 \quad \text{and} \quad 0.238 \times 500 = \underline{\underline{119 \text{ trips}}}$$

$$P_{B1} = \frac{0.299}{14.45} = 0.021 \quad \text{and} \quad 0.021 \times 500 = \underline{\underline{10 \text{ trips}}}$$

$$P_{A2} = \frac{8.365}{14.45} = 0.579 \quad \text{and} \quad 0.579 \times 500 = \underline{\underline{290 \text{ trips}}}$$

$$P_{B2} = \frac{1.132}{14.45} = 0.078 \quad \text{and} \quad 0.078 \times 500 = \underline{\underline{39 \text{ trips}}}$$

$$P_{A3} = \frac{1.195}{14.45} = 0.083 \quad \text{and} \quad 0.083 \times 500 = \underline{\underline{41 \text{ trips}}}$$

$$P_{B3} = \frac{0.022}{14.45} = 0.002 \quad \text{and} \quad 0.002 \times 500 = \underline{\underline{1 \text{ trip}}}$$

8.6   HIGHWAY ROUTE CHOICE

To summarize, the trip generation and mode-destination choice models give total highway traffic demand between a specified origin (the neighborhood from which trips originate) and a destination (the geographic area to which trips are destined), in terms of vehicles per some time period (usually vehicles per hour). With this information in hand, the final step in the sequential approach to travel demand and traffic forecasting—route choice—can be addressed. The result of the route choice decision will be traffic flow (generally in units of vehicles per hour) on specific highway routes, which is the desired output from the traffic forecasting process.

8.6.1   Highway Performance Functions

Route choice presents a classic equilibrium problem, because travelers' route choice decisions are primarily a function of route travel times, which are determined by traffic flow—itself a product of route choice decisions. This interrelationship between route choice decisions and traffic flow forms the basis of route choice theory and model development.

To begin modeling traveler route choice, a mathematical relationship between route travel time and route traffic flow is needed. Such a relationship is commonly referred to as a highway performance function. The most simplistic approach to formalizing this relationship is to assume a linear highway performance function in which travel time increases linearly with flow. An example of such a function is illustrated in Fig. 8.5. In this figure, the free-flow travel time refers to the travel time that a traveler would experience if no other vehicles were present to impede travel speed (as discussed in Chapter 5). This free-flow travel time is generally computed with the assumption that a vehicle travels at the posted speed limit of the route.

Although the linear highway performance function has the appeal of simplicity, it is not a particularly realistic representation of the travel time–traffic flow relationship. Recall that Chapter 5 presented a relationship between traffic speed and flow that is parabolic in nature, with significant reductions in travel speed occurring as the traffic flow approaches the roadway's capacity. This parabolic speed–flow relationship suggests a nonlinear highway performance function, such as that illustrated in Fig. 8.6. This figure shows route travel time increasing more quickly as traffic flow approaches capacity, which is consistent with the parabolic relationship presented in Chapter 5.

Both linear and nonlinear highway performance functions will be demonstrated through example, using two theories of travel route choice: user equilibrium and system optimization. For other theories of route choice, refer to Sheffi [1985].

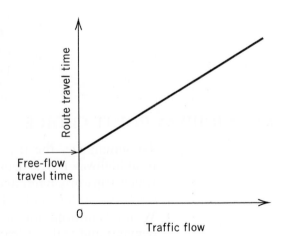

Figure 8.5 Linear travel time–traffic flow relationship.

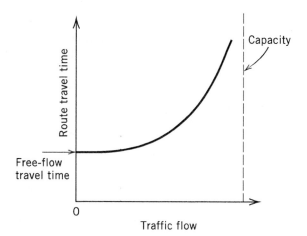

Figure 8.6 Nonlinear travel time–traffic flow relationship.

8.6.2   User Equilibrium

In developing theories of traveler route choice, two important assumptions are usually made. First, it is assumed that travelers will select routes between origins and destinations on the basis of route travel times only (they will tend to select the route with the shortest travel time). This assumption is not terribly restrictive, because travel time obviously plays the dominant role in route choice; however, other, more subtle factors that may influence route choice (scenery, pavement conditions, etc.) are not accounted for. The second assumption is that travelers know the travel times that would be encountered on all available routes between their origin and destination. This is potentially a strong assumption, because a traveler may not have actually traveled on all available routes between an origin and destination and may repeatedly (day after day) choose one route based only on the perception that travel times on alternative routes are higher. However, in support of this assumption, studies have shown that travelers' perceptions of alternative route travel times are reasonably close to actual observed travel times [Mannering, 1989].

With these assumptions, the theory of user-equilibrium route choice can be made operational. The rule of choice underlying user equilibrium is that travelers will select a route so as to minimize their personal travel time between the origin and destination. User equilibrium is said to exist when individual travelers cannot improve their travel times by unilaterally changing routes. Stated differently [Wardrop, 1952], user equilibrium can be defined as follows:

The travel time between a specified origin and destination on all used routes is the same and is less than or equal to the travel time that would be experienced by a traveler on any unused route.

EXAMPLE 8.10   BASIC USER EQUILIBRIUM

Two routes connect a city and a suburb. During the peak-hour morning commute, a total of 4500 vehicles travel from the suburb to the city. Route 1 has a 60-mi/h speed limit and is six miles in length; route 2 is three miles in length with a 45-mi/h speed limit. Studies show that the total travel time on route 1 increases two minutes for every additional 500 vehicles added. Minutes of travel time on route 2 increase with the square of the number of vehicles, expressed in thousands of vehicles per hour. Determine user-equilibrium travel times.

SOLUTION

Determining free-flow travel times, in minutes, gives

$$\text{route 1: 6 mi/(60 mi/h)} \times 60 \text{ min/h} = 6 \text{ min}$$
$$\text{route 2: 3 mi/(45 mi/h)} \times 60 \text{ min/h} = 4 \text{ min}$$

With these data, the performance functions can be written as

$$t_1 = 6 + 4x_1$$
$$t_2 = 4 + x_2^2$$

where

t_1, t_2 = average travel times on routes 1 and 2 in minutes, and
x_1, x_2 = traffic flow on routes 1 and 2 in thousands of vehicles per hour.

Also, the basic flow conservation identity is

$$q = x_1 + x_2 = 4.5$$

where q = total traffic flow between the origin and destination in thousands of vehicles per hour.

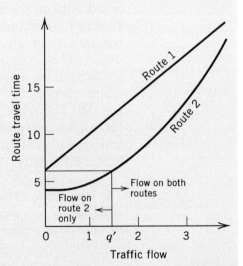

Figure 8.7 Illustration of performance curves for Example 8.10.

With Wardrop's definition of user equilibrium, it is known that the travel times on all used routes are equal. However, the first order of business is to determine whether or not both routes are used. Figure 8.7 gives a graphic representation of the two performance functions. Note that because route 2 has a lower free-flow travel time, any total origin-to-destination traffic flow less than q' (in Fig. 8.7) will result in only route 2 being used, because the travel time on route 1 would be greater even if only one vehicle used it. At flows of q' and above, route 2 is sufficiently congested, and its travel time sufficiently high, that route 1 becomes a viable alternative.

To check if the problem's flow of 4500 vehicles per hour exceeds q', the following test is conducted:

1. Assume that all traffic flow is on route 1. Substituting traffic flows of 4.5 and 0 into the performance functions gives $t_1(4.5) = 24$ min and $t_2(0) = 4$ min.
2. Assume that all traffic flow is on route 2, giving $t_1(0) = 6$ min and $t_2(4.5) = 24.25$ min.

Thus, because $t_1(4.5) > t_2(0)$ and $t_2(4.5) > t_1(0)$, both routes will be used. If $t_1(0)$ had been greater than $t_2(4.5)$, the 4500 vehicles would have been less than q' in Fig. 8.7, and only route 2 would have been used.

With both routes used, Wardrop's user-equilibrium definition gives

$$t_1 = t_2$$

or

$$6 + 4x_1 = 4 + x_2^2$$

From flow conservation, $x_1 + x_2 = 4.5$, so substituting, we get

$$6 + 4(4.5 - x_2) = 4 + x_2^2$$
$$x_2 = 2.899 \quad \text{or} \quad 2899 \text{ veh/h}$$
$$x_1 = 4.5 - x_2 = 4.5 - 2.899$$
$$= 1.601 \quad \text{or} \quad 1601 \text{ veh/h}$$

which gives average route travel times of

$$t_1 = 6 + 4(1.601) = \underline{\underline{12.4 \text{ min}}}$$
$$t_2 = 4 + (2.899)^2 = \underline{\underline{12.4 \text{ min}}}$$

EXAMPLE 8.11   USER EQUILIBRIUM—EFFECT OF CAPACITY AND TRAFFIC REDUCTION

Peak-hour traffic demand between an origin–destination pair is initially 3500 vehicles. The two routes connecting the pair have performance functions $t_1 = 2 + 3(x_1/c_1)$ and $t_2 = 4 + 2(x_2/c_2)$, where the t's are travel times in minutes, the x's are the peak-hour traffic volumes expressed in thousands, and the c's are the peak-hour route capacities expressed in thousands of vehicles per hour. Initially, the capacities of routes 1 and 2 are 2500 and 4000 veh/h, respectively. A reconstruction project reduces capacity on route 2 to 2000 veh/h. Assuming user equilibrium before and during reconstruction, what reduction in total peak-hour origin–destination traffic flow is needed to ensure that total travel times (summation of all $x_a t_a$'s, where a denotes route) during reconstruction are equal to those before reconstruction?

SOLUTION

First, focusing on the roads before reconstruction, a check to see if both routes are used gives (using performance functions)

$$t_1(3.5) = 6.2 \text{ min}, \qquad t_2(0) = 4 \text{ min}$$
$$t_1(0) = 2 \text{ min}, \qquad t_2(3.5) = 5.75 \text{ min}$$

which, because $t_1(3.5) > t_2(0)$ and $t_2(3.5) > t_1(0)$, indicates that both routes are used. Setting route travel times equal and substituting performance functions gives

$$2 + \frac{3}{2.5}(x_1) = 4 + \frac{2}{4}(x_2)$$

From conservation of flow, $x_2 = 3.5 - x_1$, so that

$$2 + 1.2x_1 = 4 + 0.5(3.5 - x_1)$$

Solving gives $x_1 = 2.206$ and $x_2 = 3.5 - 2.206 = 1.294$. For travel times,

$$t_1 = 2 + 1.2(2.206) = 4.647 \text{ min}$$
$$t_2 = 4 + 0.5(1.294) = 4.647 \text{ min}$$

The total peak-hour travel time before reconstruction will simply be the average route travel time multiplied by the number of vehicles:

$$\text{Total travel time} = 4.647(3500) = 16{,}264.5 \text{ veh-min}$$

During reconstruction, the performance function of route 1 is unchanged, but the performance function of route 2 is altered because of the reduction in capacity to

$$t_2 = 4 + \frac{2}{2}(x_2) = 4 + x_2$$

If it is assumed that both routes are used, $t_1 = t_2$. Also, it is known that the total travel time is

$$t_1(q) = t_2(q)$$
$$= 16{,}264.5 \text{ veh-min}$$

Using the performance function of route 2, we find

$$(4 + x_2)(q) = 16.2645 \text{ (using thousands of vehicles)}$$
$$q = \frac{16.2645}{4 + x_2}$$

From $t_1 = t_2$, and $x_1 = q - x_2$ (flow conservation),

$$2 + 1.2x_1 = 4 + x_2$$
$$2 + 1.2(q - x_2) = 4 + x_2$$
$$q = 1.67 + 1.83x_2$$

Equating the two expressions for q gives

$$1.67 + 1.83x_2 = \frac{16.2645}{4 + x_2}$$

$$1.83x_2^2 + 8.99x_2 - 9.5845 = 0$$

which gives $x_2 = 0.901$, $q = 1.67 + 1.83(0.901) = 3.319$, and $x_1 = 3.319 - 0.901 = 2.418$. Because flow exists on both routes, the earlier assumption that both routes would be used is valid, and a reduction of <u>181 vehicles</u> (3500 – 3319) in peak-hour flow is needed to ensure equality of total travel times.

EXAMPLE 8.12   USER EQUILIBRIUM—EFFECT OF CAPACITY REDUCTION ON TOTAL TRAVEL TIME

Two highways serve a busy corridor with a traffic demand that is fixed at 6000 vehicles during the peak hour. The performance functions for the two routes are $t_1 = 4 + 5(x_1/c_1)$ and $t_2 = 3 + 7(x_2/c_2)$, where t's are in minutes and flows (x's) and capacities (c's) are in thousands of vehicles per hour. Initially, the capacities of routes 1 and 2 are 4400 veh/h and 5200 veh/h, respectively. If a highway reconstruction project cuts the capacity of route 2 to 2200 veh/h, how many additional vehicle-hours of travel time will be added in the corridor assuming that user-equilibrium conditions hold?

SOLUTION

To determine the initial number of vehicle-hours, first check to see if both routes are used:

$$t_1(6) = 10.82 \text{ min}, \qquad t_2(0) = 3 \text{ min}$$
$$t_1(0) = 4 \text{ min}, \qquad t_2(6) = 11.08 \text{ min}$$

Both routes are used, because $t_2(6) > t_1(0)$ and $t_1(6) > t_2(0)$. At user equilibrium, $t_1 = t_2$, so substituting performance functions gives

$$4 + \frac{5}{4.4}(x_1) = 3 + \frac{7}{5.2}(x_2)$$

With flow conservation, $x_2 = 6 - x_1$, so that

$$4 + 1.136(x_1) = 3 + 1.346(6 - x_1)$$
$$x_1 = 2.85$$

and

$$x_2 = 6 - 2.85$$
$$= 3.15$$

The total travel time in hours is $(t_1 x_1 + t_2 x_2)/60$ or, by substituting,

$$\frac{\left\{ \left[4 + 1.136(2.85) \right]2850 + \left[3 + 1.346(3.15) \right]3150 \right\}}{60} = 723.88 \text{ veh-h}$$

For the reduced-capacity case, the route usage check is

$$t_1(6) = 10.82 \text{ min}, \qquad t_2(0) = 3 \text{ min}$$
$$t_1(0) = 4 \text{ min}, \qquad t_2(6) = 22.09 \text{ min}$$

Again, both routes are used [$t_2(6) > t_1(0)$ and $t_1(6) > t_2(0)$]. Equating performance functions (because travel times are equal) and using flow conservation, $x_2 = 6 - x_1$,

$$4 + \frac{5}{4.4}(x_1) = 3 + \frac{7}{2.2}(x_2)$$
$$4 + 1.136x_1 = 3 + 3.182(6 - x_1)$$
$$x_1 = 4.19$$

and

$$x_2 = 6 - 4.19 = 1.81$$

which gives a total travel time of $(t_1x_1 + t_2x_2)/60$ or, by substituting,

$$\frac{\left\{[4 + 1.136(4.19)]\, 4190 + [3 + 3.182(1.81)]\, 1810\right\}}{60} = 875.97 \text{ veh-h}$$

Thus the reduced capacity results in an additional 152.09 veh-h (875.97 − 723.88) of travel time.

8.6.3 Mathematical Programming Approach to User Equilibrium

Equating travel time on all used routes is a straightforward approach to user equilibrium, but can become cumbersome when many alternative routes are involved. The approach used to resolve this computational obstacle is to formulate the user-equilibrium problem as a mathematical program. Specifically, user-equilibrium route flows can be obtained by minimizing the following function [Sheffi, 1985]:

$$\min S(x) = \sum_h \int_0^{x_n} t_n(w)\, dw \qquad \text{(8.8)}$$

where

n = a specific route, and

$t_n(w)$ = performance function corresponding to route n (w denotes flow, x_n's).

This function is subject to the constraints that the flow on all routes is greater than or equal to zero ($x_n \geq 0$) and that flow conservation holds (the flow on all routes between an origin and destination sums to the total number of vehicles, q, traveling between the origin and destination, $q = \sum_n x_n$).

Formulating the user-equilibrium problem as a mathematical program allows an equilibrium solution to very complex highway networks (many origins and destinations) to be readily undertaken by computer. The reader is referred to Sheffi [1985] for an application of user-equilibrium principles to such a network.

EXAMPLE 8.13   USER EQUILIBRIUM—MATHEMATICAL PROGRAMMING SOLUTION

Solve Example 8.10 by formulating user-equilibrium problem as a mathematical program.

SOLUTION

From Example 8.10, the performance functions are

$$t_1 = 6 + 4x_1$$

$$t_2 = 4 + x_2^2$$

Substituting these into Eq. 8.8 gives

$$\min S(x) = \int_0^{x_1} (6 + 4w)\,dw + \int_0^{x_2} \left(4 + w^2\right)dw$$

The problem can be viewed in terms of x_2 only by noting that flow conservation implies $x_1 = 4.5 - x_2$. Substituting yields

$$S(x) = \int_0^{4.5-x_2} (6 + 4w)\,dw + \int_0^{x_2} \left(4 + w^2\right)dw$$

$$= 6w + 2w^2 \Big|_0^{4.5-x_2} + 4w + \frac{w^3}{3}\Big|_0^{x_2}$$

$$= 27 - 6x_2 + 40.5 - 18x_2 + 2x_2^2 + 4x_2 + \frac{x_2^3}{3}$$

To arrive at a minimum, the first derivative is set to zero, giving

$$\frac{dS(x)}{dx_2} = x_2^2 + 4x_2 - 20 = 0$$

which gives $x_2 = \underline{2899\ \text{veh/h}}$, the same value as found in Example 8.10. It can readily be shown that all other flows and travel times will also be the same as those computed in Example 8.10.

8.6.4   System Optimization

From an idealistic point of view, one can visualize a single route choice strategy that results in the lowest possible number of total vehicle-hours of travel for some specified origin–destination traffic flow. Such strategy is known as a system-optimal route choice and is based on the choice rule that travelers will behave such that total system travel time will be minimized, even though travelers may be able to decrease their own individual travel times by unilaterally changing routes. From this definition it is clear that system-optimal flows are not stable, because there will always be a temptation for travelers to switch to non–system-optimal routes in order to improve their travel times. Thus system-optimal flows are generally not a realistic representation of actual traffic. Nevertheless, system-optimal flows often provide useful comparisons with the more realistic user-equilibrium traffic forecasts.

The system-optimal route choice rule is made operational by the following mathematical program:

$$\min\ S(x) = \sum_n x_n t_n (x_n) \tag{8.9}$$

This program is subject to the constraints of flow conservation ($q = \sum_n x_n$) and nonnegativity ($x_n \geq 0$).

EXAMPLE 8.14   SYSTEM OPTIMIZATION

Determine the system-optimal travel time for the situation described in Example 8.10.

SOLUTION

Using Eq. 8.9 and substituting the performance functions for routes 1 and 2 yields

$$S(x) = x_1 (6 + 4x_1) + x_2 (4 + x_2^2)$$

$$= 6x_1 + 4x_1^2 + 4x_2 + x_2^3$$

From flow conservation, $x_1 = 4.5 - x_2$; therefore,

$$S(x) = 6(4.5 - x_2) + 4(4.5 - x_2)^2 + 4x_2 + x_2^3$$

$$= x_2^3 + 4x_2^2 - 38x_2 + 108$$

To find the minimum, the first derivative is set to zero, giving

$$\frac{dS(x)}{dx_2} = 3x_2^2 + 8x_2 - 38 = 0$$

which gives $x_2 = 2.467$ and $x_1 = 4.5 - 2.467 = 2.033$. For system-optimal travel times,

$$t_1 = 6 + 4(2.033) = 14.13 \text{ min}$$

$$t_2 = 4 + (2.467)^2 = 10.08 \text{ min}$$

which are not user-equilibrium travel times, because t_1 is not equal to t_2. In Example 8.10, the total user-equilibrium travel time is computed as 930 veh-h [4500(12.4)/60]. For the system-optimal total travel time [$(t_1 x_1 + t_2 x_2)/60$],

$$\frac{\left[2033(14.13) + 2467(10.08)\right]}{60} = \underline{\underline{893.2 \text{ veh-h}}}$$

Therefore, the system-optimal solution results in a systemwide travel time savings of 36.8 veh-h.

EXAMPLE 8.15   COMPARISON OF USER-EQUILIBRIUM AND SYSTEM-OPTIMAL SOLUTIONS

Two roads begin at a gate entrance to a park and take different scenic routes to a single main attraction in the park. The park managers know that 4000 vehicles arrive during the peak hour, and they distribute these vehicles among the two routes so that an equal number of vehicles take each route. The performance functions for the routes are $t_1 = 10 + x_1$ and $t_2 = 5 + 3x_2$, with the x's expressed in thousands of vehicles per hour and the t's in minutes. How many vehicle-hours would have been saved had park managers distributed the vehicular traffic so as to achieve a system-optimal solution?

SOLUTION

For the number of vehicle-hours, assuming that an equal distribution of traffic among the two routes,

$$\text{route 1: } \frac{x_1 t_1}{60} = \frac{2000[10+(2)]}{60} = 400 \text{ veh-h}$$

$$\text{route 2: } \frac{x_2 t_2}{60} = \frac{2000[5+3(2)]}{60} = 366.67 \text{ veh-h}$$

for a total of 766.67 veh-h. With the system-optimal traffic distribution, the performance functions are substituted into Eq. 8.9, giving

$$S(x) = (10 + x_1)\, x_1 + (5 + 3x_2)x_2$$

With flow conservation, $x_1 = 4.0 - x_2$, so that

$$S(x) = 4x_2^2 - 13x_2 + 56$$

Setting the first derivative equal to zero,

$$\frac{dS(x)}{dx_2} = 8x_2 - 13 = 0$$

gives $x_2 = 1.625$ and $x_1 = 4 - 1.625 = 2.375$. The total travel times are

$$\text{route 1: } \frac{x_1 t_1}{60} = \frac{2375[10+2.375]}{60} = 489.84 \text{ veh-h}$$

$$\text{route 2: } \frac{x_2 t_2}{60} = \frac{1625[5+3(1.625)]}{60} = 267.45 \text{ veh-h}$$

which gives a total system travel time of 757.27 veh-h or a savings of 9.38 veh-h (766.67 – 757.29) over the equal distribution of traffic to the two routes.

EXAMPLE 8.16   SYSTEM-OPTIMAL SOLUTION—MINIMIZING PERSON-HOURS

During the peak hour, an urban freeway segment has a traffic flow of 4000 veh/h (2000 vehicles with one occupant and 2000 vehicles with two occupants). The freeway has five lanes, four of which are unrestricted (open to all vehicles regardless of vehicle occupancy) and one that is restricted for use by vehicles with two occupants. The performance functions for the length of this freeway segment are $t_u = 4 + 0.5x_u$ for the unrestricted lanes (all four combined) and $t_r = 4 + 2x_r$ for the restricted lane (t's are in minutes and x's in thousands of vehicles per hour). Determine the distribution of traffic among the lanes such that the total number of person-hours is minimized and compare the savings in person-hours relative to a user-equilibrium solution (assume that compliance is perfect and that no single-occupant vehicles use the restricted lane).

SOLUTION

As stated in the problem, the 2000 single-occupant vehicles must use the unrestricted lanes. Begin by determining the distribution of traffic that will minimize total person-hours. Using the subscripts r for restricted lane, $u1$ for single-occupant vehicles using the unrestricted lanes, and $u2$ for two-occupant vehicles using the unrestricted lanes, total person-hours can be written as

$$S(x) = x_r t_r \times 2 + x_{u2} t_u \times 2 + x_{u1} t_u \times 1$$

where

$\quad x_r$ = flow on the restricted lane (two-occupant vehicles only),

$\quad t_r$ = travel time on the restricted lane,

$\quad x_{u2}$ = flow of two-occupant vehicles on the unrestricted lanes,

$\quad t_u$ = travel time on the unrestricted lanes, and

$\quad x_{u1}$ = flow of single-occupant vehicles on the unrestricted lanes.

It is given that $t_u = 4 + 0.5x_u$, where $x_u = x_{u1} + x_{u2}$. And because $x_{u1} = 2.0$, $t_u = 4 + 0.5(2.0 + x_{u2})$. Substituting gives

$$S(x) = 2x_r(4 + 2x_r) + 2x_{u2}(4 + 0.5(2.0 + x_{u2})) + 2(4 + 0.5(2.0 + x_{u2}))$$

$$= 8x_r + 4x_r^2 + 10x_{u2} + x_{u2}^2 + 10 + x_{u2}$$

The total number of two-occupant vehicles is 2000, so $x_r + x_{u2} = 2.0$. Substituting,

$$S(x) = 8(2 - x_{u2}) + 4(2 - x_{u2})^2 + 10x_{u2} + x_{u2}^2 + 10 + x_{u2} = 5x_{u2}^2 - 13x_{u2} + 42$$

Taking the first derivative,

$$\frac{dS(x)}{dx_{u2}} = 10x_{u2} - 13 = 0$$

which gives $x_{u2} = 1.3$, and so $x_r = 2.0 - 1.3 = 0.7$. With this, total person-hours is [with $t_r = 4 + 2(0.7) = 5.4$ and $t_u = 4 + 0.5(3.3) = 5.65$]

2[5.4(700)] + 2[5.65(1300)] + 2000(5.65) = 33,550 person-min or 559.167 person-h

For the user-equilibrium solution, with 2000 vehicles on the unrestricted lanes, t_u can be written as

$$t_u = 4 + 0.5(2.0 + x_{u2}) = 5 + 0.5x_{u2}$$

To check if both two-occupant lane choices are used by two-occupant vehicles, note that when $x_{u2} = 2$ and $x_r = 0$, $t_u = 6$ and $t_r = 4$. And when $x_{u2} = 0$ and $x_r = 2$, $t_u = 5$ and $t_r = 8$, so both lane choices might be used by two-occupant vehicles. Equating travel times ($t_u = t_r$) gives

$$5 + 0.5x_{u2} = 4 - 2x_r$$

and with $x_r = 2 - x_{u2}$,

$$5 + 0.5x_{u2} = 4 - 2(2 - x_{u2})$$

Solving gives $x_{u2} = 1.2$ and $x_r = 2 - 1.2 = 0.8$. This produces user-equilibrium travel times $t_u = t_r = 5.6$. Total person-hours for the user-equilibrium solution is then

$$2[5.6(2000)] + 5.6(2000) = 33,600 \text{ person-min or } \underline{560 \text{ person-h}}$$

So the savings is 0.833 person-h (560 − 559.167) when person-hours are minimized relative to the user-equilibrium solution.

8.7   AUTONOMOUS VEHICLES, HIGHWAY PERFORMANCE FUNCTIONS, AND SYSTEM OPTIMIZATION

Autonomous vehicles (self-driving cars), with vehicle-to-vehicle and vehicle-to-infrastructure communication, have the potential to significantly reduce congestion by allowing for closer following distances and centrally controlled vehicle routing. The speed profile of traffic would also be altered because the concept of sight-distance and reaction time would no longer apply, and stopping distances and speeds could be altered based on road-surface conditions so that the 0.35g deceleration currently assumed in highway design would likely be significantly higher in most instances (see Chapters 2 and 3).

In terms of the impact that autonomous vehicles would have on route selection and performance, one could imagine at least two possible outcomes:

1. In all likelihood, the centrally controlled vehicle routing would be system optimal to minimize total vehicle-hours or total person-hours (as demonstrated in Example 8.16). There would also be the potential to price differentially so that those wishing to travel faster would pay a fee for preferential routing.

2. The route performance function would likely change in two ways. First, the free-flow travel time would likely be reduced because vehicles will be able to travel faster and maintain a high level of safety. Second, the effect of increasing traffic demand on route travel time should be substantially less than current assumptions since vehicles will be able to maintain uniform speeds and closer following distance (the spacing between vehicles as traffic demand increases will depend on the efficiency of the technology and not on the reaction times and driving behavior of humans).

The following example provides an illustration of the possible impacts of autonomous vehicles.

EXAMPLE 8.17   EXAMPLE OF THE IMPACT OF AUTONOMOUS VEHICLES

Consider the two roads beginning at a gate entrance to a park as described in Example 8.15. As in Example 8.15, 4000 vehicles arrive during the peak hour, but the vehicles are now autonomous that allows them to be distributed among the two in a systematic way. Also, because vehicle speeds and following distances can be strictly controlled, the performance functions for the routes change (decreasing free-flow travel time and the effects of congestion) such that $t_1 = 7.5 + 0.5x_1$ and $t_2 = 3.75 + 1.5x_2$, with the x's expressed in thousands of vehicles per hour and the t's in minutes. How many vehicle-hours will be saved relative to the system-optimal solution obtained in Example 8.15 with standard human-operated vehicles?

SOLUTION

Recall from Example 8.15, it was found that the system-optimal solution gave a total travel time of 757.27 veh-h. To find the new autonomous vehicle travel time with a system-optimal route distribution and revised route performance functions, the new performance functions are substituted into Eq. 8.9, giving

$$S(x) = (7.5 + 0.5x_1)\, x_1 + (3.75 + 1.5x_2)x_2$$

With flow conservation, $x_1 = 4.0 - x_2$, so that

$$S(x) = 2x_2^2 - 7.75x_2 + 38$$

Setting the first derivative equal to zero,

$$\frac{dS(x)}{dx_2} = 4x_2 - 7.75 = 0$$

gives $x_2 = 1.938$ and $x_1 = 4 - 1.938 = 2.062$. The total travel times are

$$\text{route1}: \frac{x_1 t_1}{60} = \frac{2062\left[7.75 + 0.5(2.062)\right]}{60} = 301.77 \text{ veh-h}$$

$$\text{route1}: \frac{x_2 t_2}{60} = \frac{1938\left[3.75 + 1.5(1.938)\right]}{60} = 215.02 \text{ veh-h}$$

which gives a total autonomous vehicle system travel time of 516.79 veh-h or a savings of 240.50 veh-h (757.29 − 516.79) over the human-operated vehicle system optimization. Note the important effect that the changing route performance functions have on this solution, as well as the effect that the new performance functions have on the system-optimal route volumes.

8.8   TRAFFIC FORECASTING IN PRACTICE

With the basic procedures outlined in the previous sections of this chapter, a traffic forecasting model similar to those used in practice can be developed and implemented. Although there are many subtleties in the process that are beyond the scope of this book, the basic procedure is as follows (referring to the example network shown in Fig. 8.8):

1. The geographic region being studied is segmented into nearly homogeneous areas (see Fig. 8.8) based on similarities in land use,

socioeconomic conditions, and so on. These areas are referred to as traffic analysis zones (TAZs) and are used to determine the origins and destinations of trips (as used in the highway route choice models described in Section 8.6). The choice of the number of TAZs (often simply referred to as zones) presents a trade-off between accuracy (smaller TAZs provide more detailed forecasts) and ease of implementation (larger TAZs require less data and are easier to incorporate in the overall model system). A single point is usually chosen within the TAZ as the assumed origin/destination point of all TAZ trips. This point is referred to as the centroid (see example network in Fig. 8.8).

2. The highway network is defined to include the relevant highway segments. Highway segments are linked by using nodes (see example network in Fig. 8.8), which are usually placed at intersections or other points where highway capacity could change. Nodes permit traffic to travel from one highway segment to the next. The highway network is a representation of the actual street network and carries traffic flow between TAZs. As was the case with the size of the TAZs, a very large and detailed highway network can provide very detailed forecasts but also requires large amounts of data, and thus smaller networks are often used. Defining the highway network includes detailed information on each highway segment's performance function (see Section 8.6.1) so that traffic flows can be computed (usually by assuming user-equilibrium route choice). The performance function often used in practice to relate traffic flow with travel time was originally developed by the U.S. Bureau of Public Roads and takes the form:

$$t_n = t_{fn}[1 + \alpha(x_n/x_{cn})^{\beta}] \tag{8.10}$$

where

t_n = travel time on highway segment (route) n, usually in minutes,

t_{fn} = free-flow travel time on highway segment (route) n, usually in minutes,

x_n = traffic flow on highway segment (route) n, usually in veh/h,

x_{cn} = capacity of highway segment (route) n, usually in veh/h, and

α, β = model parameters that usually vary with respect to the capacity and speed limit of the highway segment (route). Typical values of α and β are shown in Table 8.1.

Access links are used to connect the highway network with the centroids of the TAZs (see Fig. 8.8). These links also have highway performance functions associated with them so that access/egress from the highway network to the centroids can be approximated.

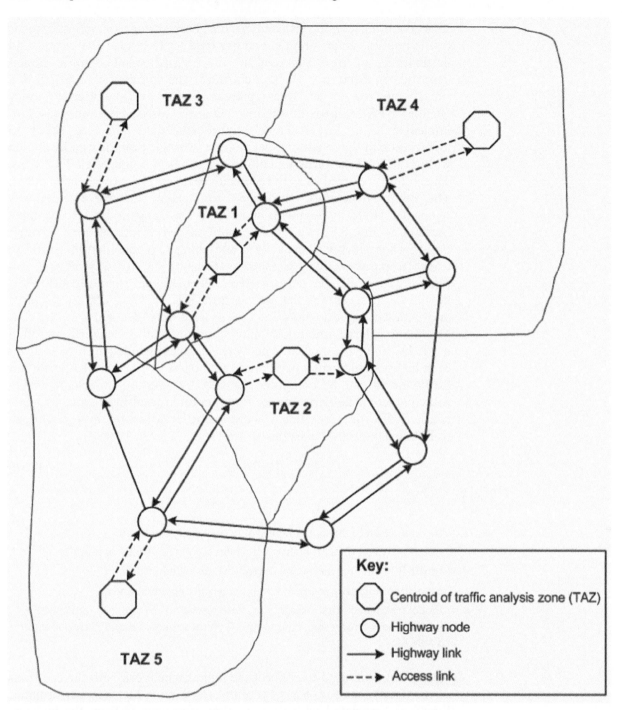

Figure 8.8 Example highway network.

Table 8.1 Typical Values of α and β for Bureau of Public Roads Highway Performance Function (see [Mannering et al., 1989])

Route speed limit (mi/h)	Route capacity, x_{cn} (veh/h)	Performance function parameters	
		α	β
< 30	< 250	0.7312	3.6596
< 30	251-499	0.6128	3.5038
< 30	500-749	0.8774	4.4613
< 30	750-999	0.6846	5.1644
< 30	1000+	1.1465	4.4239
31-40	< 499	0.6190	3.6544
31-40	500-749	0.6662	4.9432
31-40	750-999	0.6222	5.1409
31-40	1000+	1.0300	5.5226
41-50	< 750	0.6609	5.0906
41-50	750-999	0.5423	5.7894
41-50	1000+	1.0091	6.5856
> 50	< 750	0.8776	4.9287
> 50	750-999	0.7699	5.3443
> 50	1000+	1.1491	6.8677

3. Trip generation models and mode/destination choice models are then used to determine the number of vehicles traveling between all TAZs during a specified time period (usually the peak hour). The resulting vehicle trips are used to create an origin–destination matrix that gives the total number of vehicle trips going between each TAZ combination during the analysis period. Although most traffic forecasting models developed in practice have hundreds of TAZs, for illustrative purposes, Table 8.2 gives a sample vehicle origin–destination matrix with just the five TAZs shown in Fig. 8.8. If TAZs 1 and 2 are in the center of the city, the origin–destination matrix shown in Table 8.2 is what one might expect during the morning peak hour—with higher vehicular flows going from the outskirts of the region (zones 3, 4 and 5) to the city center (zones 1 and 2) and substantially lower vehicular flows going from zones 1 and 2 to zones 3, 4, and 5. As an example, Table 8.2 indicates that 3386 vehicle trips go from TAZ 5 to TAZ 1 while only 213 vehicle trips go from TAZ 1 to TAZ 5.

4. With the vehicle origin–destination trip matrix, traffic flows on each highway segment are determined, usually by assuming that user-equilibrium holds. This is achieved by solving Eq. 8.8, which requires a computer to solve the mathematical program of a network of any reasonable realistic size. An example of such a user-equilibrium computer program can be obtained from: http://swashware.com/XXE/.

Table 8.2 Example of a Peak-Hour Vehicle Origin–Destination Trips with the Five Traffic Analysis Zones (TAZs) Shown in Fig. 8.8

Origin traffic analysis zone	Destination traffic analysis zone					Total origin trips
	1	2	3	4	5	
1	–	1783	386	245	213	2627
2	2378	–	546	197	101	3222
3	4412	2232	–	745	343	7732
4	1399	1201	822	–	212	3634
5	3386	2866	1202	345	–	7799
Total destination trips	11575	8082	2956	1532	869	

8.9 THE TRADITIONAL FOUR-STEP PROCESS

The approach to travel demand and traffic forecasting presented in this chapter provides an excellent exposition of the principles underlying the problem. In practice, however, the most widely used approach to travel demand and traffic forecasting is a four-step procedure: trip generation, mode choice, destination choice, and route choice (also referred to as traffic assignment). This differs from the procedure presented in this book, which is a three-step procedure: trip generation, joint mode/destination choice, and route choice. The additional step (added by separating mode and destination choices) can make the estimation of the model less complex and was quite popular decades ago, before advances in computer estimation software became widely available. In splitting mode and destination choices, a logit formulation is still most often used for the mode choice decision. However, the destination choice is often modeled by using a gravity model [see Meyer and Miller, 2001]. The gravity model is based on the gravitational modeling principles covered in physics (the gravitational forces of planets) where the likelihood of a trip going to a destination is a function of the distance from the trip origin and some measure of attractiveness (the equivalent of mass in gravitational theory) of the destination. To implement the gravity model for trip distribution, the basic gravitational equation from Newtonian physics is appropriately modified. From physics, the basic gravity equation is

$$F_{ab} = \frac{M_a M_b}{D_{ab}^2} \tag{8.11}$$

where

F_{ab} = gravitational force between bodies a and b,
M_a = mass of body a,
M_b = mass of body b, and
D_{ab} = distance between bodies a and b.

For trip distribution, Eq. 8.11 is modified as

$$T'_{ab} = T'_a \frac{A_b f_{ab} K_{ab}}{\sum\limits_{\forall b} A_b f_{ab} K_{ab}} \tag{8.12}$$

where

T'_{ab} = total number of trips from TAZ a to TAZ b,

T'_a = total number of trips from TAZ a,

A_b = total number of trips attracted to TAZ b,

f_{ab} = distance/travel cost "friction factor," and

K_{ab} = estimated parameter to ensure results balance.

In this equation, the term for the number of trips from a TAZ (T'_a) is determined from regression techniques as described in Section 8.4, and the number of trips to a TAZ (A_b) is also determined using regression techniques. These values (trips aggregated for an entire zone) produce the origin trips or "from" trips (T''_as), which gives the last row in Table 8.2 (these from trips are sometimes referred to as trip productions), and the destination or "to" trips (A_b's) which give the last column in Table 8.2 (these to trips are sometimes referred to as trip attractions). Given these data, the intent of Eq. 8.12 is to fill in the remaining cells of Table 8.2 (given the last row and column). The other terms used to do this are the friction factor (f_{ab}), which accounts for the accessibility (cost in terms of average travel time and/or distance) between TAZs a and b, and the parameters, K_{ab}'s, which are solved iteratively to ensure that the total trips produced and attracted balance (see Table 8.2).

8.10   THE CURRENT STATE OF TRAVEL DEMAND AND TRAFFIC FORECASTING

Travel demand and traffic forecasting models, and specifically the four-step process (trip generation, trip distribution, mode choice, and route choice), have been used for more than 50 years. While they have given reasonable forecasts given the limits of the profession's understanding of travel behavior and the limits of computational tools, in the last two decades the weaknesses of this modeling approach have become obvious. Currently, the profession is shifting from the traditional four-step process to models that start at the individual traveler level and look at travel generation as an outgrowth of activity involvement (shopping, recreation, and work). Also critical in such an approach is the concept of tours, which are trips that sequentially link multiple activities (from home to exercise class, to shopping, and back home). While more complex models of traveler behavior have been appearing in the academic literature for decades, models more sophisticated than the simple four-step process have only recently begun to appear in practice. To be sure, the transition from traditional four-step models to tour-based and activity-based models presents many challenges. Included among these challenges are significant increases in required data (to be able to predict individual activity patterns) and limitations of current computer technology. However, more detailed and accurate models for travel and traffic forecasting will allow analysts to determine the impacts of many new transportation policies relating to parking controls, toll roads, congestion pricing, vehicle occupancy restrictions, reductions in energy consumption and emissions, and other emerging transportation policies.

8.11 PRACTICE PROBLEMS

PRACTICE PROBLEM 8.1

APPLICATION OF TRIP-GENERATION MODELS

Consider the Poisson trip generation model in Example 8.4. Suppose that a household has three members with an annual household income of $150,000 and lies in a neighborhood with a retail employment of 220. What is the expected number of peak-hour shopping trips, and what is the probability that the household will make more than one peak-hour shopping trip?

SOLUTION

Note: Open boxes in equations "☐" are to be completed by the reader

The Poisson trip generation model in Example 8.4 has estimated coefficients:

$BZ_i = -0.75 + 0.03$(household size)
$+ 0.004$(annual household income in thousands of dollars)
$+ 0.010$(retail employment in the household's neighborhood in hundreds)

With a household size of 5, an annual income of $150,000, and a neighborhood retail employment of 220, the expected number of peak-hour shopping trips based on this Poisson trip generation model is,

$$E[T_i] = \lambda_i = e^{BZ_i} = e^{-0.75+0.03(\boxed{})+0.004(\boxed{})+0.01(\boxed{})} = \boxed{} \text{ vehicle trips.}$$

The probability of making more than one trip is going to be equal to one minus the probability of making zero trips minus the probability of making one trip or,

$$P(T_i > 1) = 1 - P(0) - P(1).$$

Applying Eq. 8.2, the probability of making zero peak-hour shopping trips is,

$$P(0) = \frac{e^{-\boxed{}}\boxed{}^0}{\boxed{}!} = \boxed{},$$

and the probability of making one peak-hour shopping trip is,

$$P(1) = \frac{e^{-\boxed{}}\boxed{}^{\boxed{}}}{\boxed{}!} = \boxed{}.$$

So the probability of this household making more than one shopping trip during the peak hour is,

$$P(T_i > 1) = 1 - P(0) - P(1) = 1 - \boxed{} - \boxed{} = \underline{0.272}.$$

PRACTICE PROBLEM 8.2

LOGIT MODEL FOR WORK–MODE–CHOICE WITH INCREASING MODAL COSTS

Consider the conditions described in Example 8.5. If an energy crisis doubles the cost of the auto modes (drive-alone and shared-ride) and bus costs are not affected, how many workers will use each mode assuming the travel times and bus fare are unchanged?

SOLUTION

Note: Open boxes in equations "▢" are to be completed by the reader

As provided in Example 8.5, the utility functions for the three modes [automobile drive-alone (*DL*), automobile shared-ride (*SR*), and bus (*B*)] are given as, and the utility functions are estimated as (with cost is in dollars and time is in minutes),

$$U_{DL} = 2.2 - 0.2(\text{cost}_{DL}) - 0.03(\text{travel time}_{DL})$$

$$U_{SR} = 0.8 - 0.2(\text{cost}_{SR}) - 0.03(\text{travel time}_{SR})$$

$$U_{B} = -0.2(\text{cost}_{B}) - 0.01(\text{travel time}_{B}).$$

Between the residential area and an industrial complex described in Example 8.5, it is stated that 4000 workers (generating vehicle-based trips) depart for work during the peak hour. For these workers before the energy crisis, the cost of driving an automobile alone is \$6.00 with a travel time of 20 minutes, the cost of a shared ride is \$3.00 (with two travelers sharing the cost of the shared ride equally as stated in Example 8.5), and the bus fare is \$1.00 with a travel time of 25 minutes. With the energy crisis doubling the drive-alone and shared-ride costs, substitution of cost and travel time values into the utility expressions gives

$$U_{DL} = 2.2 - 0.2(\boxed{}) - 0.03(\boxed{}) = \boxed{}$$

$$U_{SR} = 0.8 - 0.2(\boxed{}) - 0.03(\boxed{}) = \boxed{}$$

$$U_{B} = -0.2(\boxed{}) - 0.01(\boxed{}) = \boxed{}$$

Substituting these values into Eq. 8.7 yields

$$P_{DL} = \frac{e^{\boxed{}}}{e^{\boxed{}} + e^{\boxed{}} + e^{\boxed{}}} = \frac{\boxed{}}{\boxed{} + \boxed{} + \boxed{}} = \frac{\boxed{}}{\boxed{}} = 0.309$$

$$P_{SR} = \frac{e^{\boxed{}}}{2.80} = \frac{\boxed{}}{2.80} = 0.253$$

$$P_{B} = \frac{e^{\boxed{}}}{2.80} = \frac{\boxed{}}{2.80} = 0.438$$

Multiplying these probabilities by 4000 (the total number of workers departing in the peak hour) gives 1235 workers driving alone (down from 2132 before the energy crisis), 1011 shared a ride (up from 956 before the energy crisis), and 1753 using a bus (up from 956 before the energy crisis). Please note that this mode shift will result in fewer vehicles being on the road, so the assumption that travel times are unchanged is not likely to be correct.

PRACTICE PROBLEM 8.3

USER EQUILIBRIUM—EFFECT OF DECREASED TRAFFIC

Two routes connect an origin and a destination, and the flow is 15,000 veh/h. Route 1 has a performance function $t_1 = 4 + 3x_1$, and route 2 has a function of $t_2 = b + 6x_2$, with the x's expressed in thousands of vehicles per hour and the t's in minutes. The user-equilibrium flow on route 1 is 9780 veh/h and it is know that both routes are used. First, determine the free-flow travel time on route 2 (the parameter b in route 2's performance function) and equilibrium travel times. Second, if population declines reduce the number of travelers at the origin, and the total origin–destination flow is reduced to 7000 veh/h, determine user-equilibrium travel times and flows.

SOLUTION

Note: Open boxes in equations "☐" are to be completed by the reader

First, using the known traffic flow (9780 veh/h) on route 1, the route-equilibrium travel time can be determined by finding the travel time on route 1 as,

$$t_1 = 4 + 3 \times \boxed{} = \boxed{}.$$

Because it is known that both routes are used and user equilibrium exists, $t_1 = t_2$. It is also know from the conservation of flow that $q = x_1 + x_2$, so,

$$x_2 = q - x_1 = \boxed{} - \boxed{} = \boxed{}.$$

With t_2 known (since $t_1 = t_2$ at user equilibrium) and x_2 known, the free-flow travel time on route 2 (the b parameter in route 2's performance function) can be determined as,

$$b = t_2 - 6x_2 = \boxed{} - 6 \times \boxed{} = \underline{2.02} \text{ min.}$$

For the second part of the problem, with the origin to destination flow reduced to 7000 veh/h, we first check that both routes are used (using performance functions),

$$t_1(7) = \boxed{} \text{ min,} \qquad t_2(0) = \boxed{} \text{ min}$$
$$t_1(0) = \boxed{} \text{ min,} \qquad t_2(7) = \boxed{} \text{ min.}$$

Because $t_1(7) > t_2(0)$ and $t_2(7) > t_1(0)$, both routes are used. Now, setting route travel times equal and substituting performance functions gives

$$4 + 3x_1 = 2.02 + 6x_2$$

From conservation of flow, $x_2 = \boxed{} - \boxed{}$, so that

$$4 + 3x_1 = 2.02 + 6(\boxed{} - \boxed{})$$

Solving gives $x_1 = \boxed{}$ and $x_2 = \boxed{} - \boxed{} = \boxed{}$, or $x_1 = \underline{4{,}447}$ veh/h and $x_2 = \underline{2{,}553}$ veh/h. And, the user-equilibrium travel times on each of the routes are computed as,

$$t_1 = 4 + 3(\boxed{}) = \underline{17.34} \text{ min}$$
$$t_2 = 2.02 + 6(\boxed{}) = \underline{17.34} \text{ min.}$$

PRACTICE
PROBLEM 8.4

THE IMPACT OF AUTONOMOUS VEHICLES
ON TOTAL TRAVEL TIME

Consider the initial conditions for the two roads in Practice Problem 8.3 (with an origin–destination traffic flow of 15,000 veh/h). Suppose all vehicles become autonomous and are centrally routed. With autonomous vehicles, the performance functions become $t_1 = 3 + 0.5x_1$ for route 1 and $t_2 = 1.6 + x_2$, for route 2 with the x's expressed in thousands of vehicles per hour and the t's in minutes. If the autonomous vehicles are assigned in a system-optimal manner to minimize vehicle-hours, how many vehicle-hours will be saved relative to the user-equilibrium solution for conventional vehicles shown in Practice Problem 8.3?

SOLUTION

Note: Open boxes in equations "**" are to be completed by the reader**

For the user-equilibrium solution in Practice Problem 8.3, using the known traffic flow (9780 veh/h) on route 1, the route-equilibrium travel time $t_1 = 33.34$ min. Since $t_1 = t_2$ at user equilibrium, the total user equilibrium is,

$$= q \times t_1 = \boxed{} \text{ veh-min or } \boxed{} \text{ veh-h.}$$

To find the new autonomous vehicle travel time with a system-optimal route distribution and revised route performance functions, the new performance functions are substituted into Eq. 8.9, giving

$$S(x) = (3 + 0.5x_1)\,x_1 + (1.6 + x_2)x_2$$

With flow conservation, $x_1 = 15 - x_2$, so that

$$S(x) = \boxed{}x_2^2 - \boxed{}x_2 + \boxed{}$$

Setting the first derivative equal to zero,

$$\frac{dS(x)}{dx_2} = \boxed{}x_2 - \boxed{} = 0$$

gives $x_2 = \boxed{}$ and $x_1 = \boxed{} - \boxed{} = \boxed{}$. The total travel times are

$$\text{route 1: } \frac{x_1 t_1}{60} = \frac{\boxed{}\left[\boxed{} + \boxed{}\left(\boxed{}\right)\right]}{60} = \boxed{} \text{ veh-h}$$

$$\text{route 2: } \frac{x_2 t_2}{60} = \frac{\boxed{}\left[\boxed{} + \boxed{}\right]}{60} = \boxed{} \text{ veh-h}$$

which gives a total autonomous-vehicle system travel time of 2291.27 veh-h or a savings of <u>6043.73 veh-h</u> (8335 − 2291.27) over the initial user-equilibrium solution shown in Practice Problem 8.3.

APPENDIX 8A   LEAST SQUARES ESTIMATION

Least squares regression is a popular method for developing mathematical relationships from empirical data. As mentioned earlier, it is a method that is well suited to the estimation of trip generation models. To illustrate the least squares approach, consider the hypothetical trip generation data presented in Table 8A.1, which could have been gathered from a typical survey of travelers.

To begin formalizing a mathematical expression, note that the objective is to predict the number of shopping trips made on a Saturday for each household, i; this number is referred to as the dependent variable (Y_i). This prediction is to be a function of the number of people in household i (z_i), which is referred to as the independent variable. A simple linear relationship between Y_i and z_i is

$$Y_i = b_0 + b_1 z_i \tag{8A.1}$$

where

$$
\begin{aligned}
Y_i &= \text{number of shopping trips made by household } i, \\
b_0, b_1 &= \text{coefficients to be determined by estimation, and} \\
Z &= \text{number of people in household } i.
\end{aligned}
$$

Ideally, one wants to determine the b's in Eq. 8A.1 that will give predictions of the number of shopping trips (Y_i's) that are as close as possible to the actual observed number of shopping trips (Y_i's, as shown in Table 8A.1). The difference or deviation between the observed and predicted number of shopping trips can be expressed mathematically as

$$\text{Deviation} = Y_i - (b_0 + b_1 z_i) \tag{8A.2}$$

Table 8A.1 Example of Shopping Trip Generation Data

Household number, i	Number of shopping trips made all day Saturday, Y_i	People in household i, z_i
1	3	4
2	1	2
3	1	3
4	5	4
5	3	2
6	2	4
7	6	8
8	4	6
9	5	6
10	2	2

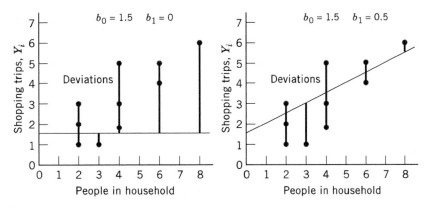

Figure 8A.1 Illustration of deviations.

Such deviations are illustrated graphically in Fig. 8A.1 for two groups of b_0 and b_1 values. In the first illustration in this figure, $b_0 = 1.5$ and $b_1 = 0$, which implies that the number of household members does not affect the number of shopping trips made. The second illustration has $b_0 = 1.5$ and $b_1 = 0.5$ and, as can readily be seen, the deviations (differences between the points representing the observed number of shopping trips and the line representing the equation $b_0 + b_1 z_i$) are reduced relative to the first illustration. These two illustrations suggest the need for some method of determining the values of b_0 and b_1 that produce the smallest possible deviations relative to observed data. Such a method can be solved by a mathematical program whose objective is to minimize the sum of the square of deviations, or

$$\min \; S\left(b_0 + b_1\right) = \sum_i \left(Y_i - b_0 - b_1 z_i\right)^2 \tag{8A.3}$$

The minimization is accomplished by setting partial derivatives equal to zero:

$$\frac{\partial S}{\partial b_0} = -2 \sum_i \left(Y_i - b_0 - b_1 z_i\right) = 0 \tag{8A.4}$$

$$\frac{\partial S}{\partial b_1} = -2 \sum_i z_i \left(Y_i - b_0 - b_1 z_i\right) = 0 \tag{8A.5}$$

Solving these equations using gives

$$\sum_i Y_i - n b_0 - b_1 \sum_i z_i = 0 \tag{8A.6}$$

$$\sum_i z_i Y_i - b_0 \sum_i X_i - b_1 \sum_i z_i^2 = 0 \tag{8A.7}$$

where

$n =$ number of households used to estimate the coefficients, and other terms are as defined previously.

Solving these equations simultaneously for b_0 and b_1 gives

$$b_1 = \frac{\sum_i (z_i - \bar{z})(Y_i - \bar{Y})}{\sum_i (z_i - \bar{z})^2} \tag{8A.8}$$

$$b_0 = \bar{Y} - b_1\bar{z} \tag{8A.9}$$

where

$\bar{Y}$ = average number of shopping trips (averaged over all households, n),

$\bar{z}$ = average household income (averaged over all households, n), and

other terms are as defined previously.

This approach to determining the values of estimable coefficients (b's) is referred to as least squares regression, and it can be shown that for the data values given in Table 8A.1, the smallest deviations between the number of predicted and actual shopping trips are given by the equation

$$Y_i = 0.33 + 0.7z_i \tag{8A.10}$$

When many coefficient values (b's) must be determined, a matrix representation of the least squares solution is appropriate:

$$B = (Z'Z)^{-1}Z'Y \tag{8A.11}$$

where

B is an $n \times 1$ vector of coefficients (with n being the number of households),

Z is an $n \times k$ matrix of variables determining Y (where k is the number of variables, such as household income, number of people in the household, etc.),

Z' is the $k \times n$ transpose matrix of Z, and

Y is an $n \times 1$ vector of the dependent variable (number of shopping trips in this example).

For additional information on least squares regression, refer to Washington et al., [2011].

APPENDIX 8B   MAXIMUM-LIKELIHOOD ESTIMATION

Maximum-likelihood estimation is used extensively in the statistical analysis of traffic data [Washington et al., 2011]. The idea underlying maximum-likelihood estimation is that different statistical distributions generate different samples, and any one sample is more likely to come from some distributions than from others.

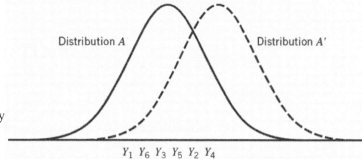

Figure 8B.1 Illustration of randomly drawn numbers and possible source distributions.

To illustrate this, suppose there is a sample of six randomly drawn numbers, $Y_1, Y_2, \ldots, Y_6$, and there are two possible distributions that could generate these numbers, as shown in Fig. 8B.1.

It is clear from Fig. 8B.1 that distribution A is much more likely to generate these six numbers than distribution A'. The objective of maximum-likelihood estimation is to estimate a coefficient vector, say B, that defines a distribution that is most likely to generate some observed data. To show how this is done, consider the Poisson regression of trip generation discussed in Section 8.4.2. The maximum-likelihood function can be written as a simple product of the probabilities of a Poisson distribution with coefficients B generating observed household trip generation. This is, for a given trip type,

$$L(B) = \prod_i P(T_i) \tag{8B.1}$$

where

B = vector of estimable coefficients,

$L(B)$ = likelihood function,

T_i = number of vehicle-based trips of a specific type (shopping, social/recreational, etc.) in some specified time period by household i, and

$P(T_i)$ = probability of household i making T trips.

Using the Poisson equation (Eq. 8.2), Eq. 8B.1 becomes

$$L(B) = \prod_i \frac{e^{-\lambda_i} \lambda_i^{T_i}}{T_i!} \tag{8B.2}$$

where

λ_i = Poisson parameter for household i, which is equal to household i's expected number of vehicle-based trips in some specified time period, $E[T_i]$, and

other terms are as defined previously.

With $\lambda_i = e^{BZ_i}$ as in Eq. 8.3,

$$L(B) = \prod_i \frac{e^{-e^{BZ_i}} \left(e^{BZ_i}\right)^{T_i}}{T_i!}$$

(8B.3)

where

  Z_i = vector of household i characteristics determining trip generation for a given trip type, and
  other terms are as defined previously.

The problem then becomes one of finding the vector B that maximizes this function (maximizes the product of probabilities as shown in Eq. 8B.1). To do this, the natural logarithm is used to transform the likelihood function into a log-likelihood function (this does not affect the maximization process). In the Poisson regression case, this log transformation of Eq. 8B.3 gives

$$LL(B) = \sum_i \left[-e^{BZ_i} + T_i BZ_i - \ln(T_i!)\right]$$

(8B.4)

where

  $LL(B)$ = log-likelihood function, and
  other terms are as defined previously.

Maximization of this expression with respect to B is undertaken by setting the first derivative to zero such that

$$\frac{\partial LL(B)}{\partial B} = \sum_i \left[-e^{BZ_i} + T_i\right] Z_i = 0$$

(8B.5)

Equation 8B.5 can be solved numerically using standard software packages [Washington et al., 2011]. Using such a software package with the data in Table 8A.1, the estimated maximum-likelihood values of B give

$$\lambda_i = e^{BZ_i} = e^{-0.206+0.2\,z_i}$$

(8B.6)

where

  z_i = people in household i (see Table 8A.1), and
  other terms are as defined previously.

Refer to Washington et al. [2011] for an extensive discussion of maximum-likelihood estimation.

NOMENCLATURE FOR CHAPTER 8

A_b	the number of trips attracted to traffic analysis zone b	T'_a	total number of trips from traffic analysis zone a
b_k	estimated coefficients	t_n	travel time on route n
B	vector of estimable coefficients	t_{fn}	free-flow travel time on route n
D_{ab}	the distance between bodies a and b	U	specifiable portion of an alternative's utility
F_{ab}	the gravitational force between bodies a and b	V	total alternative utility
f_{ab}	the distance/travel cost friction factor	w	route flow operative for x_n
K_{ab}	estimated parameter for gravity model	x_n	traffic flow on route n
$L(\cdot)$	likelihood function	x_{cn}	traffic flow capacity of route n,
$LL(\cdot)$	log-likelihood function	Y_i	dependent variable for household i
M_a	the mass of body a	z	household or alternative characteristic
P	probability of an alternative being selected	Z_i	vector of household i's characteristics
$P(T_i)$	probability of T_i trips (a nonnegative integer) being generated by household i	α	model parameter for highway performance function
q	total origin-to-destination traffic flow	β	model parameter for highway performance function
$S(\cdot)$	mathematical objective function	λ_i	Poisson parameter for household i
s	notation for the set of available alternatives	ε	unspecifiable portion of an alternative's utility (assumed to be a random variable)
T_i	number of household trips generated per unit time for household i		
T'_{ab}	total number of trips from traffic analysis zone a to traffic analysis zone b		

REFERENCES

Mannering, F., D. Garrison, and B. Sebranke. "Generation and Assessment of Incident Management Strategies, Volume IV: Seattle-Area Incident Impact Analysis: Microcomputer Traffic Simulation Results." WA-RD 204.4 and TNW90-11vol4, 1989.

Hamed, M., and F. Mannering. "Modeling Travelers' Post-Work Activity Involvement: Toward a New Methodology." *Transportation Science*, vol. 17, no. 4, 1993.

Mannering, F. "Poisson Analysis of Commuter Flexibility in Changing Route and Departure Times." *Transportation Research, Part B*, vol. 23, no. 1, 1989.

Meyer, M., and E. Miller. *Urban Transportation Planning: A Decision-Oriented Approach.* New York: McGraw-Hill, 2001.

Sheffi, Y. *Urban Transportation Networks: Equilibrium Analysis with Mathematical Programming Models.* Englewood Cliffs, NJ: Prentice-Hall, 1985.

Wardrop, F. "Some Theoretical Aspects of Road Traffic Research." *Proceedings, Institution of Civil Engineers II*, vol. 1, 1952.

Washington, S., M. Karlaftis, and F. Mannering. *Statistical and Econometric Methods for Transportation Data Analysis.* Second Edition. Boca Raton, FL: Chapman & Hall/CRC, 2011.

CHAPTER 2 PROBLEMS

SS Students solution available in interactive e-text.

Resistance, Tractive Effort, and Acceleration (Sections 2.2-2.7)

2.1 A new sports car has a drag coefficient of 0.30 and a frontal area of 21 ft^2, and is traveling at 110 mi/h. How much power is required to overcome aerodynamic drag if $\rho = 0.002378$ slugs/ft^3?

SS **2.2** For Example 2.3, how far back from the front axle would the center of gravity have to be to ensure that the maximum tractive effort developed for front- and rear-wheel–drive options is equal (assume that all other variables are unchanged)?

2.3 A vehicle manufacturer is considering an engine for a new sedan ($C_D = 0.34$, $A_f = 22$ ft^2). The car is being designed to achieve a top speed of 100 mi/h on a paved surface at sea level ($\rho = 0.002378$ slugs/ft^3). The car currently weighs 2500 lb, but the designers initially selected an underpowered engine because they did not account for aerodynamic and rolling resistances. If 2 lb of additional vehicle weight is added for each unit of horsepower needed to overcome the neglected resistance, what will be the final weight of the car if it is to achieve the 100-mi/h top speed?

2.4 A 2650-lb car is traveling at sea level at a constant speed. Its engine is running at 4500 rev/min and is producing 175 ft-lb of torque. It has a drivetrain efficiency of 90%, a drive axle slippage of 2%, 15-inch–radius wheels, and an overall gear reduction ratio of 3 to 1. If the car's frontal area is 21.5 ft^2, what is its drag coefficient?

SS **2.5** A 3000-lb car has a maximum speed (at sea level and on a level, paved surface) of 140 mi/h with 16-inch–radius wheels, a gear reduction of 3.5 to 1, and a drivetrain efficiency of 92%. It is known that at the car's top speed the engine is producing 220 ft-lb of torque. If the car's frontal area is 25 ft^2, what is its drag coefficient?

2.6 A 3200-lb car ($C_D = 0.35$, $A_f = 25$ ft^2, and $\rho = 0.002378$ slugs/ft^3) has 14-inch–radius wheels, a drivetrain efficiency of 93%, an overall gear reduction ratio of 3.2 to 1, and drive axle slippage of 3.5%. The engine develops a maximum torque of 210 ft-lb at 3600 rev/min. What is the maximum grade this vehicle could ascend, on a paved surface, while the engine is developing maximum torque?

(Assume that the available tractive effort is the engine-generated tractive effort.)

2.7 A 3400-lb car is traveling in third gear (overall gear reduction ratio of 2.5 to 1) on a level road at its top speed of 130 mi/h. The air density is 0.00206 slugs/ft^3. The car has a frontal area of 19.8 ft^2, a drag coefficient of 0.28, a wheel radius of 12.6 inches, a drive axle slippage of 3%, and a drivetrain efficiency of 88%. At this vehicle speed, what torque is the engine producing and what is the engine speed (in revolutions per minute)?

2.8 A rear-wheel–drive car weighs 2600 lb and has an 84-inch wheelbase, a center of gravity 20 inches above the roadway surface and 30 inches behind the front axle, a drivetrain efficiency of 85%, 14-inch–radius wheels, and an overall gear reduction of 7 to 1. The car's torque/engine speed curve is given by the equation $M_e = 6n_e - 0.045n_e^2$. If the car is on a paved, level roadway surface with a coefficient of adhesion of 0.75, determine its maximum acceleration from rest.

2.9 Consider the car in Problem 2.8. If it is known that the car achieves maximum speed at an overall gear reduction ratio of 2.7 to 1 with a drive axle slippage of 3.5%, how fast would the car be going if it could achieve its maximum speed when its engine is producing maximum power?

2.10 An engineer designs a rear-wheel–drive car (without an engine) that weighs 2000 lb and has a 100-inch wheelbase, drivetrain efficiency of 80%, 14-inch–radius wheels, an overall gear reduction ratio of 10 to 1, and a center of gravity (without engine) that is 22 inches above the roadway surface and 55 inches behind the front axle. An engine that weighs 3 lb for each ft-lb of developed torque is to be placed in the front portion of the car. Calculations show that for every 20 lb of engine weight added, the car's center of gravity moves 1 inch closer to the front axle (but stays at the same height above the roadway surface). If the car is starting from rest on a level paved roadway with a coefficient of adhesion of 0.8, select an engine size (weight and associated torque) that will result in the highest possible available tractive effort.

2.11 A 3000-lb car is traveling on a paved road with SS $C_D = 0.35$, $A_f = 21$ ft^2, and $\rho = 0.002378$ slugs/ft^3. Its engine is running at 3000 rev/min and is producing 250 ft-lb of torque. The car's gear reduction ratio is 4.5 to 1, drivetrain efficiency is 90%, drive axle slippage is 3.5%, and the wheel radius is 16 inches. What will the car's maximum acceleration rate be under these conditions on a level road? (Assume that the available tractive effort is the engine-generated tractive effort.)

2.12 A rear-wheel–drive car weighs 3600 lb, has 15-inch–radius wheels, a drivetrain efficiency of 95%, and an engine that develops 520 ft-lb of torque. Its wheelbase is 8.2 ft, and the center of gravity is 18 inches above the road surface and 3.3 ft behind the front axle. What is the lowest gear reduction ratio that would allow this car to achieve the highest possible acceleration from rest on good, dry pavement?

2.13 A newly designed car has a 9.0-ft wheelbase, is rear-wheel drive, and has a center of gravity 18 inches above the road and 4.3 ft behind the front axle. The car weighs 2450 lb, the mechanical efficiency of the drivetrain is 90%, and the wheel radius is 14 inches. The base engine develops 200 ft-lb of torque, and a modified version of the engine develops 240 ft-lb of torque. If the overall gear reduction ratio is 8 to 1, what is the maximum acceleration from rest for the car with the base engine and for the car with the modified engine? (It is on good, dry, and level pavement.)

2.14 A rear-wheel–drive 3000-lb drag race car has a 200-inch wheelbase and a center of gravity 20 inches above the pavement and 140 inches behind the front axle. The owners wish to achieve an initial acceleration from rest of 22 ft/s² on a level paved surface. What is the minimum coefficient of road adhesion needed to achieve this acceleration? (Assume γ_m = 1.00.)

2.15 If the race car in Problem 2.14 has a center of gravity 32 inches above the roadway and is run on a pavement with a coefficient of adhesion of 1.0, how far back from the front axle would the center of gravity have to be to develop a maximum acceleration from rest of 1.0 g (32.2 ft/s²)? (Assume γ_m = 1.00.)

SS **2.16** Consider the situation described in Example 2.5. If the vehicle is redesigned with wheels that have a 13-inch radius (assume that the mass factor is unchanged) and a center of gravity located at the same height but at the midpoint of the wheelbase, determine the acceleration for front- and rear-wheel–drive options.

Braking and Stopping Distance (Section 2.9)

2.17 If the car in Example 2.8 had C_D = 0.45 and A_f **SS** = 25 ft², what is the difference in minimum theoretical stopping distances with and without aerodynamic resistance considered (all other factors the same as in Example 2.8)?

2.18 A 3500-lb vehicle (C_D = 0.38, A_f = 26 ft², ρ = 0.002378 slugs/ft³) is driven on a surface with a coefficient of adhesion of 0.5, and the coefficient of rolling friction is approximated as 0.015 for all speeds. Assuming minimum theoretical stopping distances, if the vehicle comes to a stop 260 ft after brake application on a level surface and has a braking efficiency of 0.82, what was its initial speed (a) if aerodynamic resistance is considered and (b) if aerodynamic resistance is ignored?

2.19 A level test track has a coefficient of road adhesion of 0.80, and a car being tested has a coefficient of rolling friction that is approximated as 0.018 for all speeds. The vehicle is tested unloaded and achieves the theoretical minimum stop in 180 ft (from brake application). The initial speed was 60 mi/h. Ignoring aerodynamic resistance, what is the unloaded braking efficiency?

2.20 A driver is traveling at 90 mi/h down a 3% grade on good, wet pavement. An accident investigation team noted that braking skid marks started 410 ft before a parked car was hit at an estimated 45 mi/h. Ignoring air resistance, and using theoretical stopping distance, what was the braking efficiency of the car?

2.21 A small truck is to be driven down a 4% grade at 70 mi/h. The coefficient of road adhesion is 0.95, and it is known that the braking efficiency is 80% when the truck is empty and decreases by one percentage point for every 100 lb of cargo added. Ignoring aerodynamic resistance, if the driver wants the truck to be able to achieve a minimum theoretical stopping distance of 275 ft from the point of brake application, what is the maximum amount of cargo (in pounds) that can be carried?

2.22 Consider the conditions in Example 2.9. The **SS** car has W = 3500 lb, C_D = 0.5, A_f = 25 ft², ρ = 0.002378 slugs/ft³, and a coefficient of rolling friction approximated as 0.018 for all speed conditions. If aerodynamic resistance is considered in stopping, estimate how fast the car will be going when it strikes the object on a level and a +5% grade [all other conditions (speed, etc.) as described in Example 2.11].

2.23 A race car with a 106-inch wheelbase has its weight evenly distributed between front and rear axles. At 150 mi/h, on a race track with $\mu = 1.0$, the optimal brake force has 67.32% of the braking force on the front brakes. A new racing tire generates $\mu = 1.2$. At 150 mi/h, what percentage of the braking force should now be allocated to the front to achieve optimal braking?

2.24 A car is traveling up a 2% grade at 70 mi/h on good, wet pavement. The driver brakes to try to avoid hitting stopped traffic that is 250 ft ahead. The driver's reaction time is 0.5 s. At first, when the driver applies the brakes, a software flaw causes the antilock braking system to fail (brakes work in non-antilock mode with 80% efficiency), leaving 80 ft skid marks. After the 80 ft skid, the antilock brakes work with 100% efficiency. How fast will the driver be going when the stopped traffic is hit if the coefficient of rolling resistance is constant at 0.013? (Assume minimum theoretical stopping distance and ignore aerodynamic resistance.)

2.25 A car is traveling at 76 mi/h down a 3% grade on poor, wet pavement. The car's braking efficiency is 90%. The brakes were applied 320 ft before impacting an object. The car had an antilock braking system, but the system failed 200 ft after the brakes had been applied (wheels locked). What speed was the car traveling at just before it impacted the object? (Assume theoretical stopping distance, ignore air resistance, and let $f_{rl} = 0.015$.)

2.26 A driver traveling down a 4% grade collides with a roadside object in rainy conditions, and is issued a ticket for driving too fast for conditions. The posted speed limit is 65 mi/h. The accident investigation team determined the following: The vehicle was traveling 40 mi/h when it struck the object, braking skid marks started 205 ft before the struck object, the pavement is in good condition, and the braking efficiency of the vehicle was 93%. Using theoretical stopping distance, assuming aerodynamic resistance is negligible, and with the coefficient rolling resistance approximated as 0.015, should the driver appeal the ticket? Why or why not?

2.27 A driver is traveling 68 mi/h on a road with a -3% grade. There is a stalled car on the road 1000 ft ahead of the driver. The driver's vehicle has a braking efficiency of 90%, and it has antilock brakes. The road is in good condition and is initially dry, but it becomes wet 160 ft before the stalled car (and stays wet until the car is reached). What is the minimum distance from the stalled car at which the driver could apply the brakes and still stop before hitting it? (Assume theoretical stopping distance, ignore air resistance, and let $f_{rl} = 0.013$.)

2.28 A car is traveling at 70 mi/h on a level section of road with good, wet pavement. Its antilock braking system (ABS) only starts to work after the brakes have been locked for 100 ft. If the driver holds the brake pedal down completely, immediately locking the wheels, and keeps the pedal down during the entire process, how many feet will it take the car to stop from the point of initial brake application? (The braking efficiency is 80% with the ABS not working and 100% with the ABS working. Use theoretical stopping distance and ignore air resistance. Let $f_{rl} = 0.02$ when the brakes are locked, but compute the f_{rl} once the ABS becomes active.)

2.29 Two cars are traveling on level terrain at 60 mi/h on a road with a coefficient of adhesion of 0.8. The driver of car 1 has a 2.5-s perception/reaction time and the driver of car 2 has a 2.1-s perception/reaction time. Both cars are traveling side by side and the drivers are able to stop their respective cars in the same distance after first seeing a roadway obstacle (perception and reaction plus vehicle stopping distance). If the braking efficiency of car 2 is 0.78, determine the braking efficiency of car 1. (Assume minimum theoretical stopping distance and ignore aerodynamic resistance.)

2.30 An engineering student is driving on a level roadway and sees a construction sign 500 ft ahead in the middle of the roadway. The student strikes the sign at a speed of 25 mi/h. If the student was traveling at 55 mi/h when the sign was first spotted, what was the student's associated perception/reaction time (use practical stopping distance)?

2.31 An engineering student claims that a country road can be safely negotiated at 65 mi/h in rainy weather. Because of the winding nature of the road, one stretch of level pavement has a sight distance of only 510 ft. Assuming practical stopping distance, comment on the student's claim.

2.32 A driver is traveling at 52 mi/h on a wet road. An object is spotted on the road 415 ft ahead and the driver is able to come to a stop just before hitting the object. Assuming standard perception/reaction time and practical stopping distance, determine the grade of the road.

2.33 A test of a driver's perception/reaction time is being conducted on a special testing track with level, wet pavement and a driving speed of 50 mi/h. When the driver is sober, a stop can be made just in time to avoid hitting an object that is first visible 385 ft ahead. After a few drinks under exactly the same conditions, the driver fails to stop in time and strikes the object at a speed of 30 mi/h. Determine the driver's perception/reaction time before and after drinking. (Assume practical stopping distance.)

Acceleration and Braking (Sections 2.7 and 2.9)

2.34 On a level test track, a car with antilock brakes and 90% braking efficiency is determined to have a theoretical stopping distance (ignoring aerodynamic resistance) of 408 ft (after the brakes are applied) from 100 mi/h. The car is rear-wheel drive with a 110-inch wheelbase, weighs 3200 lb, and has a 50/50 weight distribution (front to back), a center of gravity that is 22 inches above the road surface, an engine that generates 300 ft-lb of torque, an overall gear reduction of 8.5 to 1 (in first gear), a wheel radius of 15 inches and a driveline efficiency of 95%. What is the maximum acceleration from rest of this car on this test track?

Multiple Choice Problems (Multiple Sections)

2.35 A 2500-lb vehicle has a drag coefficient of 0.35 and a frontal area of 20 ft². What is the minimum tractive effort required for this vehicle to maintain a 70 mi/h speed on a 5% upgrade through an air density of 0.002045-slugs/ft³?

   a)   242.9 lb
   b)   197.0 lb
   c)   161.9 lb
   d)   160.1 lb

2.36 A car is traveling at 20 mi/h on good, dry pavement at 5000 ft elevation. The front-wheel-drive car has a drag coefficient of 0.30, a frontal area of 20 ft² and a weight of 2500 lb. The wheelbase is 110 inches and the center of gravity is 20 inches from the ground, 50 inches behind the front axle. The engine is producing 95 ft-lb of torque and is in a gear that gives an overall gear reduction ratio of 4.5. The radius of the drive wheels is 14 inches and the mechanical efficiency of the drivetrain is 90%. What would the acceleration of the car be if the driver was accelerating quickly to avoid a collision?

   a)   3.65 ft/s²
   b)   13.26 ft/s²
   c)   15.90 ft/s²
   d)   3.48 ft/s²

2.37 A car is traveling at 60 mi/h on good, wet pavement. It has a wheelbase of 110 inches with the center of gravity 50 inches behind the front axle and at a height of 24 inches above the pavement surface. Determine the percentage of braking force that the braking system should allocate to the rear axle.

   a)   74.5%
   b)   65.4%
   c)   25.5%
   d)   34.6%

2.38 A truck traveling at 75 mi/h has a braking efficiency of 70%. The coefficient of road adhesion is 0.80. Ignoring aerodynamic resistance, determine the theoretical stopping distance on a level grade.

   a)   340.9 ft
   b)   180.6 ft
   c)   425.6 ft
   d)   338.6 ft

2.39 A child accidentally runs into the street in front of an approaching vehicle. The vehicle is traveling at 40 mi/h. Assuming the road is level, at what distance must the driver first see the child to stop just in time?

   a)   153.7 ft
   b)   300.3 ft
   c)   318.8 ft
   d)   146.7 ft

2.40 A car is traveling at sea level at 78 mi/h on a 4% upgrade before the driver sees a fallen tree in the roadway 150 ft away. The coefficient of road adhesion is 0.8. The car weighs 2700 lb, has a drag coefficient of 0.35, a frontal area of 18 ft², and a coefficient of rolling friction approximated as 0.017 for all speed conditions. The car has an antilock braking system that gives it a braking efficiency of 100%. If the driver first applies the brakes 150 ft from the tree, how fast will the car be traveling when it reaches the tree? Include the effect of aerodynamic resistance.

   a)   49.5 mi/h
   b)   48.8 mi/h
   c)   50.5 mi/h
   d)   47.7 mi/h

CHAPTER 3 PROBLEMS

Crest Vertical Curves (Section 3.3)

3.1 A 520-ft long equal-tangent crest vertical curve connects tangents that intersect at station 340 + 00 and elevation 1325 ft. The initial grade is +4.0% and the final grade is −2.5%. Determine the elevation and stationing of the high point, PVC, and PVT.

3.2 Consider Example 3.4. Solve this problem with the parabolic equation (Eq. 3.1) rather than by using offsets.

3.3 Again consider Example 3.4. Does this curve provide sufficient stopping sight distance for a speed of 60 mi/h?

3.4 An equal-tangent crest vertical curve is designed for 70 mi/h. The high point is at elevation 1011.4 ft. The initial grade is +2% and the final grade is −1%. What is the elevation of the PVT?

3.5 An equal-tangent crest curve has been designed for 70 mi/h to connect a +2% initial grade and a −1% final grade for a new vehicle that has a 3 ft driver's eye height; the curve was designed to avoid an object that is 1 ft high. Standard practical stopping distance design was used but, unlike current design standards, the vehicle was assumed to make a 0.5g stop, although driver reactions are assumed to be the same as in current highway design standards. If the PVC of the curve is at elevation 848 ft and station 43 + 48, what is the station and elevation of the high point of the curve?

3.6 A vertical curve is designed for 55 mi/h and has an initial grade of +2.5% and a final grade of −1.0%. The PVT is at station 114 + 50. It is known that a point on the curve at station 112 + 35 is at elevation 245 ft. What is the stationing and elevation of the PVC? What is the stationing and elevation of the high point on the curve?

3.7 An equal-tangent crest vertical curve is designed for 65 mi/h. The initial grade is +3.4% and the final grade is negative. What is the elevation difference between the PVC and the high point of the curve?

3.8 An equal-tangent crest vertical curve has a 50-mi/h design speed. The initial grade is +3%. The high point is at station 33 + 40.76 and the PVT is at station 37 + 24.66. What is the elevation difference between the high point and the PVT?

3.9 An equal-tangent crest curve connects a +2% initial grade with a −1% final grade and is designed for 55 mi/h. The station of the PVI is 233 + 40 with an elevation of 1203 ft. What is the elevation of the curve at station 234 + 00?

3.10 An equal-tangent vertical curve was designed in 2012 (to 2018 AASHTO guidelines) for a design speed of 70 mi/h to connect grades G_1 = +1.2% and G_2 = −2.1%. The curve is to be redesigned for a 70-mi/h design speed in the year 2025. Vehicle braking technology has advanced so that the recommended design deceleration rate is 25% greater than the 2011 value used to develop Table 3.1, but due to the higher percentage of older persons in the driving population, design reaction times have increased by 20%. Also, vehicles have become smaller so that the driver's eye height is assumed to be 3.0 ft above the pavement and roadway objects are assumed to be 1.0 ft above the pavement surface. Compute the difference in design curve lengths for the 2012 and 2025 designs.

3.11 An equal-tangent crest vertical curve is designed with a PVI at station 110 + 00 (elevation 927.2 ft) and a PVC at station 107 + 43.3 (elevation 921.55 ft). If the high point is at station 110 + 75.5, what is the design speed of the curve?

3.12 An equal-tangent crest vertical curve connects a +3.2% and a −1.1% grade. The PVI is at station 98 + 20. Due to drainage considerations, the highest point of the curve is at station 100 + 79.35. Determine the station of the PVC and PVT and the design speed of the curve.

3.13 A 1200-ft equal-tangent crest vertical curve is currently designed for 50 mi/h. A civil engineering student contends that 60 mi/h is safe in a van because of the higher driver's eye height. If all other design inputs are standard, what must the driver's eye height (in the van) be for the student's claim to be valid?

3.14 A highway reconstruction project is being undertaken to reduce crash rates. The reconstruction involves a major realignment of the highway such that a 60-mi/h design speed is attained. At one point on the highway, a 720-ft equal-tangent crest vertical curve exists. Measurements show that at 3 + 40 stations from the PVC, the vertical curve offset is 3.5 ft. Assess the adequacy of this existing curve in light of the reconstruction design speed of 60 mi/h and, if the existing curve is inadequate, compute a satisfactory curve length.

3.15 An equal-tangent crest curve connects a +1.0% and a −0.5% grade. The PVC is at station 54 + 24 and the PVI is at station 56 + 92. Is this curve long enough to provide passing sight distance for a 60-mi/h design speed?

3.16 An equal-tangent crest vertical curve connects an initial grade of +2.5% and a final grade of –0.5%. The curve is designed for 70 mi/h and the station of the *PVT* is 132+62 and is at elevation 833 ft. What is the station and elevation of the curve's high point?

3.17 An equal-tangent vertical curve connects a +2.25% initial grade and a –2.75% final grade. It is known that the highpoint of the curve is at elevation 1497.4 ft. If the *PVI* is at elevation 1500 ft, what is the design speed of the curve?

3.18 An equal-tangent vertical curve connects a +2% initial grade and a –1% final grade. It is known that the offset at 400 ft from the *PVC* is 5.298 ft and that the elevation on the curve at this point is 562 ft. What is the design speed of the curve, the elevation of the *PVT*, and the elevation of the *PVI*?

3.19 An equal-tangent vertical curve connects a +1.5% initial grade and a –3% final grade. The *PVT* is at station 120+52 and elevation 197.665 ft. If the design speed is 50 mi/h, what is the elevation of the curve at station 118+00?

Sag Vertical Curves (Section 3.3)

3.20 A 1400-ft–long sag vertical curve (equal tangent) has a *PVC* at station 115 + 00 and elevation 750 ft. The initial grade is –3.5% and the final grade is +6.5%. Determine the elevation and stationing of the low point, *PVI*, and *PVT*.

SS **3.21** An equal-tangent sag vertical curve is designed with the *PVC* at station 109 + 00 and elevation 950 ft, the *PVI* at station 110 + 77 and elevation 947.34 ft, and the low point at station 110 + 50. Determine the design speed of the curve.

3.22 An equal-tangent vertical curve connects a –2% and a +3% grade. The low point of the curve is at elevation 297.88 ft. If the *PVI* is at elevation 295 ft, what is the design speed of the curve?

3.23 An equal-tangent sag vertical curve is designed for 45 mi/h. The low point is 237 ft from the *PVC* at station 112 + 37 and the final offset at the *PVT* is 19.355 ft. If the *PVC* is at station 110 + 00, what is the elevation difference between the *PVT* and a point on the curve at station 111 + 00?

3.24 An equal-tangent vertical curve connects an initial grade of –3% and a final grade of +1% and is designed for 60 mi/h. The *PVI* is at station 250+50 and elevation 732 ft. What is the station and elevation of the lowest point on the curve?

3.25 An overpass is being built over the *PVI* of an existing equal-tangent sag curve. The sag curve has a 70-mi/h design speed and $G_1 = -5\%$, $G_2 = +3\%$. Determine the minimum necessary clearance height of the overpass and the resultant elevation of the bottom of the overpass over the *PVI*. (Ignore the cross-sectional width of the overpass.)

3.26 An existing highway-railway at-grade crossing **SS** is being redesigned as grade separated to improve traffic operations. The railway must remain at the same elevation. The highway is being reconstructed to travel under the railway. The underpass will be a sag curve that connects to 2.25% tangent sections on both ends, and the *PVI* will be centered under the railway (a symmetrical alignment). The sag curve design speed is 45 mi/h. How many feet below the railway should the curve *PVI* be located?

3.27 An existing equal-tangent sag vertical curve is designed for 60 mi/h. The initial grade is –3% and the elevation of the *PVT* is 754 ft. The *PVC* of the curve is at station 134 + 16 and the *PVI* is at 137 + 32. An overpass is being constructed directly above the *PVI*. The highway is for cars only (AASHTO minimum and recommended structure clearances do not apply) and the overpass design assumes the driver's eye height is set conservatively to 5 ft. What is the lowest possible elevation of the bottom of the overpass structure to ensure sufficient stopping sight distance at 60 mi/h?

3.28 An equal-tangent sag curve has its *PVI* at station 10 + 00 and elevation at 138 ft. Directly above the *PVI*, the bottom of an overpass structure is at elevation 162 ft. The *PVC* is at station 4 + 00. If the initial grade is –4%, what is the highest possible value of the final grade given that a 70-mi/h design speed is to be provided in daytime conditions? What is the highest possible final grade in nighttime conditions? (Note: Be careful of units of *A* and ignore the cross-sectional width of the overpass.)

3.29 An equal-tangent sag vertical curve connects a +1% and +3% initial and final grades, respectively, and is designed for 70 mi/h. The high point on the curve is at elevation 822 ft. If the *PVC* is at station 110+00, what is the elevation of the curve at station 112+12?

3.30 An equal-tangent sag vertical curve connects a –1.5% initial grade and a +1.25% final grade. The design speed is 70 mi/h and the station and elevation of the *PVT* are 138+70 and 851.5 ft, respectively. A storm drain is to be installed at the low point of the curve. What is the station and elevation of the low point?

Combined Crest and Sag Vertical Curves (Section 3.3)

SS **3.31** Consider the bridge-tunnel problem in Example 3.9. Suppose that a 70 mi/h interstate design speed is needed. If so, what would be the minimum bridge-tunnel separation distance (something higher than the current 1200 ft separation) needed to connect the elevations of the bridge and tunnel with 70 mi/h design-speed curves?

3.32 Two level sections of an east-west highway ($G = 0$) are to be connected. Currently, the two sections of highway are separated by a 4000-ft (horizontal distance), 2% grade. The westernmost section of highway is the higher of the two and is at elevation 100 ft. If the highway has a 60-mi/h design speed, determine, for the crest and sag vertical curves required, the stationing and elevation of the *PVC*s and *PVT*s given that the *PVC* of the crest curve (on the westernmost level highway section) is at station 0 + 00 and elevation 100 ft. In solving this problem, assume that the curve *PVI*s are at the intersection of $G = 0$ and the 2% grade, that is, $A = 2$.

3.33 Consider Problem 3.32. Suppose it is necessary to keep the entire alignment within the 4000 ft that currently separate the two level sections. It is determined that the crest and sag curves should be connected (the *PVT* of the crest and *PVC* of the sag) with a constant-grade section that has the lowest grade possible. Again using a 60-mi/h design speed, determine, for the crest and sag vertical curves, the stationing and elevation of the *PVC*s and *PVT*s given that the westernmost level section ends at station 0 + 00 and elevation 100 ft. (Note that *A* must now be determined and will not be equal to 2.)

3.34 Due to crashes at a railroad crossing, an overpass (with a roadway surface 26 ft above the existing road) is to be constructed on an existing level highway. The existing highway has a design speed of 50 mi/h. The overpass structure is to be level, centered above the railroad, and 180 ft long. What length of the existing level highway must be reconstructed to provide an appropriate vertical alignment?

3.35 A section of a freeway ramp has a +4.0% grade and ends at station 127 + 00 and elevation 138 ft. It must be connected to another section of the ramp (which has a 0.0% grade) that is at station 162 + 00 and elevation 97 ft. It is determined that the crest and sag curves required to connect the ramp should be connected (the *PVT* of the crest and *PVC* of the sag)

with a constant-grade section that has the lowest grade possible. Design a vertical alignment to connect between these two stations using a 50-mi/h design speed. Provide the lengths of the curves and constant-grade section.

3.36 A tangent section of highway has a −1.0% grade and ends at station 4 + 75 and elevation 82 ft. It must be connected to another section of highway that has a −1.0% grade and that begins at station 44 + 12 and elevation 131.2 ft. The connecting alignment should consist of a sag curve, constant-grade section, and crest curve, and be designed for a speed of 50 mi/h. What is the lowest grade possible for the constant-grade section that will complete this alignment?

3.37 A roadway has a design speed of 50 mi/h, and at station 105 + 00 a +3.0% grade roadway section ends and at station 125 + 00 a +2.0% grade roadway section begins. The +3.0% grade section of highway (at station 105 + 00) is at a higher elevation than the +2.0% grade section of highway (at station 125 + 00). If a −4% constant-grade section is used to connect the crest and sag vertical curves that are needed to link the +3.0 and +2.0% grade sections, what is the elevation difference between stations 105 + 00 and 125 + 00? (The entire alignment, crest and sag curves, and constant-grade section must fit between stations 105 + 00 and 125 + 00.)

3.38 A sag curve and crest curve connect a −3.5% tangent section of highway (to the west) with a +2.5% tangent section of highway (to the east). The +2.5% tangent section is at a higher elevation than the −3.5% tangent section. The two tangent sections are separated by 1150 ft of horizontal distance. If the design speed of the curves is 50 mi/h, what is the common grade between the sag and crest curves (G_2 of sag and G_1 of crest, from west to east), and what is the elevation difference between the PVC_s and PVT_c?

3.39 A level section of highway is to be connected to a section of highway with a −5% grade. The level highway section ends at station 108 + 40 (elevation 865 ft) and is to connect with the −5% section of highway at station 139 + 20 (elevation 758 ft). Using a design speed of 50 mi/h, determine the stations and elevations of the *PVC*s, *PVI*s, and *PVT*s of the two vertical curves required to connect the highway segments, as well as the length of the constant-grade section (connecting grade is to be as small as possible).

3.40 A crest and sag curve connect a 0% west highway segment (left) with a +2% east highway segment (right). The 0% west highway segment is at a higher elevation than the start of the +2% east highway segment. The two vertical curves connect with each other ($PVT_c = PVC_s$) and share a 3% common grade. If the design speed of the curves is 45 mi/h, what is the elevation difference between the two road segments?

Horizontal Curves (Section 3.4)

3.41 You are asked to design a horizontal curve for a two-lane road. The road has 12-ft lanes. Due to expensive excavation, it is determined that a maximum of 34 ft can be cleared from the road's centerline toward the inside lane to provide for stopping sight distance. Also, local guidelines dictate a maximum superelevation of 0.08 ft/ft. What is the highest possible design speed for this curve?

3.42 A horizontal curve on a two-lane highway (10-ft lanes) is designed for 50 mi/h with a 6% superelevation. The central angle of the curve is 35 degrees and the PI is at station 482 + 72. What is the station of the PT and how many feet have to be cleared from the lane's shoulder edge to provide adequate stopping sight distance?

3.43 A horizontal curve on a single-lane highway has its PC at station 123 + 70 and its PI at station 130 + 90. The curve has a superelevation of 0.06 ft/ft and is designed for 70 mi/h. What is the station of the PT?

3.44 A horizontal curve is being designed through mountainous terrain for a four-lane road with lanes that are 10 ft wide. The central angle (Δ) is known to be 40 degrees, the tangent distance is 520 ft, and the stationing of the tangent intersection (PI) is 2600 + 00. Under specified conditions and vehicle speed, the roadway surface is determined to have a coefficient of side friction of 0.08, and the curve's superelevation is 0.09 ft/ft. What is the stationing of the PC and PT and what is the safe vehicle speed?

SS **3.45** A new interstate highway is being built with a design speed of 70 mi/h. For one of the horizontal curves, the radius (measured to the innermost vehicle path) is tentatively planned as 2500 ft. What rate of superelevation is required for this curve?

3.46 On a roadway with two 12-ft lanes, a horizontal curve is designed for 35 mi/h with a 4% superelevation. It is known that $\Delta = 2\Delta_s$. The PI of the curve is at station 30 + 00. What is the station of the PT of the curve?

3.47 A developer is having a single-lane raceway constructed with a 200-mi/h design speed. A curve on the raceway has a radius of 4500 ft, a central angle of 30 degrees, and PI stationing at 1125 + 10. If the design coefficient of side friction is 0.20, determine the superelevation required at the design speed (do not ignore the normal component of the centripetal force). Also, compute the degree of curve, length of curve, and stationing of the PC and PT.

3.48 A horizontal curve is being designed for a new two-lane highway (12-ft lanes). The PI is at station 250 + 50, the design speed is 65 mi/h, and a maximum superelevation of 0.07 ft/ft is to be used. If the central angle of the curve is 38 degrees, design a curve for the highway by computing the radius and stationing of the PC and PT.

3.49 You are asked to design a horizontal curve with a 40-degree central angle ($\Delta = 40$) for a two-lane road with 11-ft lanes. The design speed is 70 mi/h and superelevation is limited to 0.06 ft/ft. Give the radius, degree of curvature, and length of curve that you would recommend.

3.50 For the horizontal curve in Problem 3.49, what **SS** distance must be cleared from the inside edge of the inside lane to provide adequate stopping sight distance?

3.51 A horizontal curve on a single-lane freeway ramp is 400 ft long, and the design speed of the ramp is 45 mi/h. If the superelevation is 10% and the station of the PC is 18 + 25, what is the station of the PI and how much distance must be cleared from the center of the lane to provide adequate stopping sight distance?

3.52 A freeway exit ramp has a single lane and consists entirely of a horizontal curve with a central angle of 90 degrees and a length of 628 ft. If the distance cleared from the centerline for sight distance is 19.4 ft, what design speed was used?

3.53 A horizontal curve on a two-lane highway (12-ft lanes) has PC at station 123 + 80 and PT at station 129 + 60. The central angle is 35 degrees, the superelevation is 0.08, and 20.6 ft is cleared (for sight distance) from the inside edge of the innermost lane. Determine a maximum safe speed (assuming current design standards) to the nearest 5 mi/h.

3.54 A horizontal curve was designed for a four-lane highway for adequate SSD. Lane widths are 12 ft, and the superelevation is 0.06 and was set assuming maximum f_s. If the necessary sight distance required 52 ft of lateral clearance from the roadway centerline, what design speed was used for the curve?

3.55 On a two-lane road with 12-ft lanes, a horizontal curve is designed for 50 mi/h with a superelevation of

10%. The *PI* of the curve is at 220+48 and the *PC* is at station 216+74. Determine the station of the *PT* and the middle ordinate for stopping sight distance.

3.56 A horizontal curve is designed for 35 mi/h with a central angle of 85 degrees. The curve has four 11-ft lanes (two in each direction) and a superelevation of 8%. If the *PI* is at station 148+40 what is the station of the *PT*?

Combined Vertical and Horizontal Curves (Section 3.5)

3.57 A section of highway has vertical and horizontal curves with the same design speed. A vertical curve on this highway connects a +1% and a +3% grade and is 420 ft long. If a horizontal curve on this highway is on a two-lane section with 12-ft lanes and has a central angle of 37 degrees and a superelevation of 6%, what is the length of the horizontal curve?

3.58 A section of a two-lane highway (12-ft lanes) is designed for 75 mi/h. At one point a vertical curve connects a −2.5% and +1.5% grade. The *PVT* of this curve is at station 36 + 50. It is known that a horizontal curve starts (has *PC*) 294 ft before the vertical curve's *PVC*. If the superelevation of the horizontal curve is 0.08 and the central angle is 38 degrees, what is the station of the *PT*?

3.59 Two straight sections of freeway cross at a right angle. At the point of crossing, the east-west highway is at elevation 150 ft and has a constant +5.0% grade (upgrade in the east direction), and the north-south highway is at elevation 125 ft and has a constant −3.0% grade (downgrade in the north direction). Design a 90-degree ramp that connects the northbound direction of travel to the eastbound direction of travel. Design the ramp for the highest design speed (to nearest 5 mi/h) with the constraint that the minimum allowable value of *D* is 8.0. (Assume that the *PC* of the horizontal curve is at station 15 + 00, and the vertical curve *PVI*s are at the *PC* and *PT*.) Give the stationing and elevations of the *PC*, *PT*, *PVC*s, and *PVT*s.

3.60 A crest vertical curve and a horizontal curve on the same highway have the same design speed. The equal-tangent vertical curve connects a +3% initial grade with a +1% final grade and has a *PVC* at 101 + 78 and a *PVT* at 106 + 72, The horizontal curve has a *PI* at 150 + 10 and a central angle of 75 degrees. If the superelevation of the horizontal curve is 8% and the road has two 12-ft lanes, what is the stationing of the *PT*?

3.61 West and east highway segments are separated by 600 ft horizontally. The west segment has a +2% grade and the east segment has a +1% grade. The west-grade segment ends at a higher elevation than the east segment and the two segments are connected by a joining sag and crest curve combination (so $PVT_s = PVC_c$). If the road is designed for 45 mi/h, what is the elevation difference between the west and east highway segments?

Multiple Choice Problems (Multiple Sections)

3.62 A 400-ft equal-tangent sag vertical curve has its *PVC* at station 100 + 00 and elevation 500 ft. The initial grade is −4.0% and the final grade is +2.5%. Determine the elevation of the lowest point of the curve.

   a)   495.077 ft
   b)   495.250 ft
   c)   485.231 ft
   d)   492.043 ft

3.63 A horizontal curve is being designed around a pond with a tangent length of 1200 ft and central angle of 0.5211 radians. If the *PI* is at station 145 + 00, determine the station of *PT*.

   a)   168 + 45.44
   b)   156 + 45.44
   c)   173 + 94.00
   d)   156 + 72.72

3.64 A car is traveling over a 1400-ft vertical curve. One of the passengers decides to calculate the current offset from the PVC. By looking at the onboard navigation device, the passenger knows that the car is 750 feet from the PVC. The initial grade is +5.5% while the final roadway grade is +3.0%. What is the current offset?

   a)   4.38 ft
   b)   17.50 ft
   c)   17.08 ft
   d)   5.02 ft

3.65 You are designing a highway to AASHTO guidelines on rolling terrain where the design speed will be 65 mi/h. At one section, a +1.25% grade and a −2.25% grade must be connected with an equal-tangent vertical curve. Determine the minimum length of curve that can be designed while meeting SSD requirements.

   a)   864.30 ft
   b)   645.00 ft
   c)   674.74 ft
   d)   673.43 ft

3.66 A car is traveling downhill on a suburban road with a grade of 4% at a speed of 35 mi/h. Determine the required stopping sight distance.

   a)   149.29 ft
   b)   245.97 ft
   c)   233.84 ft
   d)   261.26 ft

3.67 A tow truck is searching a city street at 40 mi/h for illegally parked vehicles. It travels over an equal-tangent vertical curve with an initial grade of +4.0% and final grade of −2.0%. If the height of the driver's eye is 6.0 ft and the driver spots a car 450 ft away with a height of 4.0 ft, what is the minimum length of the vertical curve for this situation?

   a)   562.94 ft
   b)   1304.15 ft
   c)   240.07 ft
   d)   306.85 ft

CHAPTER 4 PROBLEMS

Flexible-Pavement Design (Sections 4.3–4.4)

4.1 Truck A has two single axles. One axle weighs 12,000 lb and the other weighs 23,000 lb. Truck B has an 8000-lb single axle and a 43,000-lb tandem axle. On a flexible pavement with a 3-inch HMA wearing surface, a 6-inch soil-cement base, and an 8-inch crushed stone subbase, which truck will cause more pavement damage? (Assume that drainage coefficients are 1.0.)

4.2 A flexible pavement has a 4-inch HMA wearing surface, a 7-inch dense-graded crushed stone base, and a 10-inch crushed stone subbase. The pavement is on a soil with a resilient modulus of 5000 lb/in². The pavement was designed with 90% reliability, an overall standard deviation of 0.4, and a ΔPSI of 2.0 (a TSI of 2.5). The drainage coefficients are 0.9 and 0.8 for the base and subbase, respectively. How many 25-kip single-axle loads can be carried before the pavement reaches its TSI (with given reliability)?

4.3 A highway has the following pavement design daily traffic: 300 single axles at 10,000 lb each, 120 single axles at 18,000 lb each, 100 single axles at 23,000 lb each, 100 tandem axles at 32,000 lb each, 30 single axles at 32,000 lb each, and 100 triple axles at 40,000 lb each. A flexible pavement is designed to have 4 inches of sand-mix asphalt wearing surface, 6 inches of soil-cement base, and 7 inches of crushed stone subbase. The pavement has a 10-year design life, a reliability of 85%, an overall standard deviation of 0.30, drainage coefficients of 1.0, an initial PSI of 4.7, and a TSI of 2.5. What is the minimum acceptable soil resilient modulus?

4.4 Consider the conditions in Problem 4.3. Suppose the state has relaxed its truck weight limits and the impact has been to reduce the number of 18,000-lb single-axle loads from 120 to 20 and increase the number of 32,000-lb single-axle loads from 30 to 90 (all other traffic is unaffected). Under these revised daily counts, what is the minimum acceptable soil resilient modulus?

4.5 A flexible pavement was designed for the following daily traffic with a 12-year design life: 1300 single axles at 8000 lb each, 900 tandem axles at 15,000 lb each, 20 single axles at 40,000 lb each, and 200 tandem axles at 40,000 lb each. The highway was designed with 4 inches of HMA wearing surface, 4 inches of hot-mix asphaltic base, and 8 inches of crushed stone subbase. The reliability was 70%, overall standard deviation was 0.5, ΔPSI was 2.0 (with a TSI of 2.5), and all drainage coefficients were 1.0. What was the soil resilient modulus of the subgrade used in design?

4.6 A flexible pavement has a SN of 3.8 (all drainage coefficients are equal to 1.0). The initial PSI is 4.7 and the terminal serviceability is 2.5. The soil has a CBR of 9. The overall standard deviation is 0.40 and the reliability is 95%. The pavement is currently designed for 1800 equivalent 18-kip single-axle loads per day. If the number of 18-kip single-axle loads were to increase by 30%, by how many years would the pavement's design life be reduced?

4.7 An engineer plans to replace the rigid pavement in Example 4.3 with a flexible pavement. The chosen design has 6 inches of sand-mix asphalt wearing surface, 9 inches of soil-cement base, and 10 inches of crushed stone subbase. All drainage coefficients are 1.0 and the soil resilient modulus is 5000 lb/in². If the highway's traffic is the same (same axle loadings per vehicle as in Example 4.3), for how many years could you be 95% sure that this pavement will last? (Assume that any parameters not given in this problem are the same as those given in Example 4.3.)

4.8 A flexible pavement is designed with 5 inches of HMA wearing surface, 6 inches of hot-mix asphaltic base, and 10 inches of crushed stone subbase. All drainage coefficients are 1.0. Daily traffic is 200 passes of a 20-kip single axle, 200 passes of a 40-kip tandem axle, and 80 passes of a 22-kip single axle. If the initial minus the terminal PSI is 2.0 (the TSI is 2.5), the soil resilient modulus is 3000 lb/in², and the overall standard deviation is 0.6, what is the probability (reliability) that this pavement will last 20 years before reaching its terminal serviceability?

4.9 A flexible pavement is designed with 4 inches of sand-mix asphalt wearing surface, 6 inches of dense-graded crushed stone base, and 8 inches of crushed stone subbase. All drainage coefficients are 1.0. The pavement is designed for 18-kip single-axle loads (1290 per day). The initial PSI is 4.5 and the TSI is 2.5. The soil has a resilient modulus of 12,000 lb/in². If the overall standard deviation is 0.40, what is the probability that this pavement will have a PSI greater than 2.5 after 20 years?

4.10 A flexible pavement has a 4-inch sand-mix asphalt wearing surface, 10-inch soil-cement base, and a 10-inch crushed stone subbase. It is designed to withstand 400 20-kip single-axle loads and 900 35-kip tandem-axle loads per day. The subgrade CBR is 8, the overall standard deviation is 0.45, the initial PSI is 4.2, and the final PSI is 2.5. What is the probability that this pavement will have a PSI above 2.5 after 25 years? (Drainage coefficients are 1.0.)

Rigid-Pavement Design (Sections 4.5–4.6)

4.11 Consider the two trucks in Problem 4.1. Which truck will cause more pavement damage on a rigid pavement with a 10-inch slab?

4.12 You have been asked to design the pavement for an access highway to a major truck terminal. The design daily truck traffic consists of the following: 80 single axles at 22,500 lb each, 570 tandem axles at 25,000 lb each, 50 tandem axles at 39,000 lb each, and 80 triple axles at 48,000 lb each. The highway is to be designed with rigid pavement having a modulus of rupture of 600 lb/in^2 and a modulus of elasticity of 5 million lb/in^2. The reliability is to be 95%, the overall standard deviation is 0.4, the drainage coefficient is 0.9, ΔPSI is 1.7 (with a TSI of 2.5), and the load transfer coefficient is 3.2. The modulus of subgrade reaction is 200 lb/in^3. If a 20-year design life is to be used, determine the required slab thickness.

4.13 A rigid pavement is being designed with the same parameters as used in Problem 4.5. The modulus of subgrade reaction is 300 lb/in^3 and the slab thickness is determined to be 8.5 inches. The load transfer coefficient is 3.0, the drainage coefficient is 1.0, and the modulus of elasticity is 4 million lb/in^2. What is the design modulus of rupture? (Assume that any parameters not given in this problem are the same as those given in Problem 4.5.)

4.14 A rigid pavement is designed with a 10-inch slab, an E_c of 6 million lb/in^2, a concrete modulus of rupture of 432 lb/in^2, a load transfer coefficient of 3.0, an initial PSI of 4.7, and a TSI of 2.5. The overall standard deviation is 0.35, the modulus of subgrade reaction is 190 lb/in^3, and a reliability of 90% is used along with a drainage coefficient of 0.8. The pavement is designed assuming that traffic is composed entirely of trucks (100 per day). Each truck has one 20-kip single axle and one 42-kip tandem axle (the effect of all other vehicles is ignored). A section of this road is to be replaced (due to different subgrade characteristics) with a flexible pavement having a SN of 4 and is expected to last the

same number of years as the rigid pavement. What is the assumed soil resilient modulus? (Assume that all other factors are the same as for the rigid pavement.)

4.15 Consider the loading conditions in Problem 4.3. A rigid pavement is used with a modulus of subgrade reaction of 200 lb/in^3, a slab thickness of 8 inches, a load transfer coefficient of 3.2, a modulus of elasticity of 5 million lb/in^2, a modulus of rupture of 600 lb/in^2, and a drainage coefficient of 1.0. How many years is the pavement expected to last using the same reliability as in Problem 4.3? (Assume that all other factors are as in Problem 4.3.)

4.16 Consider Problem 4.15. How long would the rigid pavement be expected to last if you wanted to be 95% sure that the pavement would stay above the 2.5 TSI?

4.17 Consider the traffic conditions in Example 4.3. **SS** Suppose a 10-inch slab was used and all other parameters are as described in Example 4.3. What would the design life be if the drainage coefficient was 0.8, and what would it be if it was 0.6?

4.18 Consider the conditions in Example 4.4. Suppose all of the parameters are the same, but further soil tests found that the modulus of subgrade reaction was only 150 lb/in^3. In light of this new soil finding, how would the design life of the pavement change?

4.19 Consider the conditions in Example 4.4. Suppose all of the parameters are the same, but a quality control problem resulted in a modulus of rupture of 600 lb/in^2 instead of 800 lb/in^2. How would the design life of the pavement change?

Pavement Design with Design-Lane Traffic (Sections 4.3–4.7)

4.20 You have been asked to design a flexible pavement, and the following daily traffic is expected for design: 5000 single axles at 10,000 lb each, 400 single axles at 24,000 lb each, 1000 tandem axles at 30,000 lb each, and 100 tandem axles at 50,000 lb each. There are three lanes in the design direction (conservative design is to be used). Reliability is 90%, overall standard deviation is 0.40, ΔPSI is 1.8, and the design life is 15 years. The soil has a resilient modulus of 13,750 lb/in^2. If the TSI is 2.5, what is the required SN?

4.21 A three-lane northbound section of interstate (with the design lane conservatively designed) has rigid pavement (PCC) and was designed with a 10-inch slab, 90% reliability, 700 lb/in^2 concrete modulus of rupture, 4.5 million lb/in² modulus of elasticity, 3.0 load transfer coefficient, and an overall standard deviation of 0.35. The initial PSI is 4.6 and the TSI is 2.5. The CBR is 2 with a drainage coefficient of 1.0. The road was designed exclusively for trucks that have one 24-kip tandem axle and one 12-kip single axle. It is known from weigh-in-motion scales that there have been 13 million 18-kip–equivalent single-axle loads in the entire northbound direction of this freeway so far. If a section of flexible pavement is used to replace a section of the PCC that was removed for utility work, what SN should be used so that the PCC and flexible pavements have the same life expectancy (the new life of the flexible pavement and the remaining life of the PCC)?

4.22 A rigid pavement is designed with an 11-inch slab thickness, 90% reliability, E_c = 4 million lb/in², modulus of rupture of 600 lb/in², modulus of subgrade reaction of 150 lb/in³, a 2.8 load transfer coefficient, initial PSI of 4.8, final PSI of 2.5, overall standard deviation of 0.35, and a drainage coefficient of 0.8. The pavement has a 20-year design life. The pavement has three lanes and is conservatively designed for trucks that have one 20,000-lb single axle, one 26,000-lb tandem axle, and one 34,000-lb triple axle. What is the daily estimated truck traffic on the three lanes?

4.23 A rigid pavement is on a highway with two lanes in one direction, and the pavement is conservatively designed. The pavement has an 11-inch slab with a modulus of elasticity of 5,000,000 lb/in² and a concrete modulus of rupture of 700 lb/in², and it is on a soil with a CBR of 25. The design drainage coefficient is 1.0, the overall standard deviation is 0.3, and the load transfer coefficient is 3.0. The pavement was designed to last 20 years (initial PSI of 4.7 and a final PSI of 2.5) with 95% reliability carrying trucks with one 18-kip single axle and one 28-kip tandem axle. However, after the pavement was designed, one more lane was added in the design direction (conservative design still used), and the weight limits on the trucks were increased to a 20-kip single and a 34-kip tandem axle (the slab thickness was unchanged from the original two-lane design with lighter trucks). If climate change has caused the drainage coefficient to drop to 0.8, how long will the pavement last with the new loading and the additional lane (same volume of truck traffic)?

4.24 A four-lane northbound section of interstate has rigid pavement and was designed with an 8-inch slab, 90% reliability, a 700 lb/in² concrete modulus of rupture, a 5 million lb/in² modulus of elasticity, a 3.0 load transfer coefficient, and an overall standard deviation of 0.3. The initial PSI is 4.6 and the TSI is 2.5. The pavement was conservatively designed (assuming the upper limit of the W_{18} design lane load) to last 20 years, and the CBR is 25 with a drainage coefficient of 1.0. A design mistake was made that ignored 1000 total northbound (daily) passes of trucks with 22-kip single and 30-kip tandem axles. What slab thickness should have been used?

4.25 A flexible pavement is designed with 4 inches of an experimental asphalt wearing surface that has a structural layer coefficient of 0.575, with 4 inches of a hot-mix asphaltic base, and 10 inches of a crushed stone subbase (all drainage coefficients are 1.0). The roadway is designed to accommodate 200 delivery trucks per day and each truck has one 12-kip single front axle and one 20-kip single rear axle. The subgrade CBR is 8, the initial present serviceability index is 4.5, the TSI is 2.5, and the overall standard deviation is 0.5. How confident would you be that this pavement would last 20 years?

4.26 For the conditions in Practice Problem 4.4 **SS** (again conservatively designed), suppose that a rigid pavement was designed with a 12-inch slab (modulus of rupture of 900 lb/in², modulus of elasticity of 6.0 × 10⁶ lb/in², load transfer coefficient of 3.2, and drainage coefficient of 1.0) for the four-lane northbound direction described in Practice Problem 4.4 (all other traffic and PSI information is the same as in that problem). How confident would you be that this rigid pavement would have a PSI above 2.5 after 20 years?

4.27 A rigid pavement on a multilane highway (two lanes each direction) has been conservatively designed (for its design-lane traffic). The pavement engineer uses an 11-inch slab, E_c of 4 × 10⁶ lb/in², a concrete modulus of rupture of 695 lb/in², a load transfer coefficient of 2.8, an initial present serviceability index of 4.2, and a TSI of 2.5. The overall standard deviation is 0.6, the subgrade CBR is 20, and the drainage coefficient is 1.0. The pavement was designed for four hundred 30,000 lb triple axles, nine hundred 20,000 lb tandem axles per day, and twelve hundred 10,000 lb single axle loads per day. If the pavement was designed to last 40 years, what reliability was used?

Multiple Choice Problems (Multiple Sections)

4.28 A flexible pavement is constructed with 5 inches of sand-mix asphaltic wearing surface, 9 inches of dense-graded crushed stone base, and 10 inches of crushed stone subbase. The base has a drainage coefficient of 0.90, while the subbase drainage coefficient is 1.0. Determine the SN of the pavement.

  a) 4.47
  b) 4.31
  c) 4.76
  d) 3.98

4.29 A flexible pavement is designed to last 10 years to withstand truck traffic that consists only of trucks with two 18-kip single axles. The pavement is designed for a soil CBR of 2, an initial PSI of 5.0, a TSI of 2.5, an overall standard deviation of 0.40, and a reliability of 90%, and the SN was determined to be 6. On one section of this roadway, beneath an underpass, an engineer uses a 10-inch rigid pavement in an attempt to have it last longer before resurfacing. How many years will this rigid-pavement section last? (Given the same traffic conditions, modulus of rupture = 800 lb/in², modulus of elasticity = 5,000,000 lb/in², load transfer coefficient of 3.0, and drainage coefficient of 1.0.)

  a) 11.33
  b) 13.22
  c) 17.14
  d) 25.65

4.30 A flexible pavement at an access road to a sports stadium parking lot is designed with a 4-inch sand-mix asphaltic concrete surface, 5-inch aggregate bituminous emulsion base, and a 10-inch crushed stone subbase. There are 95 scheduled baseball and football games at the stadium per year. The access road to the parking lot is three lanes in each direction (conservatively designed). The pavement was designed for recreational vehicles with one 20-kip single axle and one 20-kip tandem axle. There are 9000 recreational vehicles estimated at each event. Given that drainage coefficients are 1.0, the overall standard deviation of traffic is 0.45, reliability is 90%, and the soil's resilient modulus is 15,000 lb/in², how many years will the access road last if the initial PSI is 4.0 and the TSI is 2.5?

  a) 7
  b) 9
  c) 12
  d) 14

4.31 A rigid pavement on a new interstate (3 lanes each direction) has been conservatively designed with a 12-inch slab, an E_c of 5.5×10^6 lb/in², a concrete modulus of rupture of 700 lb/in², a load transfer coefficient of 3.0, an initial present serviceability index of 4.5, and a TSI of 2.5. The overall standard deviation is 0.35, the subgrade CBR is 25, and the drainage coefficient is 0.9. The pavement was designed for 600 30-kip tandem axles per day and 1400 20-kip single-axle loads per day. If the desired reliability was 90%, how long was this pavement designed to last?

  a) 18
  b) 32
  c) 42
  d) 46

4.32 A flexible pavement was designed to have a 6-inch sand-mix asphaltic surface, 8-inch soil-cement base, and a 21-inch crushed stone subbase (all drainage coefficients are 1.0). The pavement was designed for 800 12-kip single axles and 1600 34-kip tandem axles per day in the design direction. The reliability used was 90%, the overall standard deviation was 0.35, initial PSI was 4.7, the TSI was 2.5, and the soil resilient modulus was 2582 lb/in². If the road has three lanes in the design direction (and was conservatively designed), for how many years was the pavement designed to last?

  a) 12
  b) 15
  c) 20
  d) 24

CHAPTER 5 PROBLEMS

Traffic Stream Parameters and Basic Traffic Stream Models (Sections 5.2–5.3)

SS **5.1** Assume that you are observing traffic in a single lane of a highway at a specific location. You measure the average headway and average spacing of passing vehicles as 3.2 seconds and 165 ft, respectively. Calculate the flow, average speed, and density of the traffic stream in this lane.

5.2 Assume that you are an observer standing at a point along a three-lane roadway. All vehicles in lane 1 are traveling at 30 mi/h, all vehicles in lane 2 are traveling at 45 mi/h, and all vehicles in lane 3 are traveling at 60 mi/h. There is also a constant spacing of 0.5 mile between vehicles. If you collect spot speed data for all vehicles as they cross your observation point, for 30 minutes, what will be the time-mean speed and space-mean speed for this traffic stream?

5.3 Four race cars are traveling on a 2.5-mile tri-oval track. The four cars are traveling at constant speeds of 195 mi/h, 190 mi/h, 185 mi/h, and 180 mi/h, respectively. Assume that you are an observer standing at a point on the track for a period of 30 minutes and are recording the instantaneous speed of each vehicle as it crosses your point. What is the time-mean speed and space-mean speed for these vehicles for this time period? (Note: Be careful with rounding.)

5.4 For Problem 5.3, calculate the space-mean speed assuming that you were given only an aerial photo of the circling race cars and the constant travel speed of each of the vehicles.

5.5 On a specific westbound section of highway, studies show that the speed–density relationship is

$$u = u_f \left[1 - \left(\frac{k}{k_f} \right)^{3.5} \right]$$

It is known that the capacity is 4200 veh/h and the jam density is 210 veh/mi. What is the space-mean speed of the traffic at capacity, and what is the free-flow speed?

5.6 A section of highway has the following flow–density relationship $q = 50k - 0.156k^2$ [with q in veh/h and k in veh/mi]. What is the capacity of the highway section, the speed at capacity, and the density when the highway is at one-quarter of its capacity?

Models of Traffic Flow (Section 5.4)

5.7 An observer has determined that the time headways between successive vehicles on a section of highway are exponentially distributed and that 65% of the headways between vehicles are 9 seconds or greater. If the observer decides to count traffic in 30-second time intervals, estimate the probability of the observer counting exactly four vehicles in an interval.

5.8 At a specified point on a highway, vehicles are known to arrive according to a Poisson process. Vehicles are counted in 20-second intervals, and vehicle counts are taken in 120 of these time intervals. It is noted that no cars arrive in 18 of these 120 intervals. Approximate the number of these 120 intervals in which exactly three cars arrive.

5.9 Consider the conditions in Practice Problem 5.2. **SS** How short would the driver reaction times of oncoming vehicles have to be for the probability of an accident to equal 0.20?

Queuing Theory and Traffic Flow Analysis (Section 5.5)

5.10 Vehicles arrive at a single toll booth beginning at 8:00 A.M. They arrive and depart according to a uniform deterministic distribution. However, the toll booth does not open until 8:10 A.M. The average arrival rate is 8 veh/min, and the average departure rate is 10 veh/min. Assuming $D/D/1$ queuing, when does the initial queue clear and what are the total delay, the average delay per vehicle, longest queue length (in vehicles), and the wait time of the 100th vehicle to arrive (assuming first-in-first-out)?

5.11 Vehicles begin to arrive at a park entrance at 7:45 A.M. at a constant rate of six per minute and at a constant rate of four vehicles per minute from 8:00 A.M. on. The park opens at 8:00 A.M. and the manager wants to set the departure rate so that the average delay per vehicle is no greater than 9 minutes (measured from the time of the first arrival until the total queue clears). Assuming $D/D/1$ queuing, what is the minimum departure rate needed to achieve this?

5.12 A toll booth on a turnpike is open from 8:00 A.M. to 12 midnight. Vehicles start arriving at 7:45 A.M. at a uniform deterministic rate of six per minute until 8:15 A.M. and from then on at two per minute. If vehicles are processed at a uniform deterministic rate of six per minute, determine when the queue will dissipate, the total delay, the maximum queue length (in vehicles), the longest vehicle delay under FIFO, and the longest vehicle delay under LIFO.

5.13 Vehicles begin to arrive at a parking lot at 6:00 A.M. at a rate of eight per minute. Due to an accident on the access highway, no vehicles arrive from 6:20 to 6:30 A.M. From 6:30 A.M. on, vehicles arrive at a rate of two per minute. The parking-lot attendant processes incoming vehicles (collects parking fees) at a rate of four per minute throughout the day. Assuming $D/D/1$ queuing, determine total vehicle delay.

5.14 Vehicles begin to arrive at a toll booth at eight vehicles per minute from 9 A.M. to 10 A.M. The booth opens at 9:10 A.M. and services at a rate of 10 vehicles per minute until 9:40 A.M. From 9:40 A.M. until 10 A.M. the service rate is six vehicles per minute. Assuming $D/D/1$ queuing, what is the total vehicle delay from 9 A.M. to 10 A.M. assuming $D/D/1$ queuing?

5.15 The arrival rate at a parking lot is 6 veh/min. Vehicles start arriving at 6:00 P.M., and when the queue reaches 36 vehicles, service begins. If company policy is that total vehicle delay should be equal to 500 veh-min, what is the departure rate? (Assume $D/D/1$ queuing and a constant service rate.)

5.16 At 8:00 A.M. there are 10 vehicles in a queue at a toll booth and vehicles are arriving at a rate of $\lambda(t) = 6.9 - 0.2t$. Beginning at 8 A.M., vehicles are being serviced at a rate of $\mu(t) = 2.1 + 0.3t$ [$\lambda(t)$ and $\mu(t)$ are in vehicles per minute and t is in minutes after 8:00 A.M.]. Assuming $D/D/1$ queuing, what is the maximum queue length, and what would the total delay be from 8:00 A.M. until the queue clears?

5.17 At the end of a sporting event, vehicles begin leaving a parking lot at $\lambda(t) = 12 - 0.25t$ and vehicles are processed at $\mu(t) = 2.5 + 0.5t$ [t is in minutes and $\lambda(t)$ and $\mu(t)$ are in vehicles per minute]. Assuming $D/D/1$ queuing, determine the total vehicle delay, longest queue, and the wait time of the 50th vehicle to arrive assuming first-in-first-out and $D/D/1$ queuing.

5.18 Vehicles arrive at a single park-entrance booth where a brochure is distributed. At 8 A.M., there are 20 vehicles in the queue and vehicles continue to arrive at the deterministic rate of $\lambda(t) = 4.2 - 0.1t$, where $\lambda(t)$ is in vehicles per minute and t is in minutes after 8:00 A.M. From 8 A.M. until 8:10 A.M., vehicles are served at a constant deterministic rate of three per minute. Starting at 8:10 A.M., another brochure-distributing person is added and the brochure-service rate increases to six per minute (still

at a single booth). Assuming $D/D/1$ queuing, determine the longest queue, the total delay from 8 A.M. until the queue dissipates; and the wait time of the 40th vehicle to arrive.

5.19 Vehicles arrive at a single toll booth beginning at 7:00 A.M. at a rate of 8 veh/min. Service also starts at 7:00 A.M. at a rate of $\mu(t) = 6 + 0.2t$ where $\mu(t)$ is in vehicles per minute and t is in minutes after 7:00 A.M. Assuming $D/D/1$ queuing, determine when the queue will clear, the total delay, and the maximum queue length in vehicles.

5.20 Vehicles begin arriving at a single toll-road booth at 8:00 A.M. at a time-dependent deterministic rate of $\lambda(t) = 2 + 0.1t$ [with $\lambda(t)$ in veh/min and t in minutes]. At 8:07 A.M., the toll booth opens and vehicles are serviced at a constant deterministic rate of 6 veh/min. Assuming $D/D/1$ queuing, what is the average delay per vehicle from 8:00 A.M. until the initial queue clears and what is the delay of the 20th vehicle to arrive?

5.21 Vehicles begin to arrive at a toll booth at 8:50 A.M. with an arrival rate of $\lambda(t) = 4.1 + 0.01t$ [with t in minutes and $\lambda(t)$ in vehicles per minute]. The toll booth opens at 9:00 A.M. and processes vehicles at a rate of 12 per minute throughout the day. Assuming $D/D/1$ queuing, when will the queue dissipate and what will be the total vehicle delay?

5.22 Vehicles begin to arrive at a toll booth at 7:50 A.M. with an arrival rate of $\lambda(t) = 5.2 - 0.01t$ [with t in minutes after 7:50 A.M. and λ in vehicles per minute]. The toll booth opens at 8:00 A.M. and serves vehicles at a rate of $\mu(t) = 3.3 + 2.4t$ (with t in minutes after 8:00 A.M. and μ in vehicles per minute). Once the service rate reaches 10 veh/min, it stays at that level for the rest of the day. If queuing is $D/D/1$, when will the queue that formed at 7:50 A.M. be cleared?

5.23 Vehicles arrive at a freeway on-ramp meter at a constant rate of six per minute starting at 6:00 A.M. Service begins at 6:00 A.M. such that $\mu(t) = 2 + 0.5t$, where $\mu(t)$ is in veh/min and t is in minutes after 6:00 A.M. What is the total delay and the maximum queue length (in vehicles)?

5.24 Vehicles arrive at a toll booth according to the function $\lambda(t) = 5.2 - 0.20t$, where $\lambda(t)$ is in vehicles per minute and t is in minutes. The toll booth operator processes one vehicle every 20 seconds. Determine total delay, maximum queue length, and the time that the 20th vehicle to arrive waits from its arrival to its departure.

5.25 There are 10 vehicles in a queue when an attendant opens a toll booth. Vehicles arrive at the booth at a rate of four per minute. The attendant opens the booth and improves the service rate over time following the function $\mu(t) = 1.1 + 0.30t$, where $\mu(t)$ is in vehicles per minute and t is in minutes. When will the queue clear, what is the total delay, and what is the maximum queue length?

SS **5.26** Vehicles begin to arrive at a parking lot at 6:00 A.M. with an arrival rate function (in vehicles per minute) of $\lambda(t) = 1.2 + 0.3t$, where t is in minutes. At 6:10 A.M., the parking lot opens and processes vehicles at a rate of 12 per minute. What is the total delay and the maximum queue length?

5.27 At a parking lot, vehicles arrive according to a Poisson process and are processed (parking fee collected) at a uniform deterministic rate at a single station. The mean arrival rate is 4.2 veh/min and the processing rate is 5 veh/min. Determine the average length of queue, the average time spent in the system, and the average waiting time in the queue.

5.28 Consider the parking lot and conditions described in Problem 5.27. If the rate at which vehicles are processed became exponentially distributed (instead of deterministic) with a mean processing rate of 5 veh/min, what would be the average length of queue, the average time spent in the system, and the average waiting time in the queue?

5.29 Vehicles arrive at a toll booth with a mean arrival rate of 3 veh/min (the time between arrivals is exponentially distributed). The toll booth operator processes vehicles (collects tolls) at a uniform deterministic rate of one every 15 seconds. What is the average length of queue, the average time spent in the system, and the average waiting time in the queue?

5.30 A business owner decides to pass out free transistor radios (along with a promotional brochure) at a booth in a parking lot. The owner begins giving the radios away at 9:15 A.M. and continues until 10:00 A.M. Vehicles start arriving for the radios at 8:45 A.M. at a uniform deterministic rate of four per minute and continue to arrive at this rate until 9:15 A.M. From 9:15 to 10:00 A.M., the arrival rate becomes 8 per minute. The radios and brochures are distributed at a uniform deterministic rate of 11 cars per minute over the 45-minute time period. Determine total delay, maximum queue length, and longest vehicle delay assuming FIFO and LIFO.

5.31 Consider the conditions described in Problem 5.30. Suppose that the owner decides to accelerate the radio-brochure distribution rate (in veh/min) so that the queue that forms will be cleared by 9:45 A.M. What would this new distribution rate be?

5.32 A ferryboat queuing lane holds 40 vehicles. If vehicles are processed (tolls collected) at a uniform deterministic rate of five vehicles per minute and processing begins when the lane reaches capacity, what is the uniform deterministic arrival rate if the vehicle queue is cleared 35 minutes after vehicles begin to arrive?

5.33 At a toll booth, vehicles arrive and are processed (tolls collected) at uniform deterministic rates λ and μ, respectively. The arrival rate is 3 veh/min. Processing begins 15 minutes after the arrival of the first vehicle, and the queue dissipates t minutes after the arrival of the first vehicle. Letting the number of vehicles that must actually wait in a queue be x, develop an expression for determining processing rates in terms of x.

5.34 Vehicles arrive at a recreational park booth at a uniform deterministic rate of 5 veh/min. If uniform deterministic processing of vehicles (collecting of fees) begins 20 minutes after the first arrival and the total delay is 3200 veh-min, how long after the arrival of the first vehicle will it take for the queue to be cleared?

5.35 Trucks begin to arrive at a truck weigh station (with a single scale) at 6:00 A.M. at a deterministic but time-varying rate of $\lambda(t) = 4.3 - 0.22t$ [$\lambda(t)$ is in veh/min and t is in minutes]. The departure rate is a constant 2 veh/min (time to weigh a truck is 30 seconds). When will the queue that forms be cleared, what will be the total delay, and what will be the maximum queue length?

5.36 Commercial trucks begin to arrive at a seaport entry plaza at 7:50 A.M., at the rate of $\lambda(t) = 6.3 - 0.25t$ [$\lambda(t)$ is in veh/min and t is in minutes]. The plaza opens at 8:00 A.M. For the first 10 minutes, one processing booth is open. After the first 10 minutes until the queue clears, two processing booths are open. Each booth processes trucks at a uniform rate of two per minute. What is the average delay per vehicle, the maximum queue length, and the average queue length?

5.37 Vehicles begin to arrive at a remote parking lot after the start of a major sporting event. They are arriving at a deterministic but time-varying rate of $\lambda(t) = 3.3 - 0.1t$ [$\lambda(t)$ is in veh/min and t is in minutes]. The parking-lot attendant processes vehicles (assigns spaces and collects fees) at a deterministic rate at a single station. A queue exceeding four vehicles will back up onto a congested street, and is to be avoided. How many vehicles per minute must the attendant process to ensure that the queue does not exceed four vehicles?

5.38 A truck weighing station has a single scale. The time between truck arrivals at the station is exponentially distributed with a mean arrival rate of 1.6 veh/min. The time it takes vehicles to be weighed is exponentially distributed with a mean rate of 2.1 veh/min. When more than five trucks are in the system, the queue backs up onto the highway and interferes with through traffic. What is the probability that the number of trucks in the system will exceed 5?

SS 5.39 Consider the convenience store described in Example 5.14. The owner is concerned about customers not finding an available parking space when they arrive during the busiest hour. How many spaces must be provided for there to be less than a 1% chance of an arriving customer not finding an open parking space?

5.40 Vehicles arrive at a toll bridge at a rate of 420 veh/h (the time between arrivals is exponentially distributed). Two toll booths are open and each can process arrivals (collect tolls) at a mean rate of 12 seconds per vehicle (the processing time is also exponentially distributed). What is the total time spent in the system by all vehicles in a 1-hour period?

5.41 Vehicles leave an airport parking facility (arrive at parking fee collection booths) at a rate of 500 veh/h (the time between arrivals is exponentially distributed). The parking facility has a policy that the average time a patron spends in a queue waiting to pay for parking is not to exceed 5 seconds. If the time required to pay for parking is exponentially distributed with a mean of 15 seconds, what is the smallest number of payment processing booths that must be open to keep the average time spent in a queue below 5 seconds?

5.42 Vehicles begin to arrive at a parking lot at 7:45 A.M. at a constant rate of 4 veh/min and continue to arrive at that rate throughout the day. The parking lot opens at 8:00 A.M. and vehicles are processed at a constant rate of one vehicle every 10 seconds. Assuming $D/D/1$ queuing, what is the longest queue, the queue at 8:15 A.M., and the average delay per vehicle from 7:45 A.M. until the queue clears?

5.43 Vehicles arrive at a toll both starting at 7:00 A.M. at a rate of $\lambda(t) = 5.1 - 0.05t$ [with $\lambda(t)$ in veh/min and t in minutes after 7:00 A.M.]. The first operator processes cars at a rate of 3 veh/minute 7:00 A.M. until 7:15 A.M when the person leaves because of illness.

From 7:15 A.M to 7:25 A.M, no one is at the toll booth but a new operator arrives at 7:25 A.M and processes at a rate of $\mu(t) = 8 + 0.3t$ [with $\mu(t)$ in veh/min and t in minutes after 7:25 A.M.]. Assuming $D/D/1$ queuing, what is the maximum queue length (in vehicles) and the average delay per vehicle?

5.44 Vehicles arrive at a toll both starting at 5:55 A.M. at a rate of $\lambda(t) = 4.0 - 0.05t$ [with $\lambda(t)$ in veh/min and t in minutes after 5:55 A.M.]. The toll booth opens at 6:00 A.M. processes vehicles at a rate of $\mu(t) = 2 + 0.2t$ [with $\mu(t)$ in veh/min and t in minutes after 6:00 A.M.] until a rate of 4.5 veh/min is reached and the service rate remains at 4.5 veh/min until the queue clears. Assuming $D/D/1$ queuing, when does the queue clear and what is the average delay per vehicle?

5.45 Vehicles begin to arrive at an amusement park entrance at 8:00 A.M. at a rate of 1000 veh/h. Some of these vehicles have electronic identifiers that allow them to enter the park immediately, beginning at 8:00 A.M., without stopping (they are billed remotely). All vehicles without such identifiers stop at a single processing booth, but they wait in line until it opens at 8:10 A.M. Once open, the operator processes vehicles at $\mu(t) = 8 + 0.5t$ [where $\mu(t)$ is in vehicles per minute and t is in minutes after 8:10 A.M.]. An observer notes that at 8:25 there are exactly 20 vehicles in the queue. What percent of arriving vehicles have electronic identifiers and what is the total delay (from the 8:00 A.M. until the queue clears) for those vehicles without the electronic identifiers (assume $D/D/1$ queuing)?

5.46 At 9:00 A.M., there are 22 vehicles in a queue at a toll booth. Starting at 9:00 A.M., vehicles are arriving at a rate of $\lambda(t) = 9.2 - 0.3t$ [where $\lambda(t)$ is in vehicles per minute and t is in minutes after 9:00 A.M.] and vehicles are being serviced at a constant rate of four vehicles per minute. What is the maximum queue length, and what would the total delay be from 9:00 A.M. until the queue clears?

5.47 A transportation engineering midterm exam starts at noon and students start completing a the exam at 1:05 P.M, and the rate they complete the exam and hand it in to the instructor at the front of the class is given by the function $\lambda(t) = 1.1 + 1.2t$ [with $\lambda(t)$ in students/minute and t is in minutes after 1:05 P.M]. The professor takes 10 seconds to process each student (record exam completion). There are 137 students in the class. Assuming $D/D/1$ queuing, when will the last student have their exam processed, what is the total student delay in handing in the exams, and what is the longest student queue?

Traffic Analysis at Highway Bottlenecks (Section 5.6)

SS **5.48** A freeway with two northbound lanes is shut down because of an accident. At the time of the accident, the traffic flow rate is 1200 vehicles per hour per lane and the flow remains at this level. The capacity of the freeway is 2200 vehicles per hour per lane when not impacted by an accident. The freeway is shut down completely for 20 minutes after the accident and then one lane is open for 20 minutes and finally both lanes are opened (40 minutes after the accident). What is the average delay per vehicle resulting from the accident (assuming $D/D/1$ queuing)?

5.49 A four-lane highway has a normal capacity of 1800 vehicles per hour per lane. In the southbound direction, a vehicle disablement on the roadway shoulder occurs at 4:30 P.M. Due to rubbernecking, the capacity in the southbound direction is reduced to 1200 veh/h/lane at this time. At 4:45 P.M., the disabled vehicle is removed from the shoulder and the capacity increases to 1500 veh/h/lane. At 5:00 P.M., the roadway capacity returns to its full value of 1800 veh/h/lane. From 4:30 P.M., until the queue clears the traffic flow rate in the southbound direction is 1600 veh/h/lane. What is the average delay per vehicle, the maximum queue length, and the average queue length in the southbound direction resulting from the incident (assuming $D/D/1$ queuing)?

Multiple Choice Problems (Multiple Sections)

5.50 Five minivans and three trucks are traveling on a 3.0 mile circular track and complete a full lap in 98.0, 108.0, 113.0, 108.0, 102.0, 101.0, 85.0, and 95 seconds, respectively. Assuming that all the vehicles are traveling at constant speeds, what is the time-mean speed of the minivans? Pay attention to rounding.

   a) 102.332 mi/h
   b) 107.417 mi/h
   c) 102.079 mi/h
   d) 102.400 mi/h

5.51 Vehicles arrive at an intersection at a rate of 400 veh/h according to a Poisson distribution. What is the probability that more than five vehicles will arrive in a 1-minute interval?

   a) 0.7944
   b) 0.6560
   c) 0.6547
   d) 0.1552

5.52 In studying of traffic flow at a highway toll booth over a course of 60 minutes, it is determined that the arrival and departure rates are deterministic, but not uniform. The arrival rate is found to vary according to the function $\lambda(t) = 1.8 + 0.25t - 0.0030t^2$. The departure rate function is $\mu(t) = 1.4 + 0.11t$. In both of these functions, t is in minutes after the beginning of the observation and $\lambda(t)$ and $\mu(t)$ are in vehicles per minute. At what time does the maximum queue length occur?

   a) 49.4 min
   b) 2.7 min
   c) 19.4 min
   d) 60.0 min

5.53 A theme park has a single entrance gate where visitors must stop and pay for parking. The average arrival rate during the peak hour is 150 veh/h and is Poisson distributed. It takes, on average, 20 seconds per vehicle (exponentially distributed) to pay for parking. What is the average waiting time for this queuing system?

   a) 4.167 min/veh
   b) 2.0 min/veh
   c) 1.667 min/veh
   d) 0.833 min/veh

5.54 At an impaired driver checkpoint, the time required to conduct the impairment test varies (according to an exponential distribution) depending on the compliance of the driver, but takes 60 seconds on average. If an average of 30 vehicles per hour arrive (according to a Poisson distribution) at the checkpoint, determine the average time spent in the system.

   a) 0.033 min/veh
   b) 1.5 min/veh
   c) 1.0 min/veh
   d) 2.0 min/veh

5.55 A toll road with three toll booths has an average arrival rate of 850 veh/h and drivers take an average of 12 seconds to pay their tolls. If the arrival and departure times are determined to be exponentially distributed, how would the probability of waiting in a queue change if a fourth toll both were opened?

   a) 0.088
   b) 0.534
   c) 0.313
   d) 0.847

CHAPTER 6 PROBLEMS

Freeways (Section 6.4)

SS **6.1** A six-lane freeway (three lanes in each direction) currently operates at maximum LOS C conditions. The lanes are 11 ft wide, the right-side shoulder is 4 ft wide, and there are two ramps within three miles upstream of the segment midpoint and one ramp within three miles downstream of the segment midpoint. The highway is on rolling terrain with 10% heavy vehicles, and the peak-hour factor is 0.90. Determine the hourly volume for these conditions.

6.2 Consider the freeway in Problem 6.1. At one point along this freeway there is a 3.5% upgrade with a directional hourly traffic volume of 5435 vehicles. The heavy vehicle split is 50% single-unit trucks/50% tractor-trailer trucks. If all other conditions are as described in Problem 6.1, how long can this grade be without the freeway LOS dropping to F?

6.3 A four-lane freeway (two lanes in each direction) is located on rolling terrain and has 12-ft lanes, no lateral obstructions within 6 ft of the pavement edges, and there are two ramps within three miles upstream of the segment midpoint and three ramps within three miles downstream of the segment midpoint. A weekday directional peak-hour volume of 1800 vehicles (familiar users) is observed, with 700 arriving in the most congested 15-min period. If a LOS no worse than C is desired, determine the maximum number of heavy vehicles that can be present in the peak-hour traffic stream.

6.4 Consider the freeway and traffic conditions described in Problem 6.3. If 180 of the 1800 vehicles observed in the peak hour were heavy vehicles (assume a 70%/30% SUT/TT split), what would the LOS of this freeway be on a 5-mi, 6% downgrade?

6.5 A six-lane freeway (three lanes in each direction) in a scenic area has a measured *FFS* of 55 mi/h. There are 7% SUTs and 7% TTs in the traffic stream. One upgrade is 5% and 0.5 mi long. An analyst has determined that the freeway is operating at capacity on this upgrade during the peak hour. If the peak-hour traffic volume is 3900 vehicles, what value for the peak-hour factor was used?

6.6 A freeway is being designed for a location in rolling terrain. The expected *FFS* is 55 mi/h. During the peak hour, it is expected that there will be a directional peak-hour volume of 2700 vehicles and 18% heavy vehicles. The *PHF* is expected to be 0.88. If a LOS no worse than D is desired, determine the necessary number of lanes.

6.7 A segment of four-lane freeway (two lanes in each direction) has a 3% upgrade that is 1500 ft long followed by a 1000-ft 4% upgrade. It has 12-ft lanes and 3-ft shoulders. The directional hourly traffic flow is 2,000 vehicles with 3% SUTs and 3% TTs. The total ramp density for this freeway segment is 2.33 ramps per mile. If the peak-hour factor is 0.90, what is the LOS of this compound-grade freeway segment?

6.8 Consider Example 6.2, in which it was determined that 1537 vehicles could be added to the peak hour before capacity is reached. Assuming level terrain as in Example 6.1, how many passenger cars could be added to the original traffic mix before peak-hour capacity is reached? (Assume only passenger cars are added and that the number of heavy vehicles originally in the traffic stream remains constant.)

6.9 An eight-lane freeway (four lanes in each direction) is on rolling terrain and has 11-ft lanes with a 4-ft right-side shoulder. The total ramp density is 1.5 ramps per mile. The directional peak-hour traffic volume is 5400 vehicles with 11% heavy vehicles. The peak-hour factor is 0.95. It has been decided that heavy vehicles will be banned from the freeway during the peak hour. What will the freeway's density and LOS be before and after the ban? (Assume that the heavy vehicles are removed and all other traffic attributes are unchanged.)

6.10 A 5% upgrade on a six-lane freeway (three **SS** lanes in each direction) is 1.25 mi long. On this segment of freeway, there is 3% SUTs and 7% TTs, and the peak-hour factor is 0.90. The lanes are 12 ft wide, there are no lateral obstructions within 6 ft of the roadway, and the total ramp density is 1.0 ramps per mile. What is the maximum directional peak-hour volume that can be accommodated without exceeding LOS C operating conditions?

6.11 A northbound two-lane (one direction) segment of freeway currently has a measured density of 32.0 pc/mi/ln and a measured *FFS* of 65 mi/h. The peak-hour factor is 0.95, the traffic stream contains 7% single-unit trucks and 3% tractor-trailer trucks. The freeway segment is on a 3.5% upgrade that is ¾ of a mile long. With the peak-hour factor unchanged, how many trucks (assuming the same ST/TT split) can be added to the peak hour before capacity is reached?

6.12 A northbound freeway segment is on a 4% upgrade from station 430+20 to 450+00 and has two 11-ft wide lanes, a 5-ft right shoulder, and has a ramp density of 1 per mile in the 3 miles before and after station 440+10. The peak-hour factor is 0.9. Northbound traffic during the peak hour is 2550 cars, 300 STs, and 300 TTs. Determine the density and LOS of the freeway segment.

6.13 A westbound freeway segment has four 12-ft wide lanes (one direction), a 4-ft right shoulder, and 4 ramps in the 3 miles before and after the midpoint of the analysis segment. The peak-hour factor is 0.85. Westbound traffic during the peak hour is 4500 passenger cars, 350 single-unit trucks, and 150 tractor-trailer trucks. The freeway segment is on a 3.5% upgrade. What is the maximum length of this grade if no worse than LOS C must be maintained?

6.14 A freeway segment currently has four 11-ft wide lanes and a 3-ft right shoulder in the southbound direction. It is operating in LOS B with a traffic density of 15.6 pc/mi/ln and a peak-hour factor of 0.85. The freeway is on level terrain with 15% heavy vehicles, and currently has no on- or off-ramps within a distance of three miles upstream and downstream of the mid-point of the analysis section. New interchanges are to be constructed that would result in 6 on- and off-ramps within the distance of three miles upstream and downstream of the mid-point of the analysis section and total traffic volume is projected to double, with the same percentage of heavy vehicles, and the peak-hour factor going to 0.95. What would the new LOS and density be after the interchange construction and new traffic conditions?

6.15 A westbound section of freeway currently has three 12-ft wide lanes, a 6-ft right shoulder, and no ramps within 3 miles upstream and downstream of the segment midpoint. It is on rolling terrain with 10% heavy vehicles and is operating at capacity with a peak-hour factor of 0.9. If the road is expanded to four 11-foot lanes with a 2-foot right shoulder, and traffic after the expansion is projected to increase by 10% with the same heavy vehicle percentage and peak-hour factor, what is the new LOS and estimated density?

6.16 A six-mile long freeway segment has 5 lanes in the northbound direction. The lanes are 10-ft wide and there is a 3-ft right shoulder. There are 8 on/off ramps in this section. What is the estimated *FFS* at the mid-point of this six-mile segment?

Multilane Highways (Section 6.5)

6.17 A multilane highway (two lanes in each direction) is on level terrain. The *FFS* has been measured at 45 mi/h. The peak-hour directional traffic flow is 1300 vehicles with 8% heavy vehicles. If the peak-hour factor is 0.85, determine the highway's LOS.

6.18 Consider the multilane highway in Problem 6.17. If the proportion of vehicle types and peak-hour factor remain constant, how many vehicles can be added to the directional traffic flow before capacity is reached?

6.19 A six-lane multilane highway (three lanes in each direction) has a peak-hour factor of 0.90, 11-ft lanes with a 4-ft right-side shoulder, and a two-way left-turn lane in the median. The directional peak-hour traffic flow is 4000 vehicles with 14% SUTs and 6% TTs. What will the LOS of this highway be on a 4% upgrade that is 1.5 miles long if the speed limit is 55 mi/h and there are 15 access points per mile?

6.20 A divided multilane highway in a recreational area has four lanes (two lanes in each direction) and is on rolling terrain. The highway has 10-ft lanes with a 6-ft right-side shoulder and a 3-ft left-side shoulder. The posted speed is 50 mi/h. Previously, there were four access points per mile, but recent development has increased the number of access points to 12 per mile. Before development, the peak-hour factor was 0.95, and the directional hourly volume was 2200 vehicles with 13% heavy vehicles. After development, the peak-hour directional flow is 2600 vehicles with the same vehicle percentages and peak-hour factor. What is the LOS before and after the development?

6.21 A multilane highway has four lanes (two lanes in each direction) and a measured *FFS* of 55 mi/h. The directional peak-hour volume is 1900 vehicles (the peak-hour factor is 0.80). One upgrade is 5% and is 0.62 mi long. Currently, heavy vehicles are not permitted on the highway, but local authorities are considering allowing heavy vehicles on this upgrade. If this is done, they estimate that 150 heavy vehicles will use the highway during the peak hour. What would be the LOS before and after the heavy vehicles are allowed on the upgrade (assuming 50% SUTs and 50% TTs.)?

SS **6.22** A four-lane undivided multilane highway (two lanes in each direction) has 11-ft lanes, 4-ft shoulders, and 10 access points per mile. It is determined that the roadway currently operates at capacity with $PHF = 0.80$. If the highway is on level terrain with 8% heavy vehicles and the speed limit is 55 mi/h, what is the directional hourly volume?

6.23 A new four-lane divided multilane highway (two lanes in each direction) is being planned with 12-ft lanes, 6-ft shoulders on both sides, and a 50-mi/h speed limit. One 3% downgrade is 4.5 mi long, and there will be four access points per mile. The peak-hour directional volume along this grade is estimated to consist of 1800 passenger cars, 140 SUTs, and 60 TTs. If the peak-hour factor is estimated to be 0.85, what LOS will this segment of highway operate under?

6.24 A six-lane divided multilane highway (three lanes in each direction) has a measured FFS of 50 mi/h. It is on rolling terrain with a traffic stream consisting of 6% heavy vehicles. One direction of the highway currently operates at maximum LOS C conditions, and it is known that the highway has $PHF = 0.90$. How many vehicles can be added to this highway before capacity is reached, assuming the proportion of vehicle types remains the same but the peak-hour factor increases to 0.95?

6.25 A four-lane undivided multilane highway (two lanes in each direction) has 11-ft lanes and 5-ft shoulders. At one point along the highway, there is a 4% upgrade that is 0.62 mi long. There are 15 access points along this grade. The peak-hour traffic volume is 2340 vehicles, with 10% single-unit trucks and 10% tractor-trailer trucks, and 620 of these vehicles arrive in the most congested 15-min period. The posted speed limit is 60 mi/h. To improve the LOS, the local transportation agency is considering reducing the number of access points by blocking some driveways and rerouting their traffic. How many of the 15 access points must be blocked to achieve LOS C?

6.26 A multilane highway, with three lanes in each direction, operates at maximum level of service D conditions during the peak hour with an observed speed of 48.0 mi/h (with 3690 veh/h observed). The highway is on rolling terrain and the traffic stream consists of cars and some percentage of heavy vehicles. The peak-hour factor is 0.85. If heavy vehicles were banned from the highway and car traffic remained the same, what would be the LOS?

6.27 An undivided multilane highway segment has two 11-ft lanes in the eastbound direction with no shoulders and a 55 mi/h speed limit. This highway segment has 17 access points on a 0.625 mile, 4% upgrade. During the highest 15 minutes of traffic flow within the peak hour, there are 720 cars, 56 single-unit trucks, and 24 tractor-trailer trucks. What are the estimated speed, density, and LOS of this upgrade?

6.28 A multilane undivided highway is on level terrain and has two lanes in the northbound direction. The posted speed limit is 55 mi/h, and there are 11-ft lanes, 3-ft right shoulders, and 8 access points per mile. The highway operates at capacity during the peak hour with 20% heavy vehicles and a peak-hour factor of 0.925. What is the peak-hour demand volume?

6.29 There are two 11-ft lanes in a northbound direction of an undivided multilane highway. The highway has a 50-mi/h speed limit with a 4-ft shoulder. During the peak hour, there are 2,475 total vehicles, with 8% heavy vehicles (even split between STs and TTs). The peak-hour factor is 0.8. One segment of this highway has a +2% grade that is ¾ of a mile long with 9 access points. What is the density and LOS of this highway segment?

Design Traffic Volumes (Section 6.7)

6.30 A four-lane freeway (two lanes in each direction) segment consists of passenger cars only, a peak-hour directional distribution of 0.70, a peak-hour factor of 0.80, and a measured FFS of 70 mi/h. Assuming Fig. 6.8 applies, if the AADT is 30,000 veh/day, determine the LOS for the 10th, 50th, and 100th highest annual hourly volumes.

6.31 A four-lane freeway (two lanes in each direction) operates at capacity during the peak hour. It has 11-ft lanes, 4-ft shoulders, and there are three ramps within three miles upstream of the segment midpoint and four ramps within three miles downstream of the segment midpoint. The freeway is on rolling terrain and has 8% heavy vehicles with a peak-hour factor of 0.85. It is known that 12% of the AADT occurs in the peak hour and that the directional factor is 0.6. What is the freeway's AADT?

6.32 A six-lane multilane highway (three lanes in each direction) currently operates at maximum LOS C conditions. The measured *FFS* is 55 mi/h. The highway is on rolling terrain with 14% SUTs and 6% TTs, and the peak-hour factor is 0.92. If 17% of all directional traffic occurs during the peak hour, determine the total directional traffic volume.

Multiple Choice Problems (Multiple Sections)

6.33 A six-lane freeway (three lanes in each direction) in rolling terrain has 10-ft lanes and obstructions 4 ft from the right edge of the traveled pavement. There are five ramps within three miles upstream of the segment midpoint and four ramps within three miles downstream of the segment midpoint. A directional peak-hour volume of 2000 vehicles is observed, with 600 vehicles arriving in the highest 15-min flow rate period. The traffic stream contains 18% heavy vehicles. What is the density of the traffic stream?

  a) 8.49 pc/mi/ln
  b) 17.15 pc/mi/ln
  c) 14.28 pc/mi/ln
  d) 14.88 pc/mi/ln

6.34 A six-lane freeway (three lanes in each direction) in level terrain has 10-ft lanes and obstructions 5 ft from the right edge. There are zero ramps within three miles upstream of the segment midpoint and one ramp within three miles downstream of the segment midpoint. The traffic stream has a peak-hour factor of 0.84, peak-hour volume of 2500 vehicles, and 4% heavy vehicles. What is the LOS?

  a) LOS A
  b) LOS B
  c) LOS C
  d) LOS D

6.35 A four-lane divided multilane highway (two lanes in each direction) in rolling terrain has five access points per mile and 11-ft lanes with a 4-ft shoulder on the right side and 2-ft shoulder on the left. The peak-hour factor is 0.84 with 13% heavy vehicles. If the analysis flow rate is 1250 pc/h/ln, what is the peak-hour volume?

  a) 1667 veh/h
  b) 1859 veh/h
  c) 1984 veh/h
  d) 3335 veh/h

6.36 A six-lane undivided multilane highway (three SS lanes in each direction) has 12-ft lanes with 2-ft shoulders on the right side. There are two access points per mile and the posted speed limit is 50 mi/h. Estimate the free-flow speed (to the nearest 1 mi/h).

  a) 50 mi/h
  b) 43 mi/h
  c) 48 mi/h
  d) 52 mi/h

6.37 You are designing a freeway as a passenger-car-only facility, and with ideal roadway characteristics. It is estimated that the freeway will have a traffic demand of 75,000 vehicles per day, a peak-hour factor of 0.88, and a directional distribution of 0.65. Determine the number of lanes (both directions) required to provide at least LOS D using the 60th highest annual hourly volume (see Fig. 6.8).

  a) 4 lanes
  b) 6 lanes
  c) 8 lanes
  d) 10 lanes

CHAPTER 7 PROBLEMS

Development of Signal Phasing and Timing Plans (Section 7.4)

SS **7.1** An intersection has a three-timing-stage signal with the movements allowed in each timing stage and corresponding analysis and saturation flow rates shown in Table 7.6. Calculate the sum of the flow ratios for the critical lane groups.

7.2 An intersection has a four-timing-stage signal with the movements allowed in each timing stage and corresponding analysis and saturation flow rates shown in Table 7.7. Calculate the sum of the flow ratios for the critical lane groups.

7.3 The minimum cycle length for an intersection is determined to be 95 seconds. The critical lane group flow ratios were calculated as 0.235, 0.250, 0.170, and 0.125 for timing stages 1–4, respectively. What X_c was used in the determination of this cycle length, assuming a lost time of 5 seconds per timing stage?

7.4 A pretimed four-timing-stage signal has critical lane group flow rates for the first three timing stages of 200, 187, and 210 veh/h (saturation flow rates are 1800 veh/h/ln for all timing stages). The lost time is known to be 4 seconds for each timing stage. If the cycle length is 60 seconds, what is the estimated effective green time of the fourth timing stage?

7.5 A four-timing-stage traffic signal has critical lane group flow ratios of 0.225, 0.175, 0.200, and 0.150. If the lost time per timing stage is 5 seconds and a critical intersection v/c of 0.85 is desired, calculate the minimum cycle length and the timing

stage effective green times such that the lane group v/c ratios are equalized.

7.6 For Problem 7.1, calculate the minimum cycle length and the effective green time for each timing stage (balancing v/c for the critical lane groups). **SS** Assume the lost time is 4 seconds per timing stage and a critical intersection v/c of 0.90 is desired.

7.7 For Problem 7.1, calculate the optimal cycle length (Webster's formulation) and the corresponding effective green times (based on lane group v/c equalization). Assume lost time is 4 seconds per timing stage.

7.8 For Problem 7.2, calculate the minimum cycle length and the effective green time for each timing stage (balancing v/c for the critical movements). Assume the lost time is 4 seconds per timing stage and a critical intersection v/c of 0.95 is desired.

7.9 For Problem 7.2, calculate the optimal cycle length (Webster's formulation) and the corresponding effective green times (based on lane group v/c equalization). Assume lost time is 4 seconds per timing stage.

7.10 Consider Example 7.3. Two additional 12-ft through lanes are added to Vine Street (the street in the intersection shown in Fig. 7.8), one lane in each direction. If the peak-hour traffic volumes are unchanged but the Vine Street left-turn saturation flow rates increase by 100 veh/h because of the added through lanes, what would the revised effective green time, yellow time, and all-red time be for each timing stage? Assume minimum cycle

Table 7.6 Data for Problem 7.1

Timing stage	1	2	3	
Allowed movements	NB L, SB L	NB T/R, SB T/R	EB L, WB L	EB T/R, WB T/R
Analysis flow rate	330, 365 veh/h	1125, 1075 veh/h	110, 80 veh/h	250, 285 veh/h
Saturation flow rate	1700, 1750 veh/h	3400, 3300 veh/h	650, 600 veh/h	1750, 1800 veh/h

Table 7.7 Data for Problem 7.2

Timing stage	1	2	3	4
Allowed movements	EB L, WB L	EB T/R, WB T/R	SB L, SB T/R	NB L, NB T/R
Analysis flow rate	245, 230 veh/h	975, 1030 veh/h	255, 235 veh/h	225, 215 veh/h
Saturation flow rate	1750, 1725 veh/h	3350, 3400 veh/h	1725, 1750 veh/h	1700, 1750 veh/h

length and a critical intersection v/c of 0.90 is desired.

7.11 Consider the intersection of Vine and Maple Streets as shown in Fig. 7.8. Suppose Vine Street's northbound and southbound approaches are both on an 8% upgrade, and the assumed vehicle approach speed is 30 mi/h. What should the yellow and all-red times be?

7.12 Consider Problem 7.10. Calculate the new minimum pedestrian green time, assuming the effective crosswalk width is 6 ft and the maximum number of crossing pedestrians in any timing stage is 20.

Analysis of Traffic at Signalized Intersections (Section 7.5)

7.13 An intersection approach has a saturation flow rate of 1500 veh/h, and vehicles arrive at the approach at the rate of 800 veh/h. The approach is controlled by a pretimed signal with a cycle length of 60 seconds and $D/D/1$ queuing holds. Local standards dictate that signals should be set such that all approach queues dissipate 10 seconds before the end of the effective green portion of the cycle. Assuming that approach capacity exceeds arrivals, determine the maximum length of effective red that will satisfy local standards.

7.14 An approach to a pretimed signal has 30 seconds of effective red, and $D/D/1$ queuing holds. The total delay at the approach is 83.33 veh-s/cycle and the saturation flow rate is 1000 veh/h. If the capacity of the approach equals the number of arrivals per cycle, determine the approach flow rate and cycle length.

SS **7.15** An approach to a pretimed signal has 25 seconds of effective green in a 60-second cycle. The approach volume is 500 veh/h and the saturation flow rate is 1400 veh/h. Calculate the average vehicle delay for the approach assuming $D/D/1$ queuing.

7.16 An observer notes that an approach to a pretimed signal has a maximum of 8 vehicles in a queue in a given cycle. If the saturation flow rate is 1440 veh/h and the effective red time is 40 seconds, how much time will it take this queue to clear after the start of the effective green (assuming that approach capacity exceeds arrivals and $D/D/1$ queuing applies)?

7.17 An approach to a pretimed signal with a 60-second cycle has 9 vehicles in the queue at the beginning of the effective green. Four of the 9

vehicles in the queue are left over from the previous cycle (at the end of the previous cycle's effective green). The saturation flow rate of the approach is 1500 veh/h, total delay for the cycle is 5.78 vehicle minutes, and at the end of the effective green there are 2 vehicles left in the queue. Determine the arrival rate assuming that it is unchanged over the duration of the observation period (from the beginning to the end of this 5.78–vehicle-minute delay cycle). (Assume $D/D/1$ queuing.)

7.18 At the beginning of an effective red, vehicles are arriving at an approach at the rate of 500 veh/h and 16 vehicles are left in the queue from the previous cycle (at the end of the previous cycle's effective green). However, due to the end of a major sporting event, the arrival rate is continuously increasing at a constant rate of 200 veh/h/min (after 1 minute the arrival rate will be 700 veh/h, after 2 minutes 900 veh/h, etc.). The saturation flow rate is 1800 veh/h, the cycle length is 60 seconds, and the effective green time is 40 seconds. Determine the total vehicle delay on this approach until complete queue clearance. (Assume $D/D/1$ queuing.)

7.19 The saturation flow rate for an intersection approach is 3600 veh/h. At the beginning of a cycle (effective red) no vehicles are queued. The signal is timed so that when the queue (from the continuously arriving vehicles) is 13 vehicles long, the effective green begins. If the queue dissipates 8 seconds before the end of the cycle and the cycle length is 60 seconds, what is the approach's arrival rate, assuming $D/D/1$ queuing?

7.20 The saturation flow rate for a pretimed signalized intersection approach is 1800 veh/h. The cycle length is 80 seconds. It is known that the arrival rate during the effective green is twice the arrival rate during the effective red. During one cycle, there are 2 vehicles in the queue at the beginning of the cycle (the beginning of the effective red) and there are 8 vehicles in the queue at the end of the effective red (the beginning of the effective green). If the queue clears exactly at the end of the effective green and $D/D/1$ queuing applies, determine the total vehicle delay on the approach in the cycle (in veh-s).

7.21 An approach to a signalized intersection has a **SS** saturation flow rate of 1800 veh/h. At the beginning of an effective red, there are 6 vehicles in the queue and vehicles arrive at 900 veh/h. The signal has a 60-second cycle with 25 seconds of effective red. What is the total vehicle delay on the approach after one cycle (assume $D/D/1$ queuing)?

7.22 An approach to a signalized intersection has a saturation flow rate of 2640 veh/h. For one cycle, the approach has 3 vehicles in queue at the beginning of an effective red, and vehicles arrive at 1064 veh/h. The signal for the approach is timed such that the effective green starts 8 seconds after the approach's vehicle queue reaches 10 vehicles, and lasts 15 seconds. What is the total vehicle delay on the approach for this signal cycle?

7.23 An approach to a signal has a saturation flow rate of 1800 veh/h. During one 80-second cycle, there are 4 vehicles queued at the beginning of the cycle (the start of the effective red) and 2 vehicles queued at the end of the cycle (the end of the effective green). At the beginning of the effective green there are 10 vehicles in the queue. The arrival rate is constant and the process is $D/D/1$. If the effective red is known to be less than 40 seconds, what is the total vehicle delay on this approach for this signal cycle?

7.24 Vehicles arrive at an approach to a pretimed signalized intersection. The arrival rate over the cycle is given by the function $v(t) = 0.22 + 0.012t$ [$v(t)$ is in veh/s and t is in seconds]. There are no vehicles in the queue when the cycle (effective red) begins. The cycle length is 60 seconds and the saturation flow rate is 3600 veh/h. Determine the effective green and red times that will allow the queue to clear exactly at the end of the cycle (the end of the effective green), and determine the total vehicle delay for this approach over the cycle (assuming $D/D/1$ queuing).

7.25 At the start of the effective red at an intersection approach to a pretimed signal, vehicles begin to arrive at a rate of 800 veh/h for the first 40 seconds and 500 veh/h from then on. The approach has a saturation flow rate of 1200 veh/h and an effective green of 20 seconds, and the cycle length is 40 seconds. What is the approach's total vehicle delay two full cycles after the 800-veh/h arrival rate begins? (Assume $D/D/1$ queuing.)

7.26 A left-turn movement has a maximum arrival rate of 200 veh/h. The saturation flow of this movement is 1400 veh/h. For this approach, the yellow time is 4 seconds, all-red time is 2 seconds, and total lost time is 3 seconds. The cycle length is 120 seconds. What minimum displayed green time must be provided to ensure that the queue in each cycle clears, and what is the movement's total delay per cycle and delay per vehicle for this green time? (Assume $D/D/1$ queuing.)

7.27 Vehicles begin to arrive at a signal approach at a rate of $v(t) = 0.2286 + 0.0008t$ [with $v(t)$ in veh/s and t in seconds] at the beginning of the cycle (the beginning of an effective red) and there are 2 vehicles already in the queue that are left over from the effective green of the previous cycle. The signal is designed so that the effective green starts when there are 10 vehicles in the queue. The saturation flow rate is 1800 veh/h. What is the approach's total vehicle delay after one cycle (cycles are 60 seconds long) and when will the effective green start in cycle #2 (or, equivalently, how long will the effective red be in cycle #2)? (Assume $D/D/1$ queuing.)

7.28 At the beginning of a signal approach's **SS** effective red there are 8 vehicles in queue. The arrival rate over the cycle is $v(t) = 0.05 + 0.001t$ [with $v(t)$ in veh/s and t in seconds]. If the saturation flow rate is 1800 veh/h and the cycle length is 80 seconds, what is the minimum effective green time needed for this cycle to have zero vehicles in the approach's queue when the effective red of the next cycle starts and what is approach's the total delay in this cycle with this green time. (Assume $D/D/1$ queuing.)

7.29 A signal approach has 20 seconds of displayed green, 4 seconds of yellow, and 3 seconds of all-red (start-up lost time and clearance times are typical). The cycle length is 60 seconds. At the beginning of an effective red there are no vehicles in queue and vehicles arrive at three-quarters of the saturation flow rate for 30 seconds. Then there is zero flow for 10 seconds, and then one-half of the saturation flow rate from 40 seconds until the end of the cycle. What is the approach's total delay for this cycle if the saturation flow rate is 1400 veh/h? (Assume $D/D/1$ queuing.)

7.30 An approach with a saturation flow rate of 1800 veh/h has 3 vehicles in queue at the start of an effective red. For the first cycle, the approach arrival rate is given by the function $v(t) = 0.5 - 0.005t$ [with $v(t)$ in veh/s and t in seconds measured from the beginning of the effective red]. From the second cycle onward (starting at the beginning of the second effective red) vehicles arrive at a fixed rate of 720 veh/h. The approach has 26 seconds of effective red and a 60 second cycle for all cycles. How many cycles will it take to have no vehicles in the queue at the start of an effective red for this approach and what would be the total delay for the approach until this happens? (Assume $D/D/1$ queuing.)

7.31 Vehicles arrive at a signal approach at a rate of $v(t) = 0.3 - 0.001t$ [with $v(t)$ in veh/s and t in seconds measured from the beginning of the effective red of the first cycle]. The signal has a 70-second cycle length with 40 seconds of effective red. The saturation flow rate of the approach is 1800 veh/h. What is the approach's total vehicle delay after two cycles (when $t = 140$ seconds) and when will the queue clear (measured from the beginning of the first cycle) during an effective green (that is, the t at which there will no longer be a queue)? (Assume $D/D/1$ queuing.)

7.32 An approach to a signalized intersection has a green time of 35 seconds, and all-red time of 2 seconds, a yellow time of 3 seconds, and a total lost time of 3 seconds. The arrival rate is $v(t) = 0.5 + 0.002t$ [with $v(t)$ in veh/s and t in seconds measured from the beginning of the effective red of the cycle]. The saturation flow rate of the approach is 3400 veh/h. How long must the cycle length be so that the queue that forms at the beginning of the cycle (effective red) dissipates exactly at the end of the cycle (end of the effective green) and what would be the average delay per vehicle over the cycle? (Assume $D/D/1$ queuing.)

7.33 Recent computations at an approach to a pretimed-signalized intersection indicate that the volume-to-capacity ratio is 0.8, the saturation flow rate is 1600 veh/h, and the effective green time is 50 seconds. If the uniform delay (assuming $D/D/1$ queuing) is 11.25 seconds per vehicle, determine the arrival flow rate (in veh/h) and the cycle length.

7.34 An approach to a pretimed signal is timed so that the effective green starts 4 seconds after 13 vehicles are in the queue. The cycle length is 70 seconds and the arrival rate for a specific cycle is $v(t) = 0.52 - 0.003t$ [with $v(t)$ in veh/s and t in seconds measured from the beginning of the effective red of the cycle]. If there are 3 vehicles in the queue at the beginning of a specific cycle (effective red) and the queue clears 4 seconds before the end of the cycle (effective green), what is the saturation flow of the approach and the approach's total delay over the cycle? (Assume $D/D/1$ queuing.)

7.35 At an intersection approach, at the beginning of an effective red there are 10 vehicles in a queue. Vehicles arrive at a constant rate of 600 veh/h and the saturation flow of the approach is 1800 veh/h. the cycle length is 60 seconds and the approach gets 30 seconds of effective green. What will the total

delay be (in veh-s) after two cycles at the approach? (Assume $D/D/1$ queuing.)

7.36 An approach at a signalized intersection with a 70-second cycle gets 25 seconds of displayed green time. Yellow time is 4 seconds and all-red is 2 seconds (lost time is to be determined from standard assumptions). At the beginning of an effective red there are 4 vehicles in the queue and the saturation flow is 818 veh/h. The arrival rate is given as $v(t) = 0.17 - 0.002t$ [with $v(t)$ in veh/s and t in seconds after the beginning of the effective red]. What is the total vehicle delay for this approach at the end of the cycle (until the next effective red) that started with the 4 vehicles queued at the beginning of the effective red? (Assume $D/D/1$ queuing.)

7.37 At the beginning of an effective red, an approach has 7 vehicles in a queue with vehicles arriving according to the function $v(t) = 0.2 + 0.00111t$ [with $v(t)$ in veh/s and t in seconds after the beginning of the effective red]. The saturation flow is 1800 veh/h, the cycle length is 60 seconds, and the effective red is 20 seconds. What is the total delay for the approach until the queue first clears? (Assume $D/D/1$ queuing.)

7.38 At one signalized intersection approach, 20 vehicles, on average, arrive during the 30-second effective green time. During the rest of the cycle, 45 vehicles, on average, arrive at the intersection from this approach. The cycle length is 90 seconds. What is the proportion of vehicles arriving on green for this approach?

7.39 Consider Problem 7.38. Assume the saturation flow rate is 8000 veh/h. Determine the average uniform delay for this approach.

7.40 Consider Problems 7.38 and 7.39. Assume that improvements were made to the signal timing such that the PVG for this approach is now 0.45. Also assume that the average number of vehicle arrivals per cycle is still 65 (but the arrivals on green and red will be different from Prob. 7.34). Determine the new average uniform delay for this approach.

7.41 An approach to a pretimed signal has 25 seconds of effective green, a saturation flow rate of 1300 veh/h, and a volume-to-capacity ratio less than 1. If the cycle length is 60 seconds and the overall delay formula (Eq. 7.27) estimates an average delay that is 34 s greater than that estimated by using just the uniform delay formula, determine the vehicle arrival rate. (Assume the signal is isolated and $d_3 = 0$.)

SS **7.42** For Problem 7.6, calculate the northbound average approach delay (using Eq. 7.27) and level of service.

7.43 For Problem 7.6, calculate the southbound average approach delay (using Eq. 7.27) and level of service.

7.44 For Problem 7.6, calculate the westbound average approach delay (using Eq. 7.27) and level of service.

7.45 For Problem 7.6, calculate the eastbound average approach delay (using Eq. 7.27) and level of service.

7.46 For Problem 7.6, calculate the overall intersection average delay (using Eqs. 7.27–7.29) and level of service.

7.47 For Problem 7.8, calculate the northbound average approach delay (using Eq. 7.27) and level of service.

7.48 For Problem 7.8, calculate the southbound average approach delay (using Eq. 7.27) and level of service.

7.49 For Problem 7.8, calculate the westbound average approach delay (using Eq. 7.27) and level of service.

7.50 For Problem 7.8, calculate the eastbound average approach delay (using Eq. 7.27) and level of service.

7.51 For Problem 7.8, calculate the overall intersection average delay (using Eqs. 7.27–7.29) and level of service.

7.52 A new shopping center opens near the intersection of Vine and Maple Streets (the intersection shown in Fig. 7.8). The net effect is to increase the approaching traffic volumes by 10%. Calculate the new level of service for the westbound approach, assuming all else remains the same. Assume other required input values are as in Table 7.5. Use Eq. 7.27 for the delay calculation.

7.53 For Problem 7.52, calculate the new level of service for the northbound approach.

7.54 Calculate the overall intersection level of service for Problem 7.10. Assume other required input values are as in Table 7.5. Use Eqs. 7.27–7.29 for the delay calculations.

7.55 Consider Problem 7.10. How much traffic volume can be added to the southbound approach (assuming the same turning movement percentage) before LOS D is reached for the approach? Use Eq. 7.27 for the delay calculation.

7.56 Consider Problem 7.10. How much traffic volume must be diverted from the eastbound approach (assuming the same turning movement percentage) to achieve LOS B for the approach? Use Eq. 7.27 for the delay calculation.

Multiple Choice Problems (Multiple Sections)

7.57 A signalized intersection has a cycle length of 70 seconds. For one traffic movement, the displayed all-red time is set to 2 seconds while the displayed yellow time is 5 seconds. The effective red time is 37 seconds and the total lost time per cycle for the movement is 4 seconds. What is the displayed green time for the traffic movement?

a) 30 s
b) 31 s
c) 33 s
d) 65 s

7.58 An isolated pretimed signalized intersection has an approach with a saturation flow rate of 1900 veh/h. For this approach, the displayed red time is 58 seconds, the displayed yellow time is 3 seconds, the all-red time is 2 seconds, the effective green time is 28 seconds, and the total lost time is 4 s/timing stage. What is the average uniform delay per vehicle when the approach flow rate is 550 veh/h?

a) 26.3 s
b) 29.7 s
c) 30.1 s
d) 413.3 s

7.59 An isolated pretimed signalized intersection has an approach with a traffic flow rate of 750 veh/h and a saturation flow rate of 3200 veh/h. This approach is allocated 32 seconds of effective green time. The cycle length is 100 seconds. Determine the average approach delay (using Eq. 7.27).

a) 4.6 s
b) 30.2 s
c) 34.8 s
d) 35.0 s

7.60 Calculate the sum of flow ratios for the critical lane groups for the three-phase timing plan, with traffic and saturation flow rates shown in the following tables.

Traffic Flow Rates (veh/h) for Problem 7.60

Timing Stage 1	Timing Stage 2	Timing Stage 3
EB L: 250	EB T/R: 1200	SB L: 75
		NB L: 100
WB L: 300	WB T/R: 1350	SB T/R: 420
		NB T/R: 425

Saturation Flow Rates (veh/h) for Problem 7.60

Timing Stage 1	Timing Stage 2	Timing Stage 3
EB L: 1800	EB T/R: 3600	SB L: 500
		NB L: 525
WB L: 1800	WB T/R: 3600	SB T/R: 1950
		NB T/R: 1950

a) 0.950
b) 0.760
c) 0.690
d) 0.622

7.61 A signalized intersection has a sum of critical flow ratios of 0.72 and a total cycle lost time of 12 seconds. Assuming a critical intersection v/c ratio of 0.9, calculate the minimum necessary cycle length.

a) 48.0 s
b) 60.0 s
c) 82.1 s
d) 42.9 s

7.62 A signalized intersection approach has an upgrade of 4%. The total width of the cross street at this intersection is 60 feet. The average vehicle length of approaching traffic is 16 feet. The speed of approaching traffic is 40 mi/h. Determine the sum of the minimum necessary change and clearance intervals.

a) 3.59 s
b) 4.96 s
c) 4.89 s
d) 2.51 s

CHAPTER 8 PROBLEMS

Trip Generation (Section 8.4)

8.1 A large retirement village has a total retail employment of 120. All 1600 of the households in this village consist of two nonworking family members with household income of $20,000. Assuming that shopping and social/recreational trip rates both peak during the same hour (for exposition purposes), predict the total number of peak-hour trips generated by this village using the trip generation models of Examples 8.1 and 8.2.

8.2 Consider the retirement village described in Problem 8.1. Determine the amount of additional retail employment (in the village) necessary to reduce the total predicted number of peak-hour shopping trips to 200.

SS 8.3 A large residential area has 1400 households with an average household income of $40,000, an average household size of 4.8, and, on average, 1.5 working members. Using the model described in Example 8.2 (assuming it was estimated using zonal averages instead of individual households), predict the change in the number of peak-hour social/recreational trips if employment in the area increases by 25% and household income by 10%.

8.4 Consider a Poisson regression model for the number of social/recreational trips generated during a peak-hour period that is estimated by (see Eq. 8.3) $BZ_i = -0.75 + 0.025(\text{household size}) + 0.008(\text{annual household income, in thousands of dollars}) + 0.10(\text{number of nonworking household members})$. Suppose a household has five members (three of whom work) and an annual income of $100,000. What is the expected number of peak-hour social/recreational trips, and what is the probability that the household will not make a peak-hour social/recreational trip?

8.5 If small express buses leave the origin described in Example 8.5 and all are filled to their capacity of 20 travelers, how many work-trip vehicles leave from origin to destination in Example 8.5 during the peak hour?

Mode and Destination Choice (Section 8.5)

8.6 It is known that 4000 automobile trips are generated in a large residential area from noon to 1:00 P.M. on Saturdays for shopping purposes. Four major shopping centers have the following characteristics:

Shopping center	Distance from residential area (mi)	Commercial floor space (thousands of ft²)
1	2.4	200
2	4.6	150
3	5.0	300
4	8.7	600

If a logit model is estimated with coefficients of -0.543 for distance and 0.0165 for commercial space (in thousands of ft²), how many shopping trips will be made to each of the four shopping centers?

8.7 Consider the shopping trip situation described in Problem 8.6. Suppose that shopping center 3 goes out of business and shopping center 2 is expanded to 450,000 ft² of commercial space. What would be the new distribution of the 4000 Saturday afternoon shopping trips?

8.8 If shopping center 3 is closed (see Problem 8.7), how much commercial floor space is needed in shopping centers 1 and 2 to ensure that each of them has the same probability of being selected as shopping center 4?

8.9 Consider the situation described in Example 8.7. **SS** If the construction of a new freeway lowers auto and transit travel times to shopping center 2 by 20%, determine the new distribution of shopping trips by destination and mode.

8.10 Consider the conditions described in Example 8.7. Heavily congested highways have caused travel times to shopping center 2 to increase by 4 min for both auto and transit modes (travel times to shopping center 1 are not affected). In order for shopping center 2 to attract as many total trips (auto and transit) as it did before the congestion, how much commercial floor space must it add (given that the total number of departing shopping trips remains at 900)?

8.11 A total of 725 auto-mode social/recreational trips are made from an origin (residential area) during the peak hour. A logit model estimation is made, and three factors were found to influence the destination choice: (1) population at the destination, in thousands (coefficient = 0.17); (2) distance from origin to destination, in miles (coefficient = −0.23); and (3) square feet of amusement floor space (movie theaters, video game centers, etc.), in thousands (coefficient $= 0.10$). Four possible destinations have the following characteristics:

	Population (thousands)	Distance from origin (mi)	Amusement space (thousands of ft^2)
Destination 1	15.5	7.5	5
Destination 2	6.0	5	10
Destination 3	0.8	2	8
Destination 4	5.0	7	15

Determine the distribution of trips among possible destinations.

8.12 Consider the situation described in Problem 8.11. If a new 6000-ft^2 arcade center is built at destination 3, determine the distribution of the 725 peak-hour social/recreational trips.

8.13 Consider the situation described in Problem 8.11. If the total number of trips remains constant, determine the amount of amusement floor space that must be added to destination 2 to attract an additional 50 social/recreational trips.

8.14 Note that with the situation described in Example 8.8, 26.6% [(110 + 23)/500] of all social/recreational trips are to destination 1. If the total number of trips remains constant, how much additional amusement floor space would have to be added to destination 1 to have it capture 35.0% of the total social/recreational trips?

Highway Route Choice (Section 8.6)

8.15 An origin–destination pair is connected by a route with a performance function $t_1 = 8 + x_1$, and another with a function $t_2 = 1 + 2x_2$ (with x's in thousands of vehicles per hour and t's in minutes). If the total origin–destination flow is 4000 veh/h, determine user-equilibrium and system-optimal route travel times, total travel time (in vehicle minutes), and route flows.

SS **8.16** Because of the great increase in vehicle-hours caused by the reconstruction project in Example 8.12, the state transportation department decides to regulate the flow of traffic on the two routes (until reconstruction is complete) to achieve a system-optimal solution. How many vehicle-hours will be saved during each peak-hour period if this strategy is implemented and travelers are not permitted to achieve a user-equilibrium solution?

8.17 For Example 8.11, what reduction in peak-hour traffic demand is needed to ensure an equality of total vehicle travel time (in vehicle minutes) assuming a system-optimal solution before and during reconstruction?

8.18 Two routes connect an origin and a destination. **SS** Routes 1 and 2 have performance functions $t_1 = 2 + x_1$ and $t_2 = 1 + x_2$, where the t's are in minutes and the x's are in thousands of vehicles per hour. The travel times on the routes are known to be in user equilibrium. If an observation for route 1 finds that the gaps between 30% of the vehicles are less than 6 seconds, estimate the volume and average travel times for the two routes. (*Hint*: Assume a Poisson distribution of vehicle arrivals, as discussed in Chapter 5.)

8.19 Three routes connect an origin and a destination with performance functions $t_1 = 8 + 0.5x_1$, $t_2 = 1 + 2x_2$, and $t_3 = 3 + 0.75x_3$, with the x's expressed in thousands of vehicles per hour and the t's expressed in minutes. If the peak-hour traffic demand is 3400 vehicles, determine user-equilibrium traffic flows.

8.20 Two routes connect a suburban area and a city, with route travel times (in minutes) given by the expressions $t_1 = 6 + 8(x_1/c_1)$ and $t_2 = 10 + 3(x_2/c_2)$, where the x's are expressed in thousands of vehicles per hour and the c's are the route capacities in thousands of vehicles per hour. Initially, the capacities of routes 1 and 2 are 4000 and 2000 veh/h, respectively. A reconstruction project on route 1 reduces the capacity to 3000 veh/h, but total traffic demand is unaffected. Observational studies note a 35.28-second increase in average travel time on route 1 and a 68.5% increase in flow on route 2 after reconstruction begins. User-equilibrium conditions exist before and during reconstruction. If both routes are always used, determine equilibrium flows and travel times before and after reconstruction begins.

8.21 Three routes connect an origin and destination with performance functions $t_1 = 2 + 0.5x_1$, $t_2 = 1 + x_2$, and $t_3 = 4 + 0.2x_3$ (with t's in minutes and x's in thousands of vehicles per hour). Determine user-equilibrium flows if the total origin-to-destination demand is (a) 10,000 veh/h and (b) 5000 veh/h.

8.22 For the routes described in Problem 8.21, what is the minimum origin-to-destination traffic demand (in vehicles per hour) that will ensure that all routes are used (assuming user-equilibrium conditions)?

8.23 A multilane highway has two northbound lanes. Each lane has a capacity of 1500 vehicles per hour.

Currently, northbound traffic consists of 3100 vehicles with 1 occupant, 600 vehicles with 2 occupants, 400 vehicles with 3 occupants, and 20 buses with 50 occupants each. The highway's performance function is $t = t_0[1 + 1.15(x/c)^{6.87}]$, where t is in minutes, t_0 is equal to 15 minutes, and x and c are volumes and capacities in vehicles per hour. An additional lane is being added (with 1500 veh/h capacity). What will the total person-hours of travel be if the lane is (a) open to all traffic, (b) open to vehicles with two or more occupants only, and (c) open to vehicles with three or more occupants only? (Assume that all qualified higher-occupancy vehicles use only the new lane, no unqualified vehicles use the new lane, and there is no mode shift.)

8.24 Consider the new lane addition in Problem 8.23. First, suppose 500 one-occupant vehicle travelers take 10 buses (50 on each bus), and the new lane is open to vehicles with two or more occupants. What would the total person-hours be? Second, referring back to part (a) of Problem 8.23, what is the minimum mode shift from one-occupant vehicles to buses (with 50 persons each) needed to ensure that the person-hours of travel time on the highway with the new lane (which is restricted to vehicles with two or more occupants) is as low as if all three lanes (the two existing lanes and the new lane) were open to all traffic? (Set up the equation, and solve to the nearest 100 one-occupant vehicles.)

8.25 Two routes connect an origin and destination with performance functions $t_1 = 5 + 3x_1$ and $t_2 = 7 + x_2$, with t's in minutes and x's in thousands of vehicles per hour. Total origin–destination demand is 7000 vehicles in the peak hour. What are user-equilibrium and system-optimal route flows and total travel times?

8.26 Consider the conditions in Problem 8.25. What is the value of the derivative of the user-equilibrium math program evaluated at the system-optimal solution with respect to x_1 (with x_1 equal to the system-optimal solution)?

8.27 Two routes connect an origin and a destination. Their performance functions are $t_1 = 3 + 1.5(x_1/c_1)^2$ and $t_2 = 5 + 4(x_2/c_2)$, with t's in minutes and x's and c's being route flows and capacities, respectively. The origin–destination demand is 6000 vehicles per hour, and c_1 and c_2 are equal to 2000 and 1500 vehicles per hour, respectively. Proposed capacity improvements will increase c_2 by 1000 vehicles per hour. It is known that the routes are currently in user equilibrium, and

it is estimated that each 1-minute reduction in route travel time will attract an additional 500 vehicles per hour (from latent travel demand and mode shifts). What will the user-equilibrium flows and total hourly origin–destination demand be after the capacity improvement?

8.28 Two routes connect an origin–destination pair with performance functions $t_1 = 5 + 4x_1$ and $t_2 = 7 + 2x_2$, with t's in minutes and x's in thousands of vehicles per hour. Assuming that both routes are used, can user-equilibrium and system-optimal solutions be equal at some feasible value of total origin-destination demand (q)? (Prove your answer.)

8.29 Three routes connect an origin–destination pair with performance functions $t_1 = 5 + 1.5x_1$, $t_2 = 12 + 3x_2$, and $t_3 = 2 + 0.2 x^2_3$ (with t's in minutes and x's in thousands of vehicles per hour). Determine user-equilibrium flows if $q = 4300$ veh/h.

8.30 Two routes connect an origin–destination pair with performance functions $t_1 = 6 + 4x_1$ and $t_2 = 2 + 0.5(x_2)^2$ (with t's in minutes and x's in thousands of vehicles per hour). The origin–destination demand is 4000 veh/h at a travel time of 2 minutes, but for each additional minute beyond these 2 minutes, 100 fewer vehicles depart. Determine user-equilibrium route flows and total vehicle travel time.

8.31 A freeway has six lanes, four of which are unrestricted (open to all vehicle traffic), and two of which are restricted lanes that can be used only by vehicles with two or more occupants. The performance function for the highway is $t = 12 + (2/NL)x$ (with t in minutes, NL being the number of lanes, and x in thousands of vehicles). During the peak hour, 3000 vehicles with one occupant and 4000 vehicles with two occupants depart for the destination. Determine the distribution of traffic between restricted and unrestricted lanes such that total person-hours are minimized.

8.32 Two routes connect an origin–destination pair with performance functions $t_1 = 5 + (x_1/2)^2$ and $t_2 = 7 + (x_2/4)^2$ (with t's in minutes and x's in thousands of vehicles per hour). It is known that at user equilibrium, 75% of the origin–destination demand takes route 1. What percentage would take route 1 if a system-optimal solution were achieved, and how much travel time would be saved?

8.33 Two routes connect an origin–destination pair with performance functions $t_1 = 5 + 3.5x_1$ and $t_2 = 1 + 0.5x^2_2$ (with t's in minutes and x's in thousands of vehicles per hour). It is known that at $x_2 = 3$, the difference between the first derivatives of the system-optimal and user-equilibrium math programs, evaluated with respect to x_2, is 7 [$dS(x)_{SO}/dx_2 - dS(x)_{UE}/dx_2 = 7$]. Determine the difference in total vehicle travel times (in vehicle minutes) between user-equilibrium and system-optimal solutions.

Multiple Choice Problems (Multiple Sections)

8.34 You are conducting a trip generation study based on Poisson regression. You estimate the following coefficients for a peak-hour shopping-trip generation model.

$BZ_i = -0.30 + 0.04$(household size)
$+ 0.005$(annual household income in thousands of dollars)
$- 0.12$(employment in the household's neighborhood in hundreds)

For a household with five members, an annual income of $95,000, and in a neighborhood with an employment of 250, what is the probability of the household making three or more peak-hour trips?

a) 0.071
b) 0.095
c) 0.905
d) 0.024

8.35 A work-mode–choice model is developed from data acquired in the field in order to determine the probabilities of individual travelers selecting various modes. The mode choices include automobile drive-alone (DL), automobile shared-ride (SR), and bus (B). The utility functions are estimated as:

$U_{DL} = 2.6 - 0.3(\text{cost}_{DL}) - 0.02(\text{travel time}_{DL})$
$U_{SR} = 0.7 - 0.3(\text{cost}_{SR}) - 0.04(\text{travel time}_{SR})$
$U_B = -0.3(\text{cost}_B) - 0.01(\text{travel time}_B)$

where cost is in dollars and time is in minutes. The cost of driving an automobile is $5.50 with a travel time of 21 minutes, while the bus fare is $1.25 with a travel time of 27 minutes. How many people will use the shared-ride mode from a community of 4500 workers, assuming the shared-ride option always consists of three individuals sharing costs equally?

a) 866
b) 2805
c) 828
d) 314

8.36 In a 1-hour period, 1100 vehicle-based shopping trips leave a large residential area for two shopping plazas. A joint shopping-trip mode-destination choice logit model is estimated, providing the following coefficients:

Variable	Auto coefficient	Bus coefficient
Constant	0.25	0.0
Travel time in minutes	–0.4	–0.5
Commercial floor space (in thousands of ft^2)	0.013	0.013

Initial travel times to shopping plazas 1 and 2 are as follows:

	By auto	By bus
Travel time to shopping plaza 1 (in minutes)	15	18
Travel time to shopping plaza 2 (in minutes)	16	19

If shopping plaza 2 has 325,000 ft^2 of commercial floor space and shopping plaza 1 has 275,000 ft^2, determine the number of bus trips to shopping plaza 2.

a) 10
b) 18
c) 597
d) 21

8.37 Two highways connect an origin–destination pair, having performance functions $t_1 = 5 + 4(x_1/c_1)$ and $t_2 = 4 + 5(x_2/c_2)$, for routes 1 and 2, respectively. For the performance functions, the t's are in minutes, and the flows (x's) and capacities (c's) are in thousands of vehicles per hour. The total traffic demand for the two highways is 4000 vehicles during the peak hour. Initially, the capacities of routes 1 and 2 are 3500 veh/h and 4600 veh/h, respectively. A construction project is planned that will cut the capacity of route 2 by 2100 veh/h. How many additional vehicle-hours of travel time will be added to the system assuming user-equilibrium conditions hold?

a) 55.3
b) 503.0
c) 77.6
d) 525.3

8.38 A study showed that during the peak-hour commute on two routes connecting a suburb with a large city, there are a total of 5500 vehicles that make the trip. Route 1 is 7 miles long with a 65-mi/h speed limit and route 2 is 4 miles long with a speed limit of 50 mi/h. The study also found that the total travel time on route 2 increases with the square of the number of vehicles, while the route 1 travel time increases two minutes for every 500 additional vehicles added. Determine the system-optimal total travel time (in veh-h).

   a) 29.9
   b) 1153.5
   c) 1364.6
   d) 1585.8

8.39 Two routes connect an origin–destination pair, with 2500 and 2000 vehicles traveling on routes 1 and 2 during the peak hour, respectively. The route performance functions are $t_1 = 12 + x_1$ and $t_2 = 7 + 2x_2$, with the x's expressed in thousands of vehicles per hour and the t's in minutes. If vehicles could be assigned to the two routes such as to achieve a system-optimal solution, how many vehicle-hours of travel time could be saved?

   a) 965.83
   b) 970.83
   c) 333.33
   d) 5.56

Index